Y0-BSD-900

ALLE ZEIT WACH
1842

Springer Series in Synergetics

Editor: Hermann Haken

Synergetics, an interdisciplinary field of research, is concerned with the cooperation of individual parts of a system that produces macroscopic spatial, temporal or functional structures. It deals with deterministic as well as stochastic processes.

1 **Synergetics** An Introduction 3rd Edition
By H. Haken

2 **Synergetics** A Workshop
Editor: H. Haken

3 **Synergetics** Far from Equilibrium
Editors: A. Pacault, C. Vidal

4 **Structural Stability in Physics**
Editors: W. Güttinger, H. Eikemeier

5 **Pattern Formation by Dynamic Systems and Pattern Recognition**
Editor: H. Haken

6 **Dynamics of Synergetic Systems**
Editor: H. Haken

7 **Problems of Biological Physics**
By L. A. Blumenfeld

8 **Stochastic Nonlinear Systems**
in Physics, Chemistry, and Biology
Editors: L. Arnold, R. Lefever

9 **Numerical Methods in the Study of Critical Phenomena**
Editors: J. Della Dora, J. Demongeot, B. Lacolle

10 **The Kinetic Theory of Electromagnetic Processes** By Yu. L. Klimontovich

11 **Chaos and Order in Nature**
Editor: H. Haken

12 **Nonlinear Phenomena in Chemical Dynamics**
Editors: C. Vidal, A. Pacault

13 **Handbook of Stochastic Methods**
for Physics, Chemistry, and the Natural Sciences
By C. W. Gardiner

14 **Concepts and Models of a Quantitative Sociology**
The Dynamics of Interacting Populations
By W. Weidlich, G. Haag

15 **Noise-Induced Transitions** Theory and Applications in Physics, Chemistry, and Biology
By W. Horsthemke, R. Lefever

16 **Physics of Bioenergetic Processes**
By L. A. Blumenfeld

17 **Evolution of Order and Chaos** in Physics, Chemistry, and Biology Editor H. Haken

18 **The Fokker-Planck Equation**
By H. Risken

19 **Chemical Oscillations, Waves, and Turbulence**
By Y. Kuramoto

20 **Advanced Synergetics**
By H. Haken

21 **Stochastic Phenomena and Chaotic Behaviour in Complex Systems**
Editor: P. Schuster

22 **Synergetics - From Microscopic to Macroscopic Order** Editor: E. Frehland

23 **Synergetics of the Brain**
Editors: E. Başar, H. Flohr, H. Haken, A. J. Mandell

24 **Chaos and Statistical Methods**
Editor: Y. Kuramoto

25 **Dynamics of Hierarchical Systems**
By J. S. Nicolis

26 **Self-Organization and Management of Social Systems**
Editors: H. Ulrich, G. J. B. Probst

27 **Non-Equilibrium Dynamics in Chemical Systems** Editors: C. Vidal, A. Pacault

28 **Self-Organization.** Autowaves and Structures Far from Equilibrium
Editor: V. I. Krinsky

Self-Organization

Autowaves and Structures Far from Equilibrium

Proceedings of an International Symposium
Pushchino, USSR, July 18–23, 1983

Editor: V. I. Krinsky

With 158 Figures

Springer-Verlag
Berlin Heidelberg New York Tokyo 1984

Professor Dr. V. I. Krinsky

Institute of Biological Physics, USSR Academy of Sciences,
SU-142292, Pushchino, USSR

Series Editor:

Professor Dr. Dr. h. c. Hermann Haken
Institut für Theoretische Physik der Universität Stuttgart, Pfaffenwaldring 57/IV,
D-7000 Stuttgart 80, Fed. Rep. of Germany

ISBN 3-540-15080-3 Springer-Verlag Berlin Heidelberg New York Tokyo
ISBN 0-387-15080-3 Springer-Verlag New York Heidelberg Berlin Tokyo

Offset printing: Beltz Offsetdruck, Hemsbach. Bookbinding: J. Schäffer OHG, Grünstadt
2153/3130-543210

Foreword

According to its definition, Synergetics is concerned with systems that produce macroscopic spatial, temporal, or functional structures. Autowaves are a specific, yet very important, case of spatio-temporal structures. The term "autowave" was coined in the Soviet Union in analogy to the term "auto-oscillator". This is a - perhaps too literal - translation of the Russian word "avto-ostsillyatory" (= self-oscillator) which in its proper translation means "self-sustained oscillator". These are oscillators, e.g., clocks, whose internal energy dissipation is compensated by a (more or less) continuous power input. Similarly, the term "autowaves" denotes propagation effects - including waves - in active media, which provide spatially distributed energy sources and thus may compensate dissipation. An example which is now famous is represented by spiral or concentric waves in a chemically active medium, undergoing the Belousov-Zhabotinsky reaction.

This book provides the reader with numerous further examples from physics, chemistry, and biology - e.g., autowaves of the heart. While the Belousov-Zhabotinsky reaction is now widely known, a number of very important results obtained in the Soviet Union are perhaps less well known. I am particularly glad that this book may help to make readers outside the Soviet Union acquainted with these important experimental and theoretical findings which are presented in a way which elucidates the common principles underlying this kind of propagation effects. Professor V. Krinsky has taken great care in editing this book to which prominent scientists from the Soviet Union and from abroad contribute and I wish to congratulate him for his successful efforts.

Hermann Haken

Preface

During recent years remarkably universal mechanisms have been found for the development of order from random distributions in active systems of quite different natures. These mechanisms are linked to the propagation of strongly nonlinear waves, the so-called autowaves which are spatio-temporal analogs of auto-oscillations. While the auto-oscillation theory is a well-developed branch of science, the study of autowaves is still only in an embryonic state.

Among interesting examples of self-organization discovered in the study of the propagation of autowaves in active media is the occurrence of dissipative wave structures in quite different fields: the morphogenesis of the simplest multicellular organisms, the optic cortex, the retina, heart tissue, the Belousov-Zhabotinsky chemical reaction. The autowave processes are also important in phase transitions, in particular for the disappearance of superconductivity, for movements of the domain walls in magnetic or magnetoelectric media, and for phenomena of critical boiling. Different types of chaos in active media, including cardiac arrhythmias, proved to be connected with the initiation and reproduction of autowave vortices. In 1983, a symposium devoted to these problems took place at the Scientific Centre of Biological Research of the USSR Academy of Sciences in Pushchino. This volume contains short reviews by invited authors, written after the symposium, with account taken of results presented there.

The camera-ready volume was prepared at the Biological Research Centre of the USSR Academie of Sciences, Institute of Biological Physics, Phushchino, USSR.

Pushchino
August 1984

V.I. Krinsky

Contents

Part I Introduction

Part II Self-Organization in Physical Systems: Autowaves and Structures Far from Equilibrium

Part V Autowaves in Biological Systems

Part VI Evolution and Self-Organization

Part I

Introduction

Synergetics - Some Basic Concepts and Recent Results

H. Haken

Institut für Theoretische Physik, Universität Stuttgart, Pfaffenwaldring 57/IV
D-7000 Stuttgart 80, Fed. Rep. of Germany

1. INTRODUCTION

In my contribution I should like to give a brief outline of some basic ideas of synergetics [1], [2] . Then I shall present some of our recent results obtained by an application of our mathematical methods. The word SYNERGETICS is composed of two greek words and means COOPERATION. What we study in this field is the cooperation of individual parts of a system so that a self-organized formation of spatial, temporal, or functional structures on macroscopic scales becomes possible. In particular we shall ask whether there are general principles which govern self-organization irrespective of the nature of the individual subsystems which may be electrons, molecules, photons, biological cells, or animals. Or, to use an idea expressed by Danilov and Kadomtsev [3], synergetics can be considered as a search for universal mathematical models (of self-organization). In particular, we wish to develop an operational approach in the sense of general systems theory. Such kind of approach has been persued in the Soviet Union by Lyapunov, Mandelstam, Andronov, Vitt, Chaikin, and many others.

2. OUTLINE OF THE GENERAL APPROACH

Let me take an example from physics. We may describe the behavior of a fluid at three different levels. At the microscopic level we deal with the motion of individual atoms or molecules. At the mesoscopic level we lump many molecules together into droplets so that we may speak of densities, temperature etc., but so that at this level no macroscopic structure is visible. At the macroscopic level we deal with the formation of structures e.g. rolls, hexagons etc. While e.g. in laser physics we directly proceed from the microscopic to the macroscopic level [4], in this lecture we shall adopt the following attitude. We assume that the transition from the microscopic to the mesoscopic level has been achieved by statistical mechanics or that adequate equations have been formulated at the mesoscopic level in a more or less phenomenological manner. An example is provided by the Navier Stokes equations, or by rate equations for chemical reactions. We then wish to study the evolution of patterns at the macroscopic level.

The state of the system is described by a set of variables $q_1 \ldots q_n$ which we lump together into a state vector q. Because in general the processes depend on space and time, q is a function of x and t also. The following list gives a number of interpretations of the various components of q

numbers or densities of atoms or molecules	fluids, solidification chemical reactions
velocity fields	flames, lasers, plasmas
electromagnetic fields	electronic devices
electrons	solid state
firing rates of neurons	neural nets
numbers of specific cells	morphogenesis
monetary flows etc.	economy
numbers of animals	ecology

The processes may take place in various geometries e.g.in the plane, in threedimensional space, but also on a sphere. For instance pattern formation on spherical shells in biology have been studied by Velarde [5] or pattern formation in the atmosphere of planets by Busse and others. Also one may think of more complicated manifolds or even evolving manifolds. The concept of approach of synergetics rests on a number of paradigms, to use a word en vogue, namely

a) evolution equations
b) instability
c) slaving
d) order parameters
e) formation of structures
f) instability hierarchies

3. A BRIEF OUTLINE OF THE MATHEMATICAL APPROACH

a) Evolution equations

These equations deal with the temporal evolution of q, i.e. we have to study $\dot{q} = N(q)$. The r.h.s. is a nonlinear function of the components q_j, e.g. q_1^2, q_1q_2 etc. The systems under consideration are dissipative i.e. they contain equations of the form

$$\dot{q}_1 = -\gamma q_1 + \ldots \tag{3.1}$$

They may contain transport terms describing

$$\begin{array}{ll} \text{convection:} & v\nabla v, \quad v\text{:velocity} \\ \text{diffusion:} & \Delta \\ \text{waves:} & \Delta \end{array} \tag{3.2}$$

The systems are controlled from the outside, e.g. by changing the energy input. This control is described by control parameters, e.g. by α in the eq.

$$\dot{q} = (\alpha - \gamma)q + \ldots \tag{3.3}$$

Finally, close to transition points of nonequilibrium phase transitions fluctuations play a decisive role. These fluctuations stem from fluctuating forces which represent the action of the microscopic "underworld" on the physical quantities q of the mesoscopic level. Lumping all the different terms together, we are led to consider coupled nonlinear stochastic partial differential equations of the type

$$dq(x,t) = N(q,\nabla,x,\alpha,t)dt + dF \tag{3.4}$$

where we may use the Stratonovich calculus. Without fluctuations the equations reduce to

$$\dot{q} = N(q, \nabla, x, \alpha, t) \tag{3.5}$$

A special case treated in chemistry has the form

$$\dot{q} = R(q) + D\nabla^2 q \tag{3.6}$$

where the first term R describes the reactions whereas the second describes diffusion processes. For sake of completeness we mention that as long as we deal with Markov processes we may also invoke other types of equations, e.g. the Chapman-Kolmogorov equation. Finally we mention that the methods we shall present below, including the slaving principle, possess a quantum mechanical analogue, where the evolution equations are replaced by Heisenberg´s operator equations which contain damping terms and fluctuating operator forces.

b. Instability

We assume that we have found a solution of the nonlinear equations for given control parameters $\alpha=\alpha_o$. In practical cases such a solution may describe, for instance, a quiescent and homogeneous state, but our treatment may also include spatially inhomogeneous and oscillatory states. We denote the corresponding solution by q_o. When we change the control parameter that solution q_o may loose its stability. To study the stability (or instability) we put

$$q(x,t,\alpha) = q_o(x,t,\alpha) + w(x,t,\alpha) \tag{3.7}$$

and insert it into (3.5). Assuming that w is a small quantity we may linearize (3.5) and study the resulting equations of the form

$$\dot{w} = L(q_o(x,t), \nabla, x, \alpha)w, \quad w = w(t) \tag{3.8}$$

If L is independent of t or depends on t periodically, or in a large class of systems in a quasiperiodic fashion, the solutions can be written in the form

$$w^{(j)}(t) = \exp(\lambda_j t) v^{(j)}(t) \tag{3.9}$$

where v(t) is bounded. Thus the global behavior of w is determined by the exponential function in (3.9). We call those solutions, whose real part of λ is positive, unstable, and those whose real part of λ is negative, stable. In order to solve the nonlinear equation (3.5) (or, more generally, its stochastic counterpart (3.4)) we make the hypothesis

$$q(x,t) = q_o(x,t,\Phi(t)) + \sum_j u_j(t) v^{(j)}(x,t,\Phi(t)) + \sum_k s_k(t) v^{(k)}(x,t,\Phi(t)) \tag{3.10*}$$

where Φ is a set of certain phase angles in case we deal with quasiperiodic motion. For details I refer the reader to my book ADVANCED SYNERGETICS. Here it may suffice to note that by inserting

* j and k run over the unstable and stable mode indices, respectively.

the hypothesis (3.10) into our original nonlinear equations (3.5) we find after some mathematical manipulations the following equations

$$\dot{u}_j = \lambda_j u_j + N_j^{(u)}(u,\Phi,t,s), \tag{3.11}$$

$$\dot{s}_k = \lambda_k s_k + N_k^{(s)}(u,\Phi,t,s), \tag{3.12}$$

$$\dot{\Phi}_l = N_l^{(\Phi)}(u,\Phi,t,s). \tag{3.13}$$

Similarly, starting from (3.4) we obtain stochastic equations for $u,s,\ \Phi$. Though in general one may not expect to simplify a problem by means of a trans-formation, the new equations (3.11)-(3.13) can be considerably simplified when a system is close to instability points, where the real parts of some λ´s change their sign from negative to positive.

c) The slaving principle

For the situations just mentioned we have derived the slaving principle for stochastic differential equations and discrete noisy maps. The slaving principle states that we may express the amplitudes s of the damped modes by means of u and Φ at the same time, so that

$$s = f(u,\Phi,t) \tag{3.14}$$

We shall call u and Φ order parameters. We have studied numerous cases of dissipative systems and have found that in practically all of them there occur only few order parameters while there are still very many slaved modes. As a consequence we achieve an enormous reduction of the degrees of freedom because we may express all damped modes s by the order parameters. In this way we obtain a closed set of equations of the form

$$\dot{u} = N(u,\Phi,t), \tag{3.15}$$

$$\dot{\Phi} = N'(u,\Phi,t). \tag{3.16}$$

Some applications of these equations will be discussed below in section 5.

4. GENERALIZED GINZBURG-LANDAU EQUATIONS

When the dimensions of continuously extended systems are large compared to the fundamental length of developing patterns, the spectrum λ is practically continuous. In such a case particular mathematical difficulties arise because it is no more possible to distinguish clearly between undamped and damped modes. A way out of this difficulty can be found when we resort to the formation of wave packets. This in turn necessitates that the order parameters, which we shall call ξ, depend not only on time but now also on space (in a slowly varying fashion). Therefore our hypothesis reads

$$q(x,t) = q_0 + \sum_{k_c} \xi_{k_c}(x,t) v_{k_c}(x) + \sum \text{slaved modes} \tag{4.1}$$

where k_c runs over a discrete set of critical wave vectors at which the instabilities occur. For simplicity let us again consider a case in which no phase angles occur and let us furtheron be satisfied with an expansion of the nonlinear terms up to third order. The order parameter equations then acquire the form

$$\xi_{k_c}(x,t) = \lambda_{k_c}(\nabla)\xi_{k_c}(x,t) + \sum_{k_1,k_2} A\ldots\ \xi_{k_1}\xi_{k_2} + \sum_{k_1,k_2,k_3} B\ldots\ \xi_{k_1}\xi_{k_2}\xi_{k_3} + F_{k_c}. \tag{4.2}$$

I have called these equations, which I derived some time ago "Generalized Ginzburg-Landau-equations", because they are strongly reminescent of the famous Ginzburg-Landau-equations. But two important distinctions should be noted. While the original Ginzburg-Landau-equations refer to a system in thermal equilibrium my Generalized Ginzburg-Landau-equations refer to systems far from thermal equilibrium. Furthermore the original Ginzburg-Landau-equations were derived in a heuristic fashion, whereas here the Generalized Ginzburg-Landau-equations have been derived rigorously. Because of the double and triple sums these equations are quite clumsy. However, under well justified assumptions these equations can be simplified as I have shown recently. To this end I define a new function

$$\Psi(x,t) = \sum_{k_c} e^{ik_c x}\xi_{k_c}(x,t). \tag{4.3}$$

After a few elementary manipulations and under specific assumptions on λ, A and B eq.(4.2) can be cast into the form

$$\dot{\Psi}(x,t) = \left(a + b(k_0^2 - \nabla^2)^2\right)\Psi + a\Psi^2 + B\Psi^3 + F, \tag{4.4}$$

where I have chosen an explicit example for $\lambda(k)$ which refers to the eigenvalues of the convection instability. We have solved this equation on a computer to study the temporal evolution of patterns. A typical result is shown in Fig. 1.

5. SOME FURTHER APPLICATIONS

By means of the mathematical methods we have outlined above my coworkers and I have treated a number of explicit cases over the recent years. I present a few of them in order to demonstrate the applicability of the mathematical method I have briefly sketched in the beginning of my lectures.

a) Pattern formation of an MHD plasma

which is heated from below and is subjected to a vertical constant magnetic field. The boundary conditions in the horizontal directions are chosen periodic. Several patterns could be found. In the single mode case rolls appear, well known from fluid dynamics. However, also two or three mode cases are possible. A typical velocity distribution is shown in Fig. 2.

b) Running waves in the positive column of a gas discharge in neon

By means of a nonlinear treatment it has been possible to derive the corresponding spatio-temporal pattern in good qualitative and semiqualitative agreement with experiments.

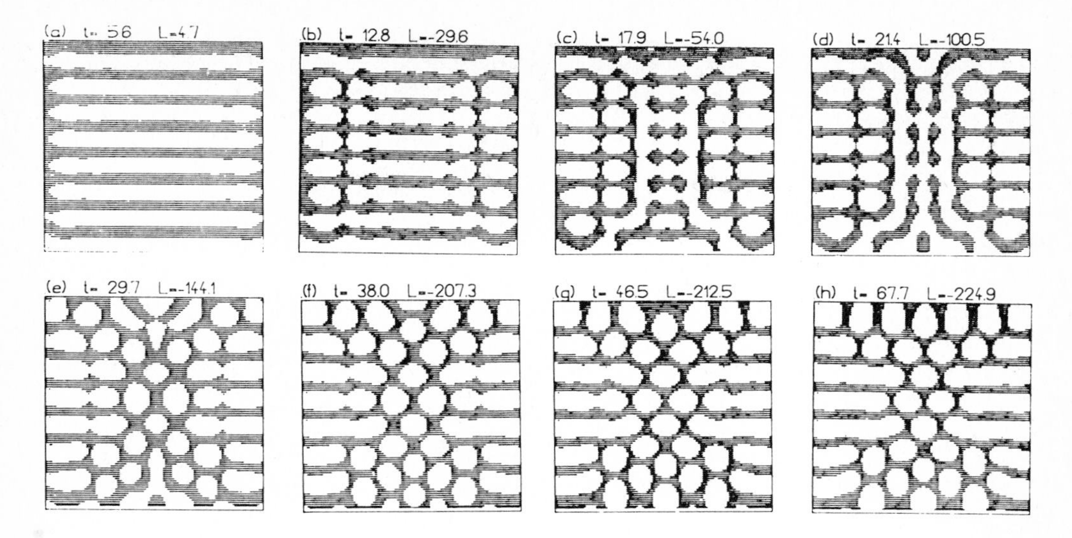

▲Fig.1. A roll pattern is prescribed but the parameter values of the equation are chosen such that hexagons should be formed. The sequence a - h shows the formation of hexagons but the final state is reached only at infinitely large time (critical slowing down)

Fig.2. Lines of constant vertical velocity in a plasma heated from below and subject to vertical constant magnetic field

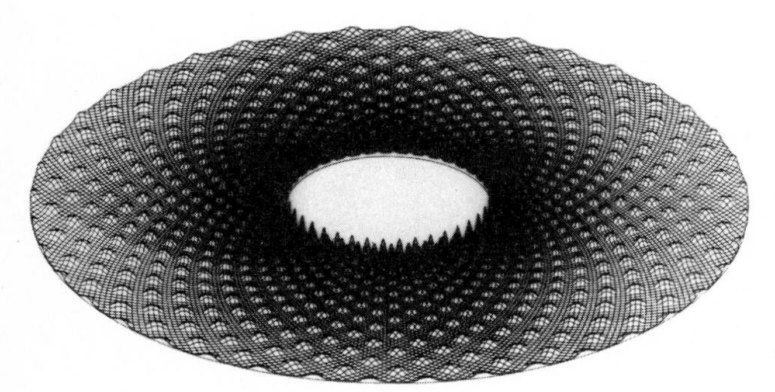

Fig.3. Embossed map on model calculation on fructification of sunflower

c) Prepattern formation of the spiral wave pattern of a sunflower head

We have adopted reaction diffusion equations of the form (cf.3.6) and have used the Gierer-Meinhardt-model for the reaction terms. Since the specific form of the reaction terms is not so important we don´t discuss them here. However, it was assumed that the diffusion is space dependent. Results of the nonlinear analysis are shown in Fig.3. They clearly exhibit two counter rotating sets of spirals in good agreement with the observed fructification of the sunflower head.

6. OUTLOOK

By means of the systematic approach of synergetics it has been possible to classify a number of spatial and temporal patterns which occur over and over again. In addition it has become possible to study the dynamics close to transition points in detail because the dynamics is governed by few order parameters only. A few words of future problems may be in order and I will list only three of them:

1) So far we have assumed that we start from a spatially homogeneous state. The whole approach works also if the original state is spatially inhomogeneous but time independent or time periodic. However, in order to solve the linearized equations and to derive the order parameter equations in most cases computer calculations may be needed.

2) When we go away from instability points, the patterns remain qualitatively the same as is known from numerous experiments. However, as it seems to me a rigorous theory far away from instability points is still lacking.

3) A rich field of further study is provided by chaos and it seems to me that we are just at the beginning of classifying and understanding chaotic motion.

In conclusion it might be worth pointing out that in the field of synergetics we need not only a further development of mathematical methods but also the corresponding experiments must be performed and a close interaction between experimentalists and theoreticians is needed. We believe that the approaches so far have not only given us fundamental insights into the way new patterns evolve at instability points but have also led to a number of practical applications by exploiting analogies between different systems. These analogies become apparent through the order parameter equations. I am sure that this will lead to a development of new devices, especially in solid state physics and quantum electronics.

References:

1. H. Haken, "Synergetics. An Introduction. Nonequilibrium Phase Transitions and Self-Organization in Physics, Chemistry and Biology". Third Revised and Enlarged Edition, Springer Verlag, Berlin, Heidelberg, New York, Tokyo 1983
2. H. Haken, "Advanced Synergetics. Instability Hierarchies of Self-Organizing Systems and Devices", Springer Verlag, Berlin, Heidelberg, New York, Tokyo 1983
3. Yu. A. Danilov, B.B. Kadomtsev, in "Nonlinear Waves, Self-Organization" edited by A.V. Gaponov-Grechov, M.I. Rabinovich, Moscow 1983
4. H. Haken "Laser Theory" Encyclopedia of Physics Vol.XXV/2c, Springer Verlag Berlin, Heidelberg, New York 1970, reprinted 1983
5. J.L. Ibanez, M.G. Velarde, J.Non-Equilib.Thermodyn. Vol.3. 63 (1978)
   Ch. Berding and H. Haken, J. Math.Biolog.14, 133 (1982)

Autowaves: Results, Problems, Outlooks

V.I. Krinsky

Institute of Biological Physics, USSR Academy of Sciences
SU-142292 Pushchino, USSR

1. The autowaves [1-3] form a type of waves which are characteristic of strongly nonlinear active media. To here belong self-sustained signals inducing local release of stored energy in an active medium, which is spent to trigger the same process in adjacent regions. The examples are provided by waves of combustion, of phase transitions, concentrational waves in chemical reactions, and also by many biological autowave processes (propagation of nerve impulses, excitation waves in heart muscle, epidemic waves in ecological communities, spreading waves in the cerebral cortex). These examples stress the importance of autowave phenomena.

Autowaves differ fundamentally from waves in traditional (conservative) media (Table). They propagate at the expense of energy taken from an active medium, and, therefore, cannot be considered as conservative systems. The shape and amplitude of autowaves remain constant during propagation, whereas the amplitude of classical waves rapidly falls with the distance and the waveform is distorted by dispersion. In the case of autowaves, there is no reflection from either the medium boundaries or inhomogeneities. Unlike waves in conservative media (solitons and soliton-like solutions) two colliding autowaves annihilate rather than penetrate one another, and, therefore, no interference takes place. It is inferred from the Table that the only property the two types of waves share is diffraction.

Table Properties of Waves and Autowaves

Properties	Waves	Autowaves
Conservation of energy	+	-
Conservation of amplitude and waveform	-	+
Reflection	+	-
Annihilation	-	+
Interference	+	-
Diffraction	+	+

In the simplest case, an active medium cannot return to the same state after the propagation of a wave - it becomes inexcitable for every wave to follow (as in the instance of combustion). Therefore, only one wave can propagate through such a medium. This simplest case has been explored exhaustively [4,5]. Of much greater interest, however, are the so called re-excitable active media, which can recover their initial state after excitation. In such media, an unlimited number of autowaves can propagate following each other. The simplest example is provided by subsequent spreading of grass-fires in a plain if considered on a long time scale (more than a year). It is obvious that the burnt grass grows again and can be reignited. Re-excitable media are characteristic of living systems (nerve fibres, heart muscle, neuron ensembles, smooth muscle cells, etc.).

2. Waves in simple active media can be described by only one equation [5]

$$\frac{\partial U}{\partial t} = f(U) + D\Delta U. \qquad (1)$$

In the case of a combustion wave, U is temperature, t time, $\dot{U}=f(U)$ a kinetic equation describing the process at a point; the term $D\Delta U$ accounts for heat diffusion from adjacent regions. For re-excitable media, a second variable (which is usually a slow one) should be added, and the resulting minimal system will have two equations. System (2) describes the behaviour of a point, and system (3) of a distributed medium

$$\left.\begin{aligned} \frac{dU}{dt} &= f(U,V) \\ \frac{dV}{dt} &= \varphi(U,V) \end{aligned}\right\} \quad (2) \qquad \left.\begin{aligned} \frac{\partial U}{\partial t} &= f(U,V) + D_U \Delta V \\ \frac{\partial V}{\partial t} &= \varphi(U,V) + D_V \Delta V \end{aligned}\right\} \quad (3)$$

For the case of a grass-fire, U is temperature and V describes the growth of the grass; for nerve impulses, U is the membrane potential of the nerve fibre and V is the current of potassium ions which resets the membrane potential; for an active chemical medium (e.g., the Belousov-Zhabotinsky (B-Z) oxidation-reduction reaction), U is the concentration of the oxidized form of the catalyst and V is the concentration of the inhibitor; for waves in heart tissue, the variables can be interpreted as in the case of nerve impulses with the only difference that the Laplacian Δ is two- or three-dimensional.

It is important to note that the laws governing the propagation of autowaves in active media of diverse nature (physical, chemical or biological) are identical. In all cases, the active medium is a two-level system. During wave propagation, the transition from a high- to low-energy level occurs and then the slow energy-pumping processes bring the medium back to the high-energy level. All re-excitable active media can be described by the same type of equations with parabolic partial derivatives (Eqs.(3)). The corresponding point system (2) has auto-oscillatory or, which is still more interesting, excitable kinetics. In the latter case, the phase plane of system (2) is of a van der Pol type (Fig.1a), U being the fast and V the slow variable. A pulse generated by the system is shown in Fig.1b,c (heart tissue).

3. The intriguing feature of active media is their ability to form local autowave sources with quite unusual properties. The autowave sources may arise in inhomogeneous active media during wave propagation [7], this results in the formation of specific structures and new types of instabilities and chaos [2,8].

Let us consider the mechanism of the initiation of the most important type of autowave sources, the reverberator. It is a rotating vortex with the waveform similar to an Archimedean spiral [9]. Reverberators naturally occur in inhomogeneous media, in which the waves can break when propagating. A wave in a two-dimensional medium is schematically shown in Fig.1d, and a wavebreak, occurring at an inhomogeneity, is shown in Fig.2. The broken wave (wave 2) propagates

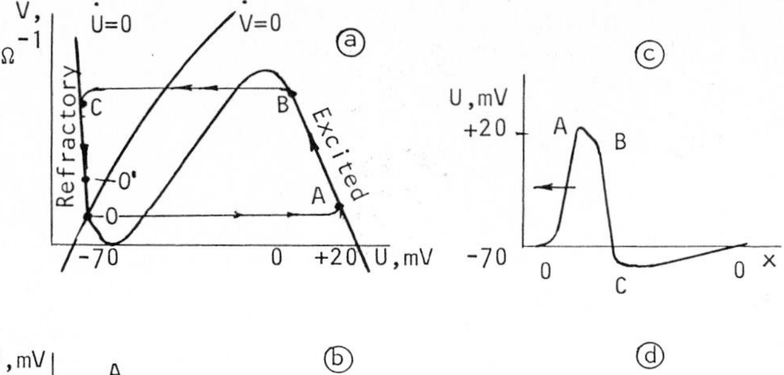

Fig. 1 Pulses and waves in system (3). a - the phase plane, b - a pulse generated by system (2). Trajectories OA and BC are the pulse fronts (fast motions), AB and CO the plateaus (slow motions), c- a pulse in a one-dimensional medium described by system (3), d - a scheme of wave propagating in a two-dimensional medium. The regions in the vicinity of AB are shaded ("excited state"), those in the vicinity of CO are dashed ("refractory state") and those in the vicinity of OO^1 ("resting state") are blank

more slowly than the preceding one, and the wavebreak comes to the region in which the recovery has already been completed, and, therefore, the excitation wave can propagate through it. From Huyghens' construction of the consecutive wavefront positions it is seen that the wavebreak begins to curl into a spiral (Fig.3) [10].

Fig.4 shows the reverberator in different active media: chemically active medium, culture of social amoeba *Dictyostelium discoideum*, retina, heart tissue, and in computational experiments based on Eqs.(3).

Along with rotating vortices, other types of autowave sources are also possible, in particular, various types of leading centres which emit concentric waves. If in an active medium several autowave sources exist simultaneously, the waves emitted by them annihilate when colliding (see Table), and the source with the highest frequency suppresses all others [7]. Among known local autowave sources the reverberator has the highest frequency and therefore synchronizes the whole medium by overwhelming the sources of concentric waves (Fig.5).

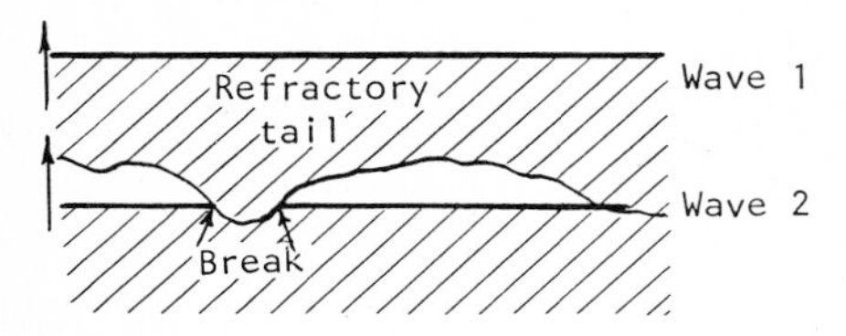

Fig. 2 Occurrence of a wavebreak in an inhomogeneous medium. Two waves propagating upwards are shown. The length of the refractory tail (dashed) of wave 1 is somewhat longer in the middle of the picture than it is at its edges. Wave 2 cannot propagate through the dashed region, this results in a wavebreak.

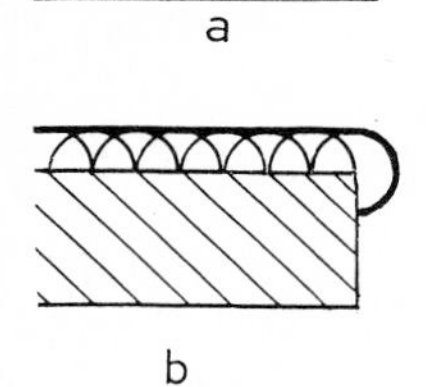

Fig. 3 Development of the wavebreak into a spiral wave. The thin line shows the initial position of the wave. The wave position after a short time interval dt (solid line) is constructed, according to Huyghens' principle, as an envelope of circumferences of radius V.dt, where V is the wave velocity

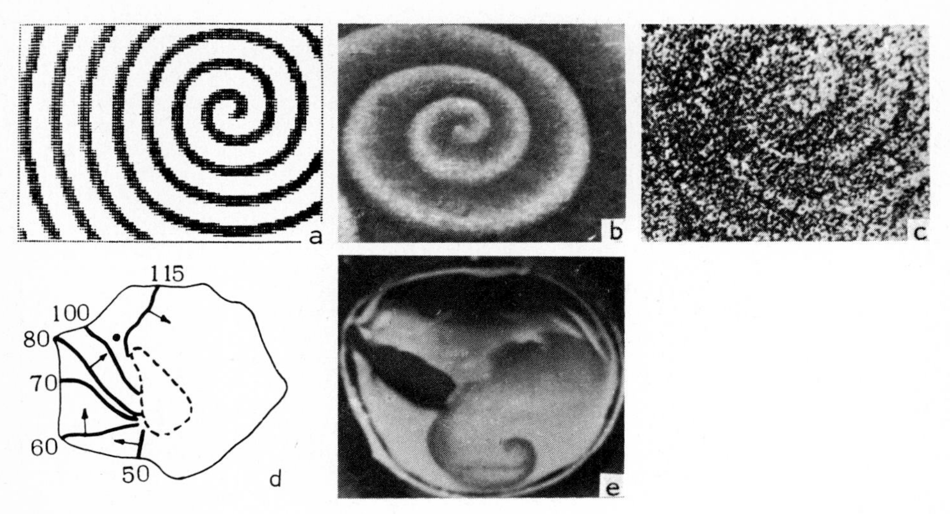

Fig.4 Rotating spiral waves (reverberators) in different active media; a - active medium given by system (3) (numerical calculations [17]); b - chemical active medium [11]; c - culture of social amoeba *Dictyostelium discoideum*[12]; d - rabbit atrium [13]; e - retina [14]

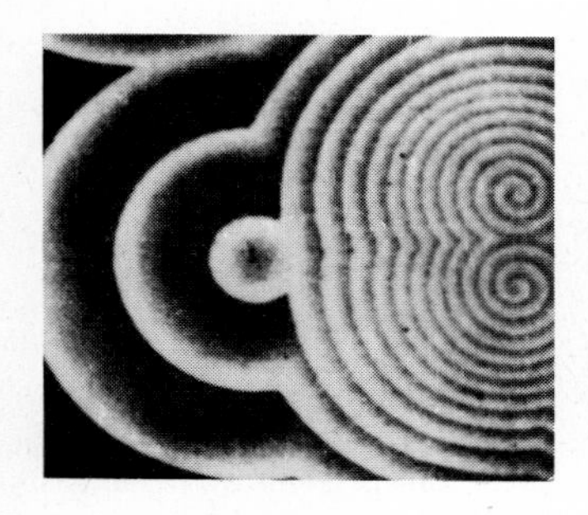

Fig.5 The reverberator supresses a source of concentric waves in a chemical active medium [2]

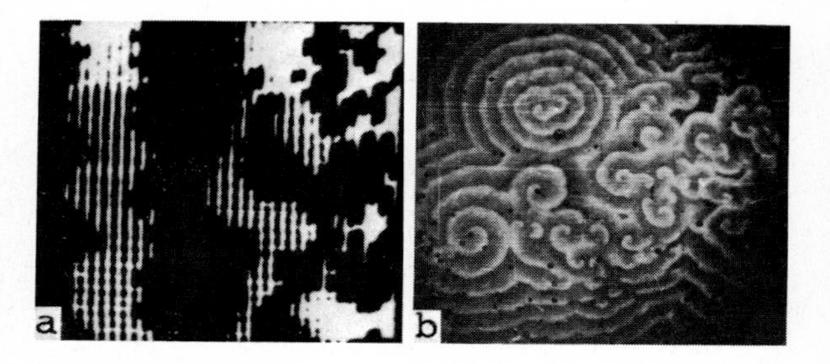

Fig.6 Wave pattern at the onset of the development of chaos in active media; a - computer calculations (inhomogeneous case) based on a Wiener type model [3]; b - chemical active medium [2]

The most important property of rotating vortices is their ability for reproduction [7,8]. In an inhomogeneous medium, the waves emitted by a rotating reverberator can produce new reverberators when breaking at inhomogeneities, and so on. The chain reaction of the reproduction of reverberators leads to a decrease in the characteristic scales of the wave pattern, and the whole medium turns out to be filled with fragments of rotating vortices. The resulting chaotic regime resembles a small-scale turbulence (Fig.6), though the underlying mechanisms differ significantly (Fig.7).

4. The above properties of vortices in active media (their ability to occur at inhomogeneities during wave propagation, to synchronize other wave sources and to reproduce themselves) give an insight into

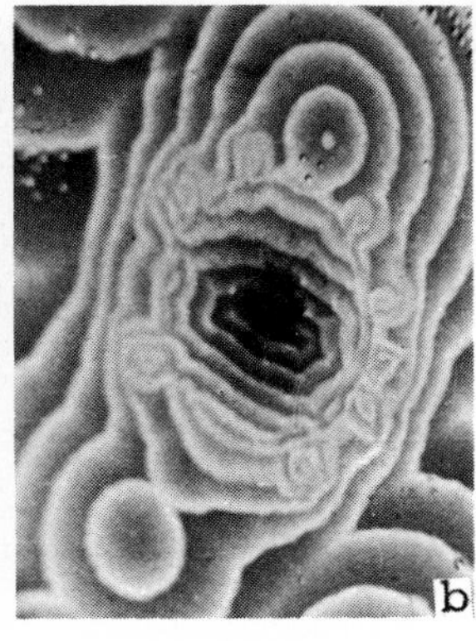

Fig.7 Reproduction of reverberators at an inhomogeneity (chemical active medium) [2]

the fundamental mechanisms of a number of serious heart disorders [8,18]. A vortex-like wave source (reverberator) occurring in heart muscle suppresses the normal heart pacemaker (SA-node) with a resulting abrupt increase in the cardiac rate and dramatic impairment of hemodynamics. This is just the case with paroxysmal tachycardia (severe cardiac arrhythmia), as shown in experiments on animals [13].

The chain reaction of reverberator production by inhomogeneities in heart tissue is the cause of another serious heart disease, ventricular fibrillation. During fibrillation, the myocardial cells contract asynchronously and the living pump turns out to be a chaotically pulsing muscle bag uncapable of pumping blood. This is the dramatic way a turbulent autowave regime manifests itself in the heart.

5. Note that the present-day knowledge of autowave sources and their properties originates from attempts to understand the mechanisms of cardiac muscle malfunction. The pioneering mathematical work in the field is due to N. Wiener and A. Rosenblueth [9]. The active media were further investigated by members of I.M.Gelfand's seminar in Moscow in 1960-1966, who described the properties and the fundamental mechanisms of the generation of autowave sources [7,10] using formalized Wiener type models. These theoretical studies together with experimental investigations of the oscillatory chemical reaction [6] discovered by Belousov in 1951 (see [1], p.176) resulted in the creation of a chemically active medium by Zaikin and Zhabotinsky [19] and Winfree [20].

6. The chemically active medium is a thin (1 mm) layer of the solution in which the Belousov periodic oxidation-reduction reaction runs. In this reaction, about 1/100 of the substrate present is oxidized in one period. Therefore, the oxidation wave can propagate through the medium about 100 times.

The chemical active medium is the first known case of re-excitable media. It turned out to be very convenient for both demonstration and investigation of autowaves: oxidation wave here is clearly visible (blue against the bright red background), the system is simple and easy to operate. All the theoretically predicted properties of autowaves and their sources (Table) could be demonstrated using this medium.

However, biologically important active media suitable for studying the autowave phenomena, have not been found until later (see Fig.4).

The photos of spiral rotating waves organizing the morphogenetic process in social amoeba *Dictyostelium discoideum* or the maps of rotating vortices in rabbit atrium during the experimentally induced paroxysmal tachycardia are now widely known and travel from book to book. Here may be added the recent splendid result of the Czechoslovak electrophysiologist Bureš who has demonstrated, together with his colleagues from Moscow, the initiation of a reverberator in the retina in experiments with spreading depression [14]. The fact that the wave of spreading depression can produce vortices may have direct relations to the mechanisms of epilepsy.

7. One of the major problems in this new and rapidly expanding branch of science is the search for new autowave regimes. For practical purposes it is very important to find effective methods to control the autowave sources (to prevent their spontaneous generation, to lower the rate of their reproduction, to regulate their position in active medium and their interaction with each other, etc. (see Fig.5)).

To solve these problems, the following approaches may be used: 1. Analytical investigation of the solutions of simplified versions of Eqs.(3); 2.Computer simulation; 3. Experimental modelling of autowave media.

For a relatively simple case of a weakly nonlinear medium there are well-developed analytical methods, which give quasiharmonic solutions [21-27]. However, the major problems are related to strongly nonlinear media [7,8]. Note the problem of wave sources in three-dimensional active media, where the centre of a vortex is, in contrast to two-dimensional cases, a filament rather than a point. The topology of filament arrangement has been studied in [28]. The basic experimental findings by Winfree on vortex evolution in 3-dimensions [20] can be explained by the simple hypothesis that the filament is resilient. It is noteworthy that elasticity is well-established for vortices in conservative media (He^4, superconductors, etc.), as a consequence of the minimum energy principle. This principle does not hold for active media where energy is not conserved. However, the computational experiments show that a toroidal vortex in relaxation systems contracts because of "springiness" of the filament and, in addition, it drifts along the symmetry axis [29]. The effect can be studied analytically in the case of weakly nonlinear systems. In particular, the evolution of the radius R of a toroidal vortex is described by the equation [30]

$$dR/dt \sim -D/R$$

(for a λ-ω system without dispersion), from which follows that the compression velocity of the filament is proportional to the vortex curvature. Similar problems for relaxation systems remain as yet unsolved.

8. An important class of problems arises from studying the dependence of the characteristics of an autowave source on the medium parameters. Fig.8 demonstrates experimental means to change the characteristic size of reverberators. It can be seen that the wavelength and the size of the reverberator core may be increased by almost one order of magnitude by variation of the parameters responsible for the decrease in the excitability (increase in the threshold) of the active medium.

Fig.9 presents data on the reverberator period which were obtained for three types of active media: in numerical experiments using the Fitz-Hugh - Nagumo model, in heart tissue and chemical active medium.

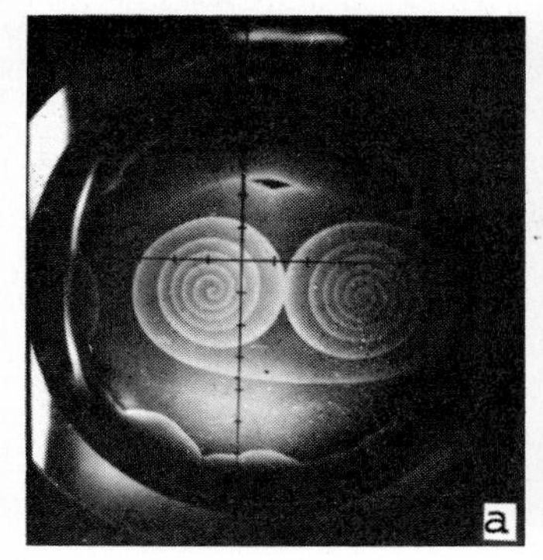

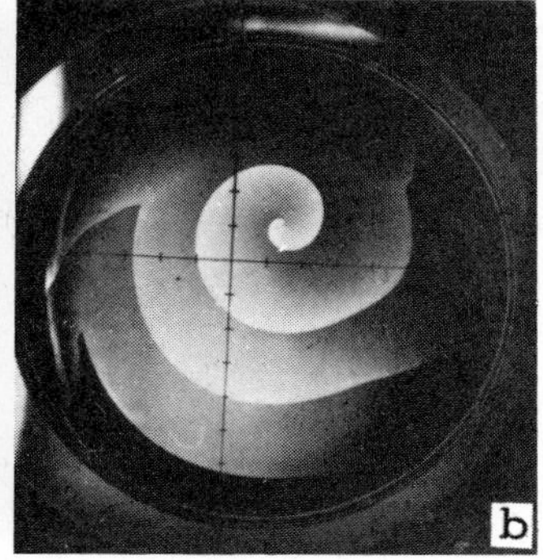

Fig.8 Increase in the reverberator wavelength after the suppression of excitability of a chemical medium. a - $[H_2SO_4]$= 0.7 M; b - $[H_2SO_4]$ = 0.15 M [15]

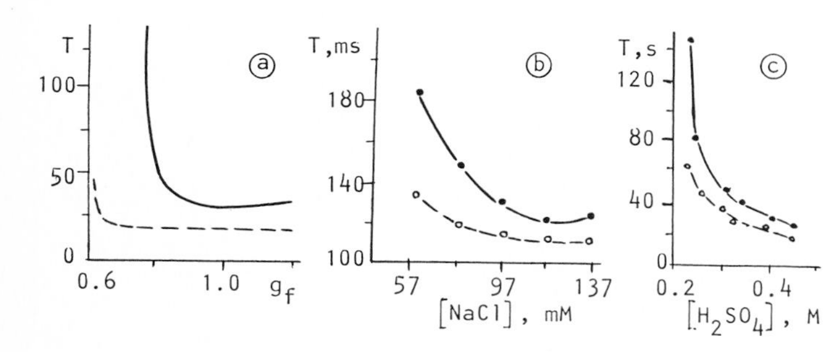

Fig. 9 The dependence of the reverberator period on the excitability of an active medium. a - the Fitz-Hugh - Nagumo model [17]; b - the chemical active medium [15]; c -the heart [16]

As each medium has its own adjustable parameter (the x-axis), these were chosen to represent the same integral characteristics of the medium, its excitability. The excitability decreases when the parameters are changed in the direction indicated by the arrow. It is seen that in all the cases these changes are qualitatively the same: the period increases as the excitability decreases.

Note that the study of the autowave source characteristics (in particular, the change in the period under the action of various physiological drugs) may be useful in the elaboration of methods for differential diagnostics of cardiac arrhythmia [16].

9. During recent years, a new autowave regime has been discovered in two-dimensional active media, the so called anomalous reverberator [17]. The normal reverberator is known to have a core, the region where induced oscillations take place. Their amplitude goes down smoothly from unity (the amplitude of autowaves) to zero, the zero value can be recorded at only one point, which is the centre of the core. For the anomalous reverberator the oscillation amplitude is zero in the entire core region, which is rather large (Fig.10c). In the vicinity of the central (quiescent) region, the wave curvature is critical, and therefore the wave cannot penetrate into the region. However, the medium remains excitable in this region (unlike the normal reverberator core). This can be readily seen if an excitation is applied from outside during the rotation of the revernerator. In this case a propagating wave occurs which involves the entire central region, though the reverberator failed to penetrate into it. The anomalous reverberator can be initiated by changing any of the para-

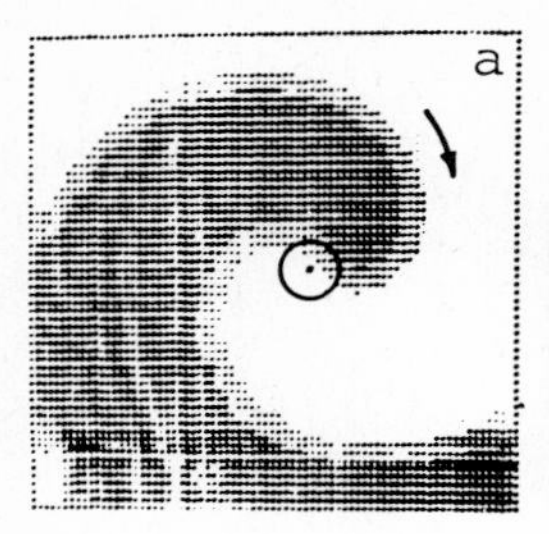

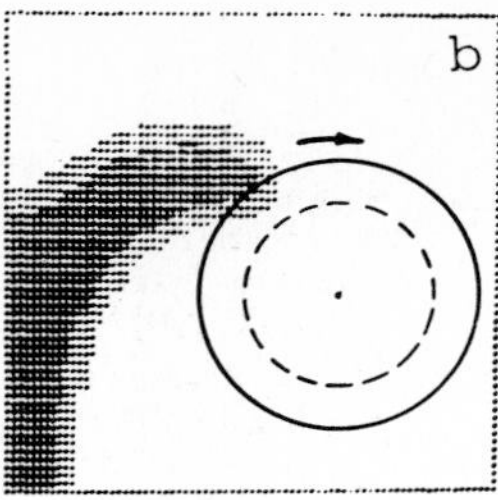

Fig. 10 Normal (a) and anomalous (b) reverberators (schematic). The successive wavefront positions are shown. In b, dashed is the inexcitable region [17]

meters leading towards a decrease in the excitability of the medium (increase in the threshold).

10. As already mentioned, analytical investigations of reverberators in strongly nonlinear (relaxatory) media present a serious problem. However, for media in which the size of the reverberator core exceeds the diffusion length, an approximate theory was constructed [31] using Greenberg's transformations [32]. This theory explains the dependence of the basic characteristics of a reverberator (period, core size) on the medium parameters and the transformation of the normal reverberator to the anomalous.

In what follows some autowave regimes will be demonstrated, which are still a challenging problem for theoreticians. In Fig.11 is shown a cinegram of a rotating vortex in a chemically active medium with the topological charge N=3 [33]. It is remarkable that these multi-armed vortices proved to be stable and observable for more than half an hour, very much unlike the multi-armed species in conservative media (He^4, superconductors), which are always unstable and degrade to simple vortices. The stability of multi-armed vortices has been analysed theoretically, but only for the simple and unrealistic case of a weakly nonlinear active medium without dispersion (in such media vortices are no more than rotating straight lines). Note also that the now-familiar solutions of reaction-diffusion equations for the vortex with a topological charge N suggest rigid rotation of the vortex as a whole

$$U = F(N\theta - \omega t, r) \tag{4}$$

(θ and r are polar coordinates). However, the experiments reveal a much more complicated wave pattern (see Fig.11) than described by

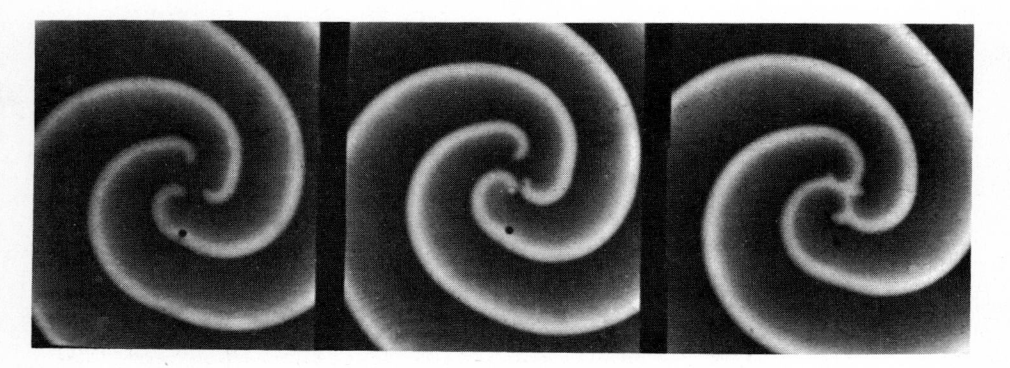

Fig. 11 Reverberator with the topological charge N=3 in a chemical active medium [33]

Eq. (4): the waves in the vortex core are intermittently connected to each other (in a, connected are 1st and 2nd, in b, 2nd and 3rd, in c all the three waves, etc.).

An interesting phenomenon - the induced drift of a vortex - has been discovered in experiments with chemically active media [34]. It has been shown that if a vortex is exposed to external waves with a frequency (f_e) higher than its own rotational frequency (f_r), it drifts away slowly from the applied wave source. The velocity of drifting, described by the empirical formula

$$V_d \approx (1 - f_r/f_e) \qquad (5)$$

is small as compared to the velocity V of wave propagation in the medium and increases with f_e. By induced drift the position of the vortex in the medium can be regulated, which opens new ways to attack the problem of cardiac arrhythmias. Note that no precise theory (based on the solutions of equations (3)) for this effect is available at present.

12. As pointed out, the most intriguing autowave effects are associated with the occurrence in re-excitable media of autowave sources (in particular, vortices) and their interaction with each other. The propagation of autowaves in the simplest active media, in which only one wave can propagate, has been studied already for a long time in connection with the problems of spreading of flames or biological populations [4,5]. Also in these studies new trends are now followed, which may lead to surprising results and important practical applications.

A curious example of autowaves in a chemical system has been provided by Goldansky and co-workers [35], who described chemical reactions proceeding at extremely low temperatures with rates significantly higher than predicted by Arrhenius' law. The mechanism of such superhigh reaction rates proved to be linked with the generation of autowaves. A chemical reaction in the solid phase initiates local cracking of the specimen and at the new-formed surface of the crack (in nonequilibrium conditions !) a reaction can occur, which would be impossible at equilibrium. The liberated energy provides further growth of the crack and so on. The reaction can take place because the chemical and mechanical processes proceed tightly coupled as autowave (at a rate of about 10 cm/s). Neither of the processes can run separately.

The next example concerns a boiling liquid on heat-generating elements. The transition from the nucleate to the film boiling was also found to have the characteristics of autowaves. This allows the autowave theory to be used to control the rate and direction of this autowave process and thus to gain more safety as well as quality of heat-generating elements [36].

Another practically important example is the autowave transition of a catalyst from inactive into active state, which was studied for the case of platinum catalysts [37]. Based upon these studies, an approach is being developed to reduce catalyst expenditures, which is of particular interest in the case of expensive catalysts.

13. In conclusion, I would like to note that the autowave mechanisms of structure formation and evolution and of chaos development are only a small part of what is presently being studied as nonequilibrium, self-organized, synergistic systems [38-45]. As the above examples show, it is however a very important part of this rapidly advancing branch of science with its own problems on the frontiers of the theories for oscillations, strongly nonlinear waves and structure formation.

References

1. Autowave Processes in Systems with Diffusion (ed. M.T.Grekhova), Gorky, Acad. Sci. USSR (1981)
2. G.R.Ivanitsky, V.I.Krinsky, A.N.Zaikin, A.M.Zhabotinsky, Sov. Sci. Rev. D2, 280 (198o)
3. G.R.Ivanitsky, V.I.Krinsky, E.E.Sel'kov. Mathematical Biophysics of the Cell, M., "Nauka" Publ. (1978) chapters 6-8
4. A.N.Kolmogorov, I.E.Petrovsky, N.S.Piskunov. Vestnik Moscovskogo Univ., Ser. 1, Mathematika, Mekhanika, 1, 1 (1937)
5. Ya.B.Zeldovich, G.I.Barenblat, V.B.Librovich, G.M.Makhviladze. Mathematical Theory of Combustion and Explosion (in Russian), M., "Nauka" Publ. (1980)
6. A.M.Zhabotinsky, Biophizika, 9, 306 (1964)
7. V.I.Krinsky, Biofizika, 11, 676 (1966)
8. V.I.Krinsky. Problemy kibernetiki, 20, 59 (1968)
9. N.Wiener, A.Rosenblueth. Arch. Inst. Cardiologia de Mex., 16, 205 (1946)
10. N.S.Balakhovsky, Biofizika, 10, 1063 (1965)
11. A.M.Zhabotinsky, A.N.Zaikin, in: Oscillatory Processes in Biological and Chemical Systems, 2, Pushchino, Acad. Sci. USSR, (1971) p. 279
12. A.J.Durston. Dev. Biol, 37, 221 (1974)
13. M.Allessie, F.Bonke, S.Shopman. Circ. Res. 33, 54 (1973)
14. J.Bures, N.A.Goralova, J. Neurobiol, 14, 353 (1983)
15. K.I.Agladze. Preprint, Pushchino (1983)
16. A.M.Pertsov, A.K.Grenadier, this volume
17. A.M.Pertsov, A.V.Panfilov., in: Autowave Processes in Systems with Diffusion, Gorky, Acad. Sci. USSR, (1981) p.77
18. V.I.Krinsky, in: Internat. Encycl. of Pharm. and Therapeutics, Pergamon Press, London, p.105 (1981)
19. A.N.Zaikin, A.M.Zhabotinsky, Nature, 225, 535 (1970)
20. A.T.Winfree, Science, 181, 937 (1973)
21. D.C.Cohen, T.C.Neu, R.R.Rosales, SIAM J. Appl. Math. 35, 536 (1978)
22. Koppel M., Howard L.N., Stud.Appl. Math. 52, 291 (1973); 56, 95 (1977)
23. Y.Kuramoto, T.Tsuzuki, Progr. Theor. Phys. 55, 356 (1976)
24. T.Erneux, Herschkowitz-Kaufman. J. Chem. Phys. 66, 248 (1977)
25. A.C.Mikhailov, I.V.Uporov. Doklady Akad. Nauk SSSR, 249, 733 (1979)
26. Ya.B.Zeldovich, B.A.Malomed. Doklady Akad. Nauk SSSR, 254, 93 (1980)
27. V.I.Krindky, B.A.Malomed. Physica 9D, 81 (1983)
28. A.T.Winfree, S.H.Strogatz. Physica 8D, 35 (1983); 65 (1983); Physica 9D, 333 (1983)
29. A.V.Panfilov, A.M.Pertsov. Doklady Acad. Nauk SSSR, N 6 (1984)
30. L.V.Yakushevitz. Studia Biophysica (1984) (in press)
31. A.S.Mikhailov, V.I.Krinsky. Physica 9D, 346 (1983)
32. J.M.Greenberg. SIAM J. Appl. Math. 30, 199 (1976)
33. K.I.Agladze, V.I.Krinsky. Nature 296, 424 (1982); 308, 834 (1984)
34. V.I.Krinsky, K.I.Agladze. Physica 8D, 50 (1983)
35. A.M.Zanin, D.P.Kiryukhin, I.M.Barkalov, V.V.Barelko, V.I.Goldansky Doklady Akad. Nauk SSSR 260, N 6, 1397 (1981)
36. S.A.Zhukov, V.V.Barelko, L.F.Bokova in: Autowave Processes in Systems with Diffusion. Gorky, Acad. Sci. USSR, (1981), p.149
37. V.V.Barelko. Problems of Cybernetics and Catalysis, 18 (1981)
38. A.M.Zhabotinsky. Concentrational Auto-Oscillations (in Russian) M. "Nauka" Publ. (1974)
39. V.A.Vasiljev, Yu.M.Romanovsky, V.G.Yakhno, Uspekhi fizitcheskikh nauk , 128, 625 (1979)
40. A.T.Winfree, Geometry of Biological Time. Springer Verlag (1980)
41. V.I.Krinsky, A.S.Mikhailov. Autowaves. Znanie, Publ.Moskow (1984)
42. L.S.Polak, A.S.Mikhailov. Self-Organization in Nonequilibrium Physico-Chemical Systems (in Russian), M., "Nauka" Publ. (1983)

43. G.Nicolis, I.Prigogine. Self-Organization in Non-Equilibrium Systems; From Dissipative Structures to Order through Fluctuations. N.Y., Wiley (1977)
44. H.Haken, Synergetics. An Introduction. Nonequilibrium Phase Transitions and Self-Organization in Physics, Chemistry and Biology (3rd edition). Springer Verlag (1983)
45. H.Haken. Advanced Synergetics. Instability Hierarchies of Self-Organizing Systems and Devices. Springer Verlag (1983)

Part II

Self-Organization in Physical Systems: Autowaves and Structures Far from Equlibrium

The Microscopic Theory of Irreversible Processes

I. Prigogine

Faculté des Sciences Université Libre de Bruxelles, B-Brussels, Belgium
and
Center for Studies in Statistical Mechanics University of Texas at Austin
Austin, TX 78712, USA

1. Introduction

I am very happy to present this lecture in Pushchino. The work done here had wide repercussions on the development of nonequilibrium physics. The study of the Belousov-Zhabotinskii reaction has led to a new way of looking on macroscopic physics. We see now that matter in nonequilibrium conditions can acquire strikingly new properties which nobody could have predicted two decades before. However, I have decided to devote my lecture to the microscopic theory of irreversible processes, as this still remains a somewhat controversial subject. Indeed, in all fundamental theories (be it classical dynamics, quantum mechanics or relativity theory), entropy is conserved as a result of the unitary (or measure preserving) character of the evolution, in flagrant contradiction to the formulation of the second law of thermodynamics. As a result, the second law has usually been regarded as an approximation or as even being subjective in character. However, precisely because of the striking new developments in the phenomenological theory of irreversible processes, such an attitude becomes more difficult to accept today. For this reason, in the approach to the problem of irreversibility as developed by us, the law of entropy increase, and therefore the existence of an arrow of time, is taken as a fundamental fact [1,2,3]. The task of a satisfactory theory of irreversibility is thus conceived as the study of the fundamental change of the conceptual structure of dynamics which the law of entropy increase implies.

Let us make here two preliminary remarks. First, the second law of thermodynamics is not universal in the sense that there exist dynamical systems, such as the frictionless harmonic oscillator or two-body planetary motion, to which it does not apply. Therefore, the elucidation of the microscopic basis of the second law is closely related to the study of the specific conditions which have to be satisfied in order that we can speak at all about entropy and entropy increase. There is also a second aspect: irreversibility in the macroscopic theory means that not all transformations compatible with other principles of physics are possible. There is the well-known dissymmetry between transforming work into heat or heat into work. There are other situations in physics where some impossibilities arise, such as the impossibility to transmit signals with a velocity which would be larger than the velocity of light in the vacuum, or such as the impossibility to determine simultaneously momenta and coordinate as a result of Planck's constant. Every time such an "impossibility" has been identified, the structure of physical theories changed radically. We have therefore also to expect that if we take the second law of thermodynamics seriously, and I really do not know how we could avoid it, we will come out with a basic change in the structure of classical or quantum theory. In this lecture we want to present a few qualitative remarks concerning this change. Further details can be studied in the original articles. In short, irreversibility in classical theory requires a nonabelian formulation of classical mechanics quite similar to quantum theory. Similarly, in quantum theory we have to go from the ordinary Hilbert space formalism to a superspace formalism involving nonfactorizable superoperators [4,5]. We shall limit ourselves in this presentation to the case of classical mechanics.

2. Transformations in Dynamical Systems

An abstract dynamical system is characterized by a phase space Γ, a point transformation S_t, and an invariant measure μ. We have two alternative descriptions of the evolution of a dynamical system. One is in terms of trajectories corresponding to the point transformation S_t; the other is in terms of unitary operators U_t acting on distribution functions ρ in phase space. The unitary operator U_t satisfies the group relation:

$$U_t U_s = U_{t+s} \qquad t,s \gtrless 0 \tag{2.1}$$

For continuous time transformations we may write:

$$U_t = e^{-iLt} \tag{2.2}$$

where L is the Liouville operator. We may therefore present classical mechanics in terms of operators. However, they are not genuine operators, as they can be expressed in terms of point transformations in phase space. For example,

$$\rho_t(\omega) = (U_t\rho)(\omega) = (S_{-t}\omega) \tag{2.3}$$

In contrast, a probabilistic process such as a Markov chain may be generated by an operator W_t satisfying a semigroup property

$$W_t W_s = W_{t+s} \qquad t,s \geqslant 0 \tag{2.4}$$

The basic quantity of Markov chains is the transition probability $p(t,\omega,\Delta)$, which leads us at time t from point ω to the domain Δ. This quantity is related to the basic operator W_t through (see [6])

$$W_t\phi_\Delta(\omega) = p(t,\omega,\Delta) \tag{2.5}$$

where $\phi_\Delta(\omega)$ is the characteristic function of domain Δ. Obviously the Markov process operator cannot be reduced to a point transform (except in the trivial case in which the transition probabilities take only the values 0 or 1). In other words, a probabilistic process generated in the phase space is necessarily nonlocal.

Following the pioneering work of Boltzmann, we want to relate entropy to a probabilistic process. To do so the basic step we have to make is to associate to each distribution function ρ_t a new distribution $\hat{\rho}_t = \Lambda\rho_t$ using a necessarily nonunitary transformation Λ such that entropy

$$-\int \hat{\rho}_t \log \hat{\rho}_t \, d\mu \tag{2.6}$$

of the transformed states has the desired property of monotonic increase. The transformation $\rho_t \to \hat{\rho}_t$ corresponds to transcribing the physical states in terms of new elementary entities different from the entities in terms of which the initial dynamical description is formulated. It is in terms of this new description that we can speak of irreversible processes, be it on the level of elementary particles or on the level of cosmological evolution. The desired property of Λ follows most naturally if it satisfies an intertwining relation

$$\Lambda U_t = W_t \Lambda \qquad (t \geqslant 0) \tag{2.7}$$

where W_t is associated as stated before with a dissipative semigroup, while U_t is a unitary transform corresponding to a measure-preserving dynamical process. If we can introduce such a transformation operator Λ, the transformed states $\hat{\rho}_t$ will indeed evolve under the Markov process corresponding to W_t. As we have noticed, U_t is a local operator which can be expressed in terms of a point transformation, while W_t is not. Therefore the transformation operator Λ has itself to be a nonlocal operator. However, the full content of the second law is not expressed by the mere existence of a suitable transformation Λ satisfying (2.7). As a consequence of the time reversal symmetry of classical dynamics, there also exists another transformation Λ' satisfying

$$\Lambda' U_t = W'_t \Lambda' \qquad (t \geq 0) \tag{2.8}$$

so that the new transformed states $\Lambda' \rho_t \equiv \hat{\rho}'_t$ will evolve under a Markov process corresponding to W'_t for $t \leq 0$ and will thus exhibit monotonic increase of entropy in the "negative direction" of time.

The dynamical formulation of the second law of thermodynamics therefore consists of two statements: (1) first, it is the statement that the dynamical system admits two distinct Λ and Λ' so that the dynamical group U_t is transformed into two distinct Markov semigroups W_t and W'_t corresponding to the two directions of time. Misra and I have called this property intrinsic randomness. (2) In addition, there is a selection principle according to which only one of these transformations gives rise to physically realizable states and physical evolution. This is a stronger property which we have called intrinsic irreversibility.

The essential element which permits the construction of nonlocal transformations Λ or Λ' is that of a suitable degree of instability (such as in K systems [7]). Let us now indicate briefly how, in such types of systems, a nonabelian algebra can be constructed and how this is related to the construction of the transformation operator Λ.

3. The Internal Time and the Operator Formulation of Classical Mechanics

When a classical dynamical system presents a sufficient degree of instability we may associate to it a "time operator" T such that its commutator with the Liouville operator is given by (see i.e.[6])

$$-i[L,T] = 1 \tag{3.1}$$

This implies the relation

$$U_t^{+} \, T \, U_t = T + t.1 \tag{3.2}$$

between the time operator T, and the unitary operator U_t. This operator T is nonlocal. In other words, it does not commute in general with phase functions defined on the phase space Γ. Let us illustrate this statement in terms of the simple baker transformation.

The baker transformation may be described as the transformation B of the unit square on to itself which is the combined result of two successive operations: (1) first the unit square is squeezed in the vertical direction to half its width and is at the same time elongated in horizontal directions to double the length; (2) next, the resulting rectangle is cut in the middle and the right half is stacked on the left half. The iterates B^n of B may be considered to model dynamical evolution of a system at unit interval of time. The unitary operation U^n induced from B^n:

$$(U^n \rho)(\omega) \equiv \rho(B^{-n}\omega) \tag{3.3}$$

describes the evolution of distribution functions on the unit square. In the case of the baker transformation, a complete set of orthogonal eigen functions of T can be constructed as follows: Let X_o be the function which assumes the value -1 on the left half of the square and +1 on the right half. Define

$$X_n = U^n X_o \, . \tag{3.4}$$

A few of these functions are represented in the figure below:

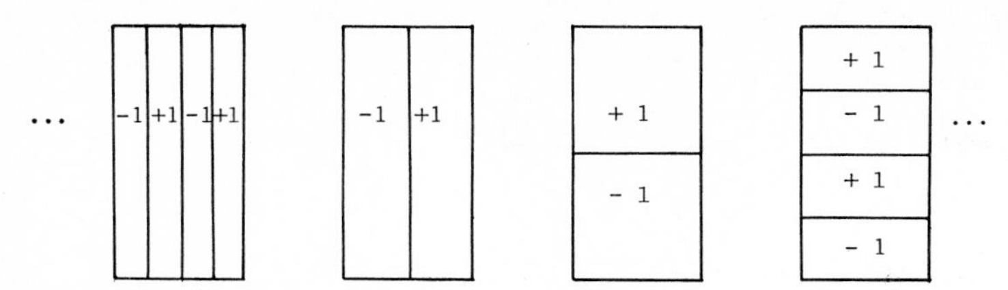

Fig.1 Partition associated to the baker transformation

A complete set of eigen functions of T is obtained by taking all possible finite products of X_n. Such a product belongs to the eigen value m of T when m is the maximum of the indices n of X_n appearing in the product. For example $X_{-5}X_3$, $X_{-1}X_2X_3$ and X_3 etc. are all eigenvectors of T corresponding to eigenvalue +3. We shall denote by $\phi_{n,i}$ a complete set of eigenvectors of T; the index i labels the additional degeneracy. The eigenfunctions $\phi_{n,i}$ together with the constant function Λ form a complete set of orthogonal functions.

The eigenvalue problem for T is therefore solved by the functions

$$T\,\phi_{n,i} = n\phi_{n,i} \qquad \text{(n integer)} \tag{3.5}$$

and these functions (together with the constant function over Γ) forms a complete and orthonormal set. Each distribution function ρ (or better, its excess over its constant value, $\bar{\rho}$) can be expanded in terms of the eigenfunctions $\phi_{n,i}$. Moreover, we can calculate averages $\langle T^m\rangle_{\bar{\rho}}$ using the formula well known from quantum mechanics:

$$\langle T^m\rangle_{\bar{\rho}} = \frac{\langle\bar{\rho}, T^m\bar{\rho}\rangle}{\langle\bar{\rho},\bar{\rho}\rangle} \tag{3.6}$$

Obviously, there is a complementarity in the usual sense between the local description in terms of the phase point and the topological description in terms of the eigenfunctions of T. Suppose we know that the distribution function has age 0, and that it is represented by X_0 (see figure 1). The distribution function $\rho = 1 + \bar{\rho}$ is then nonvanishing in the right half of the phase space, but we have no further information about the localization of the system. Inversely, if we assume the localization to be given (the distribution function is then represented by the δ function), then it can be easily shown using formula (3.6) that the dispersion

$$\langle\Delta T^2\rangle \to \infty \tag{3.7}$$

is infinite.

The introduction of the nonlocal operator T leads therefore for unstable classical systems to a new kind of complementarity formally quite analogous to the Heisenberg uncertainty relations.

4.

Let us now turn to the construction of the transformation operator Λ. As we have shown in recent papers [2,3,4,6] this can be performed by considering Λ as an operator function of T. In other words, to obtain the transition from the unitary transformation to the contraction semigroup for t>0, we have to write

$$\Lambda = \sum_{-\infty}^{+\infty} \lambda_n E_n + 1\rangle\langle 1 \tag{4.1}$$

where the set of E_n corresponds to the projection operator associated to the internal time operator T. Moreover, to satisfy all the necessary requirements (such as positivity, preservation and so on), it has been shown that the set of number λ_n have to be positive$(\lambda_n \leq 1)$ such that

$$\lambda_n \to 0 \qquad n \to +\infty \tag{4.2}$$

as well as

$$\lambda_{n+1}/\lambda_n \to 0 \qquad n \to +\infty \tag{4.3}$$

These two condition can be satisfied, for example, by the choice

$$\lambda_n = (1 + e^n)^{-1} \tag{4.4}$$

By definition we have (dropping the degeneracy index i)

$$\Lambda\phi_n = \lambda_n \phi_n \tag{4.5}$$

Let us now compare the transformation induced by the unitary operator U to the transformation induced by the Markov semigroup. As we have seen

$$U\phi_n = \phi_{n+1} \tag{4.6}$$

In contrast,

$$W \phi_n = \Lambda U \Lambda^{-1} \phi_n = \frac{\lambda_{n+1}}{\lambda_n} \phi_{n+1} \tag{4.7}$$

It is essential to notice that because of the property (4.3)

$$\frac{\lambda_{n+1}}{\lambda_n} < 1 \tag{4.8}$$

the Markov transformation is indeed a contracting one.

Let us comment briefly on the second law of thermodynamics as a selection principle. We may now require that the only measures which can be observed or prepared are the ones which are in the domain of Λ. In the case of the baker transformation, it may be shown that this excludes contracting fibers but keeps dilating fibers (see figures).[3,8]

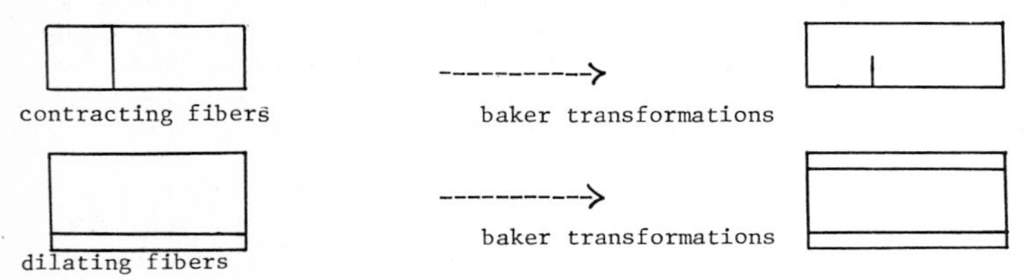

Fig. 4.1 Distinction between "contracting" and "dilating" fibers for the baker model.

Of course, the example of the baker transformation is somewhat artificial. However, there are physical situations corresponding to highly unstable systems, such as the Lorentz gas or systems of hard spheres. The dilating fiber would correspond to a set of hard spheres which in the distant past had parallel velocities, while in the distant future, after a large number of collisions, they have randomly distributed velocities. On the contrary, a contracting fiber would correspond to the velocity inverted situation in which in the distant past the velocities of the hard spheres had been randomly distributed but become all parallel in the distant future. The selection of the dilating fiber with the exclusion of the contracting fiber takes here a very intuitive meaning. Moreover, we have shown that there is an infinite entropy barrier between the states which are accepted and the states which are rejected by the second law. Intuitively this is clear as you would need an infinite information to produce a contracting fiber. This leads us to what I believe is really the crux of the matter in the problem of irreversibility.

5.

In classical physics there is a strong distinction made between initial conditions which are given arbitrarily and the laws of motion. However, this is an idealization which is not always applicable. In fact, the initial condition is the outcome of a previous evolution and will be propagated again by the subsequent dynamical evolution of the system. We have therefore to link the type of initial condition with the law of evolution in contrast with what is the traditional attitude. Let us consider a distribution function ρ which we expand in terms of the complete set in the phase space ϕ_n. We may write it

$$\rho = \sum_{-\infty}^{+\infty} c_n \phi_n + 1 \tag{5.1}$$

It is clear that the form of this function will be preserved both when we go to the future or to the past using the unitary operator U_t. Indeed, as an example, when we refer once more to the baker transformation we have

$$V\rho = \sum_{-\infty}^{+\infty} c_n \phi_{n+1} + 1 = \sum_{-\infty}^{+\infty} c_{n-1} \phi_n + 1 \tag{5.2}$$

Let us now consider the transformed distribution function:

$$\hat{\rho} = \Lambda\rho = \sum_{-\infty}^{+\infty} c_n \lambda_n \phi_n + 1 \tag{5.3}$$

The characteristic feature of this family of functions is that they have an in-built arrow of time. The partitions ϕ_n corresponding to the far distant future have a decreasing statistical weight (as $\lambda_n \rightarrow 0$ when $n \rightarrow \infty$). This characteristic property is preserved by the contracting semigroup as

$$W\hat{\rho} = \sum_{-\infty}^{+\infty} c_n \lambda_{n+1} \phi_{n+1} + 1 = \sum_{-\infty}^{+\infty} c_{n-1} \lambda_n \phi_n + 1 \tag{5.4}$$

Therefore, in the case of highly unstable systems, we may construct different families of distribution functions: distribution functions having no intrinsic arrow of time or distribution functions "oriented towards the future" such as (5.3) or oriented towards the past. As the time at which we consider the "initial" state is arbitrary, the self-preservation of the class of distribution functions implies itself the type of dynamical evolution. In other words, the existence of the second law of thermodynamics means that on the microscopic level the distribution functions must themselves have a broken time symmetry which is then propagated in time. What seems to me so remarkable is that the concepts of "state" (as expressed by ρ) and of dynamical evolution are so closely linked. States which have a broken line symmetry emerge as the result of dynamical laws having themselves a broken time symmetry and are transformed by these laws into states preserving the broken time symmetry.

6.

It is obvious that living systems such as human beings have an arrow of time. An amusing experiment which everybody can perform is to hear an acoustical impression in two ways corresponding to time inversion. We can start with a given sound very weak and increase its intensity, or we can start with the same sound very strong and decrease its intensity over the same period. The acoustical impression is obviously very different, which means that we who hear this sound have an inbred direction of time. At which level does this arrow of time appear? In the classical concept it was considered to be an artifact or, as Einstein once wrote an illusion. In the description I have tried to outline to you, it appears to have an objective meaning for the class of dynamical systems for which nonlocal operators can be constructed. Of course, in the model I presented here the decreasing sequence of the number λ_n remained arbitrary. To calculate it from dynamics we need supplementary considerations which are connected to kinetic theory [1,8]. Also, as I said in the beginning, the theory can be extended to quantum systems; there also it requires the giving up of locality in the Hilbert space.

The validity of the second law on the microscopic level is in this way reduced to a precise question which has to be further investigated both from the theoretical point of view testing its internal consistency and from the experimental point of view through experiments which would make explicit this microscopic arrow of time.

It is interesting to note that from the classical point of view time played a rather subordinate role. Through the existence of the velocity of light, time could in a sense be reduced to space. On the contrary, here we see that time as irreversibility has striking consequences on the structure of space as it involves a form of nonlocality. It also leads to a redefinition of states and observables as a consequence of the nonunitary transformation Λ. It is hardly an exaggeration to say that ireversibility not only leads to new chapters of macroscopic physics but also to a new way of conceiving the fundamental laws of nature.

I would like to acknowledge my colleagues in both Brussels and Austin with special thanks to Dr. B. Misra for his inspiring discussions. This work was sponsored by the Solvay Institute (Brussels) and the Robert A. Welch Foundation (Houston, Texas).

REFERENCES

1. I. Prigogine, C. George, F. Henin, L. Rosenfeld, Chem. Scripta 4, 5-32 (1973).

2. B. Misra, I. Prigogine, in Long Time Predictions in Dyn. Systems,(eds. C.W. Horton, L. E. Reichl and A. Szebehely; Wiley 1983).

3. M. Courbage, I. Prigogine, Proc. Nat. Acad. Sc. U.S.A., 80, 2412-2416 (1983).

4. B. Misra, I. Prigogine, M. Courbage, Proc. Nat. Acad. Sc., U.S.A. 76, 4768-4772 (1979).

5. I. Prigogine, C. George, Proc. Natl. Acad. Sci. U.S.A. 80 4590 (1983).

6. B. Misra, I. Prigogine, M. Courbage, Physica A 98 4590 (1983).

7. V. Arnold and A. Avez, Erogdic Problems of Classical Mechanics, New York, Amsterdam , W.A. Benjamin, Inc. (1968).

8. C. George, F. Mayne, I. Prigogine, to appear in Physica.

Coherent Structures in Plasmas

B.B. Kadomtsev

I.V. Kurchatov Institute of Atomic Energy, SU-Moscow, USSR

Plasma collective phenomena, characterized by spontaneous generation and development of nonlinear structures, are reviewed.

1. Introduction

Both laboratory and cosmic plasmas are extremely non-equilibrium media. This fact is especially evident for laboratory plasmas, since the ionized gas cannot be in equilibrium with the cold walls of a vessel in which plasma is sustained. (We do not consider here the case of a weakly ionized thermal gas within a hot-walled vessel). To maintain plasma conditions it is necessary to pump energy through the system. In this sense plasma is an open system. Moreover,plasma instabilities of various kinds are easily excited due to the interaction between charged particles via long distance forces and to weakness of dissipation. In other words, plasma is an open dissipative system and also it is an active medium.

The fact that instabilities of different kinds can spontaneously develop in active plasmas was known long ago. Also known is the term "collective phenomena". It includes a wide range of phenomena, connected with the development of instabilities up to finite amplitudes and with the feedback influence of waves with finite amplitudes upon the global plasma parameters.

Early theoretical studies of various instabilities were carried out to a linear approximation. Even for the simplest case of completely ionized, two-component (electron-ion) plasmas a rich variety of collective spontaneously excited degrees of freedom has been discovered. Later on, some theoretical approaches have been developed to study nonlinear phenomena of both regular (coherent) and chaotic character. The chaotic (or turbulent) phenomena are characterized by excitation of a number of degrees of freedom.

In distinction to liquids, a series of weakly-turbulent collective processes can appear in plasmas. These processes are connected with waves, weakly interacting with particles and with each other. The theory of weakly-turbulent plasmas, developed mainly by Soviet theorists, was found to be appropriate and convenient for the theoretical treatment of a wide range of nonlinear plasma processes. Naturally, this theory is not universal, since high amplitude perturbations can arise, resulting in either a small-scale, strong turbulence with many degrees of freedom or large-scale coherent structures. The phenomena of this kind are more difficult for theoretical consideration but, at the same time, they are very rich in manifestations and very interesting from the view-point of self-organization phenomena, which can be gathered under the general umbrella of synergetics.

In the present paper the nonlinear plasma phenomena, having the character of large-scale plasma structures (sometimes in combination with a small-scale turbulence) are briefly reviewed. These coherent structures are of interest from the view-point of a general approach to self-organization not only in physics, but also in other branches of science.

2. Strata

Stratification of a positive glow-discharge column was the first structure observed in laboratory plasmas. Such a one-dimensional modulation of the initially uniform plasma is the simplest type of structures. Stratification of a positive column is wellknown to be of ionization-diffusive origin, it is provided by the non-linear dependence of the ionization rate on electron temperature and density. To the present time, a fairly deep insight into the nature of strata has been gained. The phenomenon has adequately been described both qualitatively and quantitatively (see, e.g., review [1] and references therein). Stratifications of a more complicated form appear in discharges with crossed electric and magnetic fields [2].

3. Current-Convective Instability

If a positive column is put into a sufficiently strong longitudinal magnetic field, the luminous cord bends into a helical line. Usually, this screw-shaped configuration rapidly rotates, so that a stroboscope is needed to observe it. Initially this phenomenon was detected by its secondary manifestation, by anomalous fast transport of charged particles across the magnetic field [3]. Only when the theory has been built [4], which explains this effect by helical bending of the plasma cord, the direct observations of helical waves in plasma have been performed. In this case the plasma transport to the walls is due to convection and not to diffusion.Therefore, this instability is usually referred to as current-convective instability (see review [5]). It is a dissipative modification of the more strong helical instability of the completely ionized current-carrying plasma column in a strong longitudinal magnetic field. The current-convective instability was also detected and studied in semiconductor plasmas [6].

4. Plasma Focus

Complex phenomena of collective nature can arise in the so-called high current pinch discharge compressed by its own magnetic field. In particular, this refers to what is called plasma focus, the discharge without cylindrical symmetry [7]. In such a discharge plasma collapses into a compact body, which becomes a source of X-ray and neutron emission. Generation of a plasma focus is a complicated dynamical process with elements of self-organization, viz.spontaneous development of a small-scale structure. In the initial stage of the discharge, filamentation occurs due to the ionization (overheating) instability. At the final compression, a complicated scenario of the events is being displayed, including field reconnection in the filaments, plasma escape from the local necks, moderation in the evolution of the necks due to plasma being a two-liquid medium (the field is frozen only into the electron component), etc. The processes with strong energy radiation, resulting in radiative collapse and formation of micropinches [8], can arise in discharges with admixtures of heavy gases or metal vapours. These phenomena are being extensively investigated.

5. Solar Prominences

At first sight, solar prominences appear to be a phenomenon of completely different nature and scale. But in fact we again come across the dynamics of plasmas in the magnetic field with participation of radiation. The most popular model of quiet solar prominences describes them as plasma sheathes, suspended upon magnetic force lines and produced by condensation of coronal plasma because of thermal instability (colder prominence plasma provides stronger radiation). However, there is a variety of prominences, and their description requires a deeper knowledge of the subject (see, e.g., [9]).

6. Tokamak Plasma

Tokamak is the most advanced concept of high temperature plasma confinement in the magnetic field, aimed at realization of controlled thermonuclear fusion. The idea of tokamak is very simple. It is a toroidal, high current discharge stabilized by a strong longitudinal, i.e. toroidal, magnetic field.

Tokamak plasma is a system with a high degree of self-organization, characterized by a variety of self-consistent nonlinear processes. The most interesting tokamak regimes are near the margin of stability, where, therefore, the MHD-activity spontaneously develops. This activity controls the profiles of temperature and density distributions in plasma. The specific activity arises at the center of a plasma column as saw-tooth oscillations, which represent a periodic development of helical perturbations with the subsequent reconnection of the field lines. Other helical modes arise at the periphery. They form what is known as an island structure of magnetic surfaces. This structure seems to give rise to a small-scale flatter of magnetic surfaces, which results in a weak stochasticity of the field lines and an enhanced heat transport via electrons. All these processes are overimposed by high frequency oscillations, excited by so-called run-away electrons. The still more complicated set of phenomena takes place in plasma heated with high power sources other than the ohmic current. All these processes are being carefully studied, using a great number of experimental facilities (e.g., see [10]).

7. Drift Waves and Rossby Solitons

One type of oscillations of inhomogeneous plasma in the magnetic field was called drift waves. In these oscillations, electrons are in equilibrium along the magnetic field lines, but ions drift in the crossed electric and magnetic fields. It has been found out that the nonlinear drift waves can look like solitary waves, i.e. solitons. Equations for these waves are similar to those for gravitational waves in a shallow rotating liquid. V.Petviashvili has shown [11] that these equations have the solutions of two-dimensional soliton type. This allows a hypothesis to be proposed that the Jupiter's Great Red Spot is just a two-dimensional soliton in its atmosphere.

To verify this hypothesis, M.Nezlin and his co-workers [12] carried out experiments on the excitation of solitons in a shallow rotating water. In these experiments, long-lasting local vortices were detected, the vortices being anticyclonic in complete accordance with the theory. The localized perturbations of cyclonic type rapidly disappeared, probably, due to dispersion. Thus, the Great Red Spot is, most likely, a soliton in the rotating atmosphere. Solitons of this type were named the Rossby solitons after the Swedish scientist who studied the waves in the rotating atmosphere to a linear approximation.

8. Aurora and Double Layers

The aurora polaris with its inimitable beauty is one of the most impressive natural phenomena. In its essence, this is, without any doubt, a process of plasma physics, or, to be more precise, a chain of strongly non-linear, tightly bound processes. In the final analysis, the aurora is an electric discharge produced by currents passing through the atmosphere. These currents are due to convective streams in the magnetospheric plasma of the Earth at a distance of 6 to 10 Earth's radii. The currents pass through rather rarefied plasma and thereby a specific nonlinear process is being developed, namely, a double electric layer is being formed at the altitude of some thousand kilometers above the Earth. The double layer accelerates electrons passing through it and the very electrons, exciting atoms and molecules, produce a unique picture of the aurora polaris.

Conclusion

There is no doubt that the above examples of nonlinear processes do not complete the list of nonlinear collective plasma phenomena, which exhibit the features of self-organization. Actually, all the plasma physics is a manifestation of complex nonlinear processes in the ionized non-equilibrium gases.

Some plasma phenomena are very instructive, showing how complicated can be the behavior of an ensemble of simple identical particles. This is particularly true of fully-ionized plasmas where two species of particles bound by the long-range forces expose many complicated cooperative phenomena. Having got acquainted with the plasma physics, one can understand Mr. F. Hoyle, who prefers a Black Cloud in his fiction "The black cloud" to be a representative of living beings completely dissimilar to the known biological ones. Of course, plasma does not have as many inner degrees of freedom as has the organic world of molecules in a living organism, but it demonstrates the great opportunities of self-organization in Nature.

The studies of self-organization phenomena in such a simple medium as plasma could help understanding the phenomena of self-organization in more complicated systems.

References

1. P.S.Landa, N.A.Miskinova, Yu.V.Ponomarev. Uspekhi Fiz. Nauk, 132, (1980), 601
2. A.V.Nedospasov. Uspekhi Fiz. Nauk, 94, (1968)
3. B.Lehnert. Proc. of the 2nd UN Intern. Conf. of PUAE, 32 (1958), 349
4. B.B.Kadomtsev, A.V.Nedospasov. J. Nucl. Ener., Cl, (1960), 2301
5. A.V.Nedospasov. Uspekhi Fiz. Nauk, 116 (1975) 643
6. V.V.Vladimirov. Uspekhi Fiz. Nauk, 115 (1975) 73
7. N.V.Filippov, T.I.Filippova, V.P.Vinogradov. Nuclear Fusion, pt.2 Supp., (1962) 577
8. E.D.Korop, B.E.Meyerovich, Yu.V.Sidel'nikov, S.T.Sukhorukov. Uspekhi Fiz. Nauk, 129 (1979) 87
9. S.V.Pikel'ner. "Foundations of cosmic electrodynamics" (in Russian), Fizmatgiz, Moscow, 1961
10. B.B.Kadomtsev, V.D.Shafranov. Uspekhi Fiz. Nauk, 139 (1983) 399
11. V.I.Petviashvili. Pis'ma ZhETF, 32 (1980) 632
12. S.V.Antipov, M.V.Nezlin, E.N.Snezhkin, A.S.Trubnikov. Pis"ma ZhETF, 33 (1981) 368. Ibid, 35 (1982) 521

Structures in the Universe

Ya.B. Zeldovich

Institute of Physical Problems, The USSR Academy of Sciences
SU-Moscow, USSR

Introduction

The universe offers to the investigators a peculiar combination of local structures and global statistical homogeneity.

On an astronomically small scale, we deal with stars and galaxies. These objects are autonomous, and they stand out sharply against the surrounding background. Stars and galaxies are stationary almost at all times. The characteristic size of a galaxy is up to 100-200 thousand light years. On the other hand, observation of volumes of the order of one milliard light years reveals that they do not differ appreciably in quantity and quality of the substance they consist of and in radiation they emit. In the range of sizes between 10^5 and 10^9 light years some peculiar structures have recently been discovered. These are the giant globular clusters of galaxies; the superclusters, in which galaxies and clusters of galaxies are located beside surfaces or, more frequently, beside lines; and, finally, vast empty regions which practically do not contain galaxies [1,2].

The origin of these structures is quite clear [3]. In the course of evolution, the universe had gone through the stage of hot plasma. According to A.A.Friedman's theory and the ideas of G.A.Gamov, expansion and cooling of the plasma had taken place. As a result, the nonrelativistic substance, i.e. neutral atoms and, apparently, some elementary particles, e.g. neutrino with nonzero resting mass, had become dominant. A substance is called nonrelativistic if its particles move with a velocity which is very small compared to the speed of light. Accordingly, the pressure of the substance is small compared to the energy density ρc^2.

Small perturbations of homogeneity, which occur in the initial state of the hot plasma, grow under the action of gravitational instability in the nonrelativistic substance. At the final stage of perturbation growth, the pressure of the substance may be neglected. In the last few years, we have determined what types of structures develop at the nonlinear stage [3,4,5].

In hot plasma, short-wave perturbations are damped by dissipative processes. Therefore, the initial spectrum of perturbations of a nonrelativistic substance contains no short waves. In other words, the initial density and velocity distributions are smooth, they contain neither jumps nor fractures. However, it turns out that with growing perturbation there first arise characteristic local singularities, and then a global structure appears.

Approximate Theory of Growing Perturbations

The linear theory of gravitational instability of the quiescent medium yields the dispersion equation

$$\gamma = \sqrt{4\pi G\rho_o - k^2c^2}$$

for a plane wave

$$\frac{\delta\rho}{\rho_o} = Ae^{\gamma t+ikx},$$

where γ is the instability growth rate, k the wave vector, G the Newton gravity constant, ρ_o the mean density, $\delta\rho$ the perturbation of the density, and c the sound velocity. If the pressure is small, the sound velocity is also small, and one may assume

$$\gamma = \gamma_o = \sqrt{4\pi G\rho_o}.$$

Independence of the instability growth rate on the wavelength implies that there exist factorized solutions

$$\frac{\delta\rho}{\rho} = \Phi(\vec{x})e^{\gamma t}$$

with an arbitrary function Φ. The only required condition is smoothness of the function, i.e. the absence of high harmonics in its Fourier expansion. This condition being met, the Fourier expansion becomes unnecessary.

For the expanding world, whose scale varies as $t^{2/3}$, it turns out that the factorized solution of a linear problem has the form

$$\frac{\delta\rho}{\rho} = t^{2/3}\Phi\left(\frac{\vec{x}}{t^{2/3}}\right).$$

The mean density varies as $\bar{\rho} = (6\pi Gt^2)^{-1}$, i.e. inversely with the cube of the scale. However, it is obvious that the solution of the linear problem for density perturbations cannot be extrapolated to the range of large perturbations, otherwise ranges with $(\delta\rho/\rho)<-1$ will appear, where $\rho=\bar{\rho}+\delta\rho$ is negative. Extrapolation of the expression for displacements of particles proves to be a good approximation [6]. In other words, the solutions should be written in the form of the Euler coordinates ($\vec{x}$) of particles depending on their initial position, i.e. their Lagrange coordinates ($\vec{\xi}$)

$$\vec{x} = t^{2/3}\vec{\xi} + t^{4/3}\vec{\Phi}(\vec{\xi}).$$

The first term in the right-hand side describes the general "Hubble" expansion of the universe, while the second concerns the growing perturbations.

Given the solution $\vec{x}\,(t,\vec{\xi})$, the density can be calculated exactly. Note that choosing the new variables $\vec{r} = \vec{x}/t^{2/3}$, $\tau = t^{2/3}$, one obtains

$$\vec{r} = \vec{\xi} + \tau\vec{\Phi}(\vec{\xi}).$$

The problem is reduced to free motion with a constant velocity $\vec{\Phi}(\vec{\xi})$ of particles which were initially located at the point ξ.

However, the fact that we indeed deal with perturbations growing under the action of gravity is contained in $\vec{\Phi}(\vec{\xi})$ being a potential function. The scalar $\varphi(\vec{\xi})$ can be chosen as a potential, then $(\vec{\xi}=\xi_1\xi_2\xi_3)$

$$\Phi_x(\vec{\xi}) = -\frac{\partial\varphi}{\partial\xi_1} = -\varphi,_1$$

$$\Phi_y(\vec{\xi}) = -\varphi,_2, \qquad \Phi_z(\vec{\xi}) = -\varphi,_3.$$

The problem is reduced to the Lagrange $\vec{\xi}$ into $\vec{r}$ map depending on the parameter τ.

The density is given by the formula

$$\rho=\bar{\rho}\,|\delta_{ik} - \tau\varphi_{ik}|^{-1},$$

where | | are determinant bars. The tensor φ_{ik} is symmetric and has three real eigenvalues to be denoted as α, β, γ. Assuming $\alpha>\beta>\gamma$,

$$\rho=\bar{\rho}\,/(1-\alpha\tau)(1-\beta\tau)(1-\gamma\tau).$$

Local Catastrophes (Caustics) and "Pancakes"

The above solution contains a hint at a catastrophe: if $\alpha>0$, the density becomes infinite at a certain finite value of τ [6]. The nature of the catastrophe is obvious: the trajectories of neighbouring particles may converge akin to optical beams which converge to form a caustic. Neglecting the pressure, such a behaviour of neighbouring trajectories is quite natural. It is important that in the general case $\alpha\neq\beta$. The caustic is formed due to contraction along one axis, which corresponds to the eigenvalue α of the deformation tensor determined by the matrix φ_{ik}. The point where $\alpha=\alpha_m$ is maximum, can be found in the ξ-space. The caustic intersection of trajectories will first occur at this point at time $\tau=\alpha_m^{-1}$. Then the isosurfaces α=const can be constructed. They are shown for a two-dimensional case in Fig.1 $(\alpha_m>\alpha_1>\alpha_2)$. These isolines are determined by the particles which will undergo the caustic contraction at the subsequent instants of time τ_1 and τ_2.

They are ellipsoids (ellipses) in agreement with the fact that α_m is the maximum and there are no first derivatives in the expansion of α in the powers of the distance from the maximum point [4].

To proceed from the $\vec{\xi}$- to the $\vec{r}$-space it is essential that the map which becomes singular along one axis should be employed. This means that in the $\vec{r}$-space the ellipsoid (the ellipse in two dimensions) turns inside out and forms a figure, the plane projection of which is shown in Fig.2. Note that D and C have exchanged their positions, and the caustic line has return-points at A and B. The density is

Fig. 1

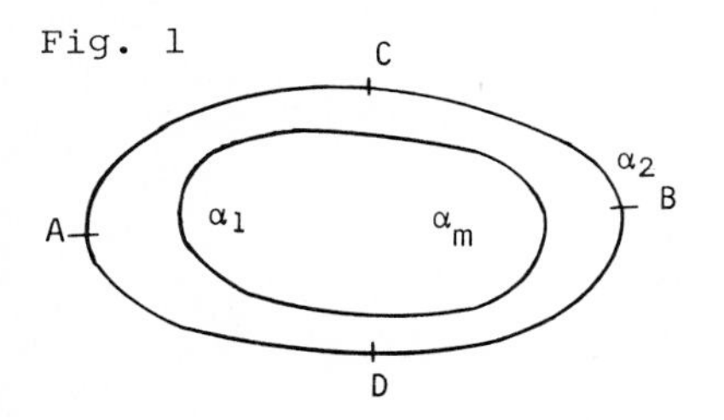

Fig. 2

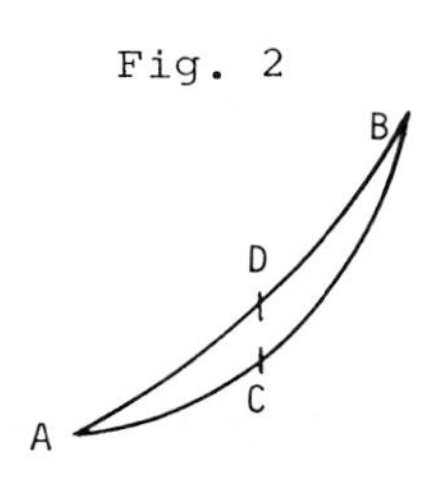

infinite on the line, and the three rays reach any point inside the figure. Omitting many interesting details, we shall list only the basic methodical conclusions.

1. The nonlinear growth of perturbations, as well as free motion of particles, result in the appearance of singularities, i.e. caustic lines and surfaces of infinite density.

2. In terms of the Fourier expansion of the density, the appearance of singularities implies a catastrophic rise of high harmonics corresponding to short wavelengths. It will be recalled that these harmonics were absent at the linear stage. The particular shape of the singular surfaces means that the high harmonics are strictly cophasal. The spectral approach in the random phase approximation is absolutely irrelevant in the case of catastrophes.

Global Picture of the Phenomenon

Once arisen, the caustic singularity does not disappear. The high density regions bounded by caustics grow both normal to the caustic surface and sideways, resulting in an increase of the pancake area. At a certain instant of time, separate "pancakes" merge to form a common global structure. The question may be raised as to the topological or percolational properties of this structure [7,8].

Let us consider the two- (2D) and three-dimensional (3D) cases separately. On a plane (2D), the high density regions are initially separated. In the course of time, the mass of contracted substance and the portion of the $\vec{x}$-surface occupied by the contracted substance (briefly, the "dense" surface) increase. However, the portion of the dense surface remains less than one half for a long period of time. How is the dense portion disposed? Will it occupy separate areas surrounded by the nondense(rarefied) substance?

The numerical calculations suggest that this is not the case. The dense portion forms thin walls which separate large cells with rarefied matter. This situation can be explained if we come back to the ξ-plane. In this plane the density is constant by definition, the portion of mass and the portion of area coincide.

It is evident that, due to the asymmetric condition $\alpha>\beta$, the region where $\alpha>0$ occupies more than half the area. The exact figures for the random function are: $\alpha>\beta>0$, 20%; $\alpha>0>\beta$, 60%; $0>\alpha>\beta$,20%. It means that for a sufficiently long time, 60% of the substance undergo the caustic contraction. The expansion along one axis does not prevent the appearance of infinite density, if the contraction along the other axis takes place.

As already mentioned, the portion of area equals the portion of mass in the $\vec{\xi}$-plane. As the isolines of α are smooth in the $\vec{\xi}$-plane, there is no doubt that the dense regions (80% of the area in ξ) form a single domain surrounding the insulated nondense regions (20% of the mass). This situation can actually occur even earlier, at a certain value of $\alpha'>0$ and at $\tau'=(\alpha')^{-1}$ when the portion of mass and the portion of area in the dense region do not considerably exceed 50%.

The next important remark is that the map of the $\vec{\xi}$ into $\vec{x}$ transition does not change the topology and percolation [7]. The continuous region with $\alpha>\alpha'$ in $\vec{\xi}$ remains continuous in $\vec{x}$. A closed curve drawn in

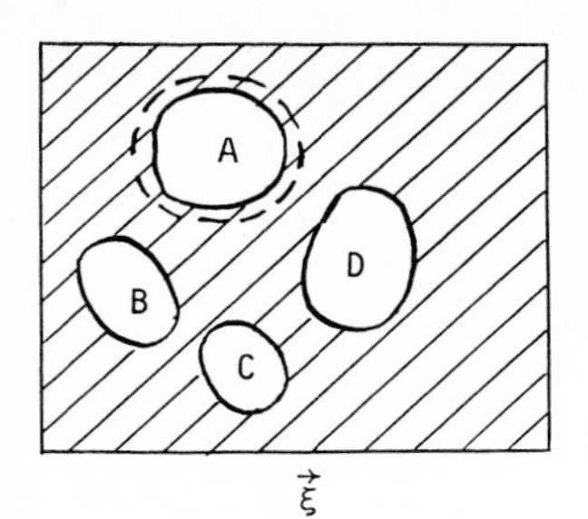

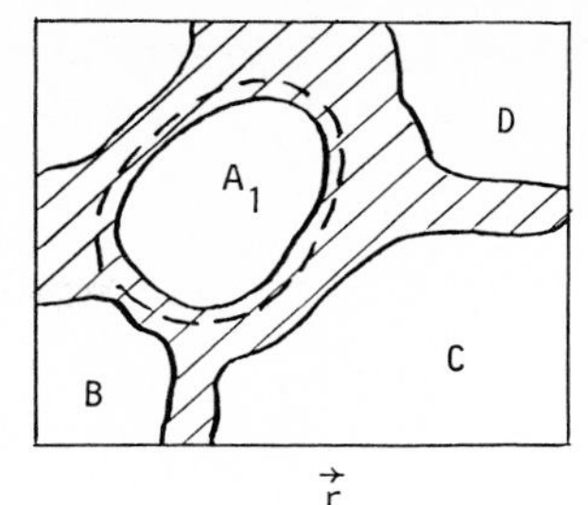

Fig. 3

the dense region and surrounding the waste one remains also closed in $\vec{x}$. The situation is illustrated by Fig.3.
The two-dimensional maps occur in the problem of reflection or refraction of a beam of parallel solar rays by the water surface with random choppiness. The time coordinate is substituted here by the Z-coordinate which is the distance to the surface. In a certain interval of Z, a network-like structure can easily be observed, e.g., in a basin or on the downward vault of a bridge.

In the case of optics the rays diverge after intersection. With an increase in Z, the distinct network-like structure turns into a multiple intersection pattern. In the gravitation problems, the dense regions retain the particles which found themselves therein. Here the network-like structure survives for a longer time and, according to numerical calculations, it decays due to the motion along the dense surface (in 2D, along the line).

The real 3D-situation which occurs in astronomy is more complicated. There is a wide interval of portions of mass where the dense and nondense regions interlace, and percolation takes place along both the regions.

The numerical calculations show [5] that in the real astronomical problem the specific heat processes, radiation transfer, formation of separate galaxies and stars from the dense gas are essential at the later stage. Naturally, all these problems are not discussed here. However, the general conclusion is that the gravitational growth of perturbations, as well as free inertial motion and refraction of light rays, result in a nontrivial network-like or cellular structure of the dense regions. The simple idea that small portions of area are always arranged as separate formations (islands, colonies) is sharply inconsistent with reality [1,2]. At the same time, the specific structures exist, for some reasons, as certain intermediate asymptotics.

We must be happy if it is confirmed that the universe which surrounds us is just at this remarkable stage of development.

1. J.Dort. Ann. Rev. Astron. Astrophys. (1983)
2. Ya.B.Zeldovich, J.Einasto, S.F.Shandarin. Nature 300, 407 (1982)
3. S.F.Shandarin, A.G.Doroshkevich, Ya.B.Zeldovich. Usp. Fiz. Nauk 139, 83 (1983)
4. V.I.Arnold, S.F.Shandarin, Ya.B.Zeldovich. Geophys. Astrophys. Fluid Dynamics 20, 111 (1982)
5. A.A.Klypin, S.F.Shandarin. Mon. Not. R. Astr. Soc. 204, 891 (1983)
6. Ya.B.Zeldovich. Astrophysica (Russian) 6, 119 (1982)
7. Ya.B.Zeldovich. Pisma Astron. Zhurn (Russian) 8, 102 (1982)
8. S.F.Shandarin, Ya.B.Zeldovich. Comments on Astrophys. 10, 33 (1983)
9. S.F.Shandarin, Ya.B.Zeldovich. Phys. Rev. Lett. (1984)
10. Ya.B.Zeldovich, V.A.Manaev, S.F.Shandarin. Usp.Fiz.Nauk 139, 153 (1983)

Interfacial Instability in Fluid Layers Under Thermal Constraints

Manuel G. Velarde
U.N.E.D.-Ciencias, Apdo. Correos 50.487, E-Madrid, Spain

1. Introduction

Although reports on interfacial convection were published before the beginning of our century and their relevance had been discussed by several authors, a systematic study of the phenomenology did not really occur until the experimental work conducted by Henri BENARD /1/. This author posed himself the task of providing a quantitative description of the (steady) flows arising in a horizontal thin liquid layer heated from below and open to the ambient air. There is some evidence that Bénard perceived the relevant role of surface tension in his problem but, however, he did not really address himself the question of interest to us here: how and how much the liquid-air interface affected or was affected by the observed convective flows. It took some fifty years until the right experimental (BLOCK /2/, KOSCHMIEDER /3/) and theoretical questions (PEARSON /4/) were asked and, to a first approximation, unambiguously answered (BIRIKH /5/, NIELD /6/, STERNLING and SCRIVEN /7/).Yet, today we do not dispose of a complete theory of the interfacial phenomena involved in Bénard convection. However,interfacial convection is so relevant to chemical engineering, materials sciences, crystal growth (LANGER /8/, OSTRACH /9/, SCHWABE and SCHARMANN/10/) that many aspects of the problems have already been elucidated.

An interesting feature of Bénard convection is that the flow pattern does not depend on the liquid used and depends little on the geometry and, moreover, may not rely on any buoyancy-assisted motor (BLOCK / 2 /, DAUZERE /11/, VOLKOVISKY /12/ . Buoyancy and thermocapillary were not brought together to account for Bénard's experiments until several decades after a seminal paper by LORD RAYLEIGH /13/ on buoyancy-driven flows. He predicted that a motionless horizontal liquid layer heated from below would be convectively unstable only past a certain threshold in the thermal gradient along the vertical. He also gave a quantitative estimate of the expected flow patterns. Rayleigh's theory was thought for many decades to be the theory of Bénard convection which it is not, to a major extent (LOW and BRUNT /14/). Rayleigh's masterly analysis of the stability of fluid layers opened the way to a fertile understanding of natural convection (VELARDE and NORMAND /15/). Influenced by evidence of polygonally patterned convection arising in the drying of paint films, PEARSON /4/ developed a theory for the onset of thermocapillary (steady) convection in thin liquid layers heated from below. No buoyancy effects were incorporated in his analysis. Pearson considered a highly idealized interface liquid-air: indeformable (Crispation number $Cr = \mu\kappa/\sigma h = 0$, with μ the dynamic viscosity, κ the thermal diffusivity, σ the surface tension and h the vertical liquid layer thickness), inviscid and perfectly elastic. Interfacial tension depended only on temperature leading to interfacial tractions ($Ma = -(\partial\sigma/\partial T)\Delta T\, h/\mu\kappa \neq 0$ with ΔT the temperature difference (it could be any other constraint as shown in Fig.1). For standard fluids Ma is positive when heating the layer from below. Pearson predicted that a motionless horizontal liquid layer heated from below would be convectively unstable at and above a non-vanishing value of Ma. This value was shown to be independent of material parameters provided the liquid layer was like in Rayleigh's analysis a Newtonian-Boussinesquian fluid (PEREZ-CORDON and VELARDE /16,17/). We shall disregard here all non-Boussinesquian effects. A more complete theory of interfacial convection did not appear until the work by SCRIVEN and STERNLING /7,18,19/. These authors focussed their attention on the stability of an interface separating two infinitely extended, both horizontally and vertically, liquid layers.

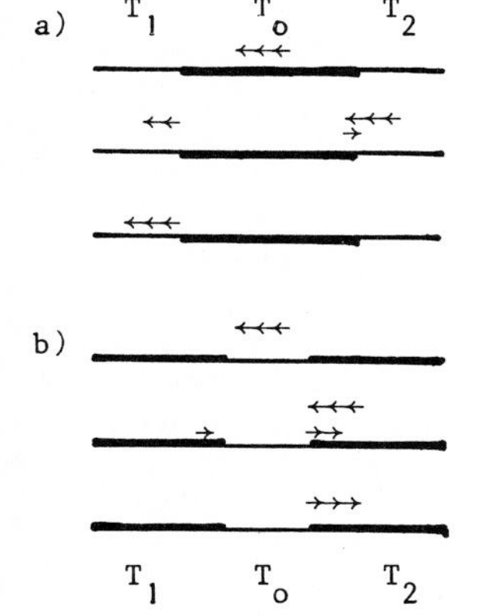

Fig.1.*THERMOCAPILLARY INSTABILITY*
A portion of liquid is drawn away from its initial position due to the temperature gradient as the net resulting force proceeds from the hotter to the cooler region. Case a) corresponds to an unstable state and thus relaxes through convection, whereas case b) corresponds to a stable state and thus the disturbance imposed by the initial motion cannot be sustained and the system returns to its initial stable state. Arrows(→) indicate motion of the fluid. In case a) $T_1 < T_2 < T_o$ whereas in case b , $T_o < T_1 < T_2$.

STERNLING and SCRIVEN /7/ assumed that transfer of a solute existed between two bulk phases, A and B (see Fig.2). If solute is diffusing from A to B and by any chance there is a substantial increase of solute at points in the interface thus creating a non-homogeneous distribution of solute concentration (say higher in '1' than in '2') interfacial traction may eventually lead to convection. Whether or not such a disturbance is sustained can rely on various properties and transport phenomena in the phases A and B. The motion initiated at the interface penetrates into the bulk phases. The flow in A and B develops according to properties like the kinematic viscosity. The greater its value the more vigorous the flow is expected. On the other hand, solute transport to or from the interface also depends on the relative solutal diffusivities in the bulk. The lower the

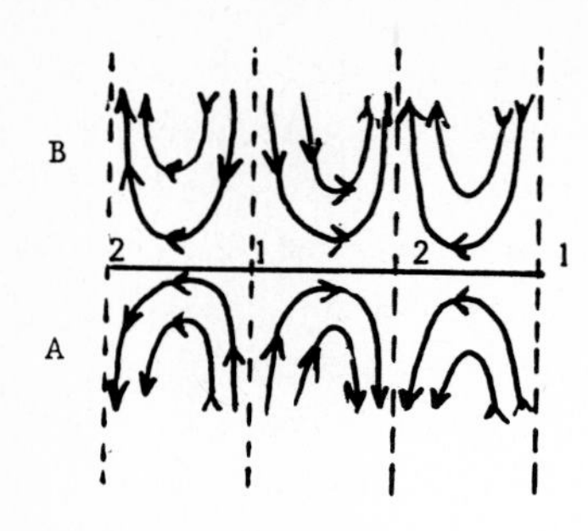

Fig.2. *INTERFACIAL CONVECTION WITH TWO FLUIDS* The figure shows a cross-section of two dimensional cellular convection when there is an interface 1-2 -1-2 between two fluids A and B of properties $\nu_A > \nu_B$, $D_A > D_B$. Solute is assumed to diffuse from A to B and the tension is lower at the higher solute concentration $\sigma(1) < \sigma(2)$.

diffusivity the weaker its influence is on the convective transport. Hence, the bulk phase of higher diffusivity and lower kinematic viscosity would affect less dramatically the stability of the interface.

If interfacial tension decreases with increasing solute concentration and there is increase of solute at '1' then the interface stretches there (Fig.1,& 2) and contracts at '2'. Fluid motion develops from the former to the latter point. In the case of figure 2 we have $\nu_A > \nu_B$ and $D_A < D_B$. If, however, the interfacial tension increases with increasing solutal concentration no motion is expected to be sustained. Note that if the viscosity and the diffusivity are both higher in the same phase one cannot infer the flow direction from the heuristic argument.The competition between the two transport phenomena may lead to oscillations and then whether or not we take into account other phenomena at the interface (material properties, dissipative processes, deformation, reactive processes, evaporation , adsorption) becomes crucial. On the other hand, it appears that for single component standard liquids no (buoyancy-thermocapillary) convective instability is to be expected in a layer cooled from below and open to the ambient air! Air has vanishingly small dynamic viscosity and a negligible traction exists above the interface. Under the simplifying assumptions used by STERNLING and SCRIVEN /7/ and PEARSON/4/ no convective instability develops when the liquid at the bottom does not provide support for the convection. Lastly, we should recall that two other surface transport phenomena considered by STERNLING and SCRIVEN, the surface shear and dilational viscosities, tend also to damp out interfacial convection.

When both, energy and mass transport are involved in the problem, generally, the interfacial tension always decreases with increasing temperature whereas either sign is possible for increasing solute concentration, as this depends on the specific tensioactive 'solute' added to a given liquid (for instance, salt tends to increase the tension in water whereas a detergent rather does the opposite). When kinematic viscosity and solute diffusivity are lower in one of the bulk phases there is a possibility for interfacial convective instability with either sign of the solute transfer between the bulk phases (A to B or vice versa). Quite a spectacular experimental verification of the predictions given by STERNLING and SCRIVEN has been provided by the work of ORELL and WESTWATER /20,21/, WANN-

GARD /22/ and PANTALONI et al. /23,24/. Bénard's convection was indeed thermocapillary-driven to a major extent. In thermocapillary flows, as reported already by Bénard and later determined by various authors, the liquid-air interface is depleted where ascending (hot) currents exist. The opposite structure is predicted for buoyancy-driven flows when surface tension phenomena are negligible (HERSHEY /25/, HICKMAN /26/, JEFFREYS /27/). Other predictions made by SCRIVEN and STERNLING /19/ correspond to a thin layer at non-vanishing crispation and surface viscosity coefficients. The latter two have a stabilizing role for short wavelength disturbances while the former $(Cr \neq 0)$ is needed for a realistic situation at large wavelengths. At finite crispation number, interfacial instability is predicted for all values of Marangoni number, i.e., without threshold unless gravity waves are incorporated in the description of the dynamics of the liquid-air interface. The latter problem was studied by SMITH /28/ and BENTWICH /29/.

Reactive processes, adsorption-desorption phenomena (BERG and ACRIVOS /30/ , HICKMAN /26/, evaporation (BERG et al. /31/, BOSE and PALMER/32/),deformation(DAVIS and SEGEL /32/, LUDVIKSSON and LIGHTFOOT /34/, SCANLON and SEGEL /35/) are mechanisms that can induce surface tension inhomogeneities thus leading to interfacial instability and convection. On occasions coupling of phenomena at an interface suffices regardless of whether or not each of the processes is destabilizing by itself. An origin for instability could be a large disparity in time or space scales in the coupled processes. In the remaining we shall focus attention to the coupling of buoyancy, however small, to interfacial tractions leading to interfacial instability in a fluid layer with an open deformable interface. Results have been obtained for both finite and linear stability analyses.

2. Parameters and equations for a binary fluid layer

The theory here presented applies to single and two-component fluid layers. The equations, however, will be introduced for the case of a binary mixture. The second component can always be thought as an impurity in the liquid layer open to ambient air.

There are some interesting phenomena in binary fluid mixtures. Salt-fingering is the kind of instability that develops at the interface of a warm, salty water overlying a layer of cold and fresh fluid. It is originated by the large separation between two diffusion scales: heat diffusivity ($\kappa \sim 0.001\ cm^2\ s^{-1}$) is about two orders of magnitude different from mass diffusivity ($D \sim 0.000\,01\ cm^2 s^{-1}$) for standard liquid mixtures.

When the fresh and cold layer rather sits on top of a salty warmer water there is no tendency to fingering but to the appearance of oscillations in the thermal and hydrodynamic variables and once more a simple explanation originates from the same large separation of diffusion scales. Thus knowledge of the actual value in an experiment of the Lewis number (or inverse Lewis number, according to the author), $Le = D/\kappa$, is crucial for the understanding of the evolution of the fluid layer under varying thermal constraints. The other relevant parameters in a Rayleigh-Bénard geometry are the thermal Rayleigh number and the solutal Rayleigh number $Ra = \alpha g d^3 \Delta T/\nu\kappa$ and $Rs = \gamma g d^3 \Delta N/\nu D$, respectively, where α and γ are the thermal and solutal expansion coefficients, g is the gravitational acceleration, ΔT is the thermal gradient across the layer of thickness d, and ν is the kinematic viscosity of the mixture. $\Delta N/d$ denotes a concentration gradient. With an open interface several other parameters enter the problem. Surface tension tractions must be considered if there is variation of surface tension with either temperature or solute (an impurity). This is accounted with the inclussion of the thermal and solutal Marangoni numbers (the latter is usually called the elasticity), $M = -(\partial\sigma/\partial T)d\Delta T/\mu\kappa$ and $E = -(\partial\sigma/ N_1)d\Delta N_1/\mu D$, respectively. Here σ is the surface tension (liquid-air, say) and σ_0 is some mean value. N_1 is the mass-fraction of the impurity, which for convenience is here considered the heavier component of the binary mixture. Consideration of surface tension tractions does not necessarily force the consideration of the deformation of the interface for they may be operating even if the interface remains level. If, however, the deformation is to be considered at least two more dimensionless groups appear, the Bond number (Bo or G) and the capillary (or crispation) number already defined in the Introduction. We have $G = \rho g d^2/\sigma_0$ and $C = \mu\kappa/\sigma_0 d$, respectively, where ρ accounts for the density of the fluid mixture. $\mu = \nu\rho$ is the dynamic viscosity.

The Bond number estimates the strength of the gravitational forces with respect to the surface tension and thus large values of G correspond to rather flat and level interfaces whereas low values of the Bond number exist when interfaces tend to be spherically shaped (at thermodynamic equilibrium this means minimization of free energy). Rather low Bond numbers appear in experiments aboard space crafts where gravity might decrease in four or six orders of magnitude the value on Earth. Then capillary lengths may reach the order of the meter and for this reason there is no need of a container for the handling of liquids at very low Bond numbers. The capillary number compares dissipation to surface tension forces. To a first approximation dissipation tends to damp out all inhomogeneities whereas, as before, surface tension tends to bend interfaces. For standard fluids(out of the critical point) and standard modes of operation C varies between 10^{-7} and 10^{-2}. Large values of C appear, however, when the operation takes place near a critical point (in temperature for a single component liquid layer, the consolutal/ demixion point) where the interfacial tension goes to zero or when we handle extremely thin films.

It is of some interest to read the definition of the capillary number from another perspective. Consider an interface where a capillary wave may develop. Let

ξ be an estimate of its length. The actual value could eventually be d, the cell gap in a Rayleigh-Bénard (Marangoni) experiment. Then a quantity, the (mechanical) time constant of such a disturbance upon the interface (the interface may be likened to a stretched membrane) can be defined through the relation $\tau_\xi^2 = \rho\xi^3/\sigma$ (from now on $\xi = d$). With heat and mass diffusion the other two time constants are $\tau_V = d^2/\nu$ and $\tau_T = d^2/\kappa$. We have $C = \tau_d^2/\tau_V\,\tau_T$ and thus large values of the capillary number correspond to the case of disturbances that decay so fast on the thermal and momentum dissipation scales that this happens before the mechanical disturbance, the "wave" decays, i.e., before the interface returns to the level position. Restoring mechanical forces (potential energy) make the interface overshoot the level position thus leading to interfacial oscillations. It is this potential energy that provides in a fluid layer the possibility of reversing an initially given fluid motion. Note that the above given argument can be extended to the case of an interface where some chemical reaction takes place. It suffices to replace the heat diffusivity by the appropriate chemical constant and apply the argument just sketched.

Still two more parameters are needed in the problem considered here: the Prandtl number, $P = \nu/\kappa$ and the Schmidt number ν/D. Low Prandtl number fluids also tend to show oscillatory instabilities as there inertial terms tend to dominate dissipation. Finally the combination $A = (Ra + LeRs)\,C/G$ is a quantity that estimates the validity of the Boussinesquian approximation (for a single component $A = \alpha\Delta T$). In the following we shall consider values of A and C small with emphasis however, on the role of C when the Rayleigh and Marangoni numbers compete or cooperate for the onset of instability. All results will show the dependence on P, Le and G and, for illustration, cases with vanishing Rayleigh numbers are also discussed.

We use the following conventions and notation (Fig.3): d is the mean distance between two infinitely extended surfaces: the lower is a rigid, heat conducting plate held at constantly controlled temperature. The upper surface is the one open to the ambient air, for simplicity considered adiabatic (poor heat conductor). The fluid enclosed between these two surfaces is an incompressible binary mixture and we restrict consideration to a two-dimensional problem. Thus x and z denote, respectively, the horizontal and vertical coordinates. The ambient air is assumed to have negligible density and dynamic viscosity.

For universality in the description the following units (scales) are introduced: d for length, d^2/κ for time, κ/d for velocity, ΔT and ΔN_1 for temper-

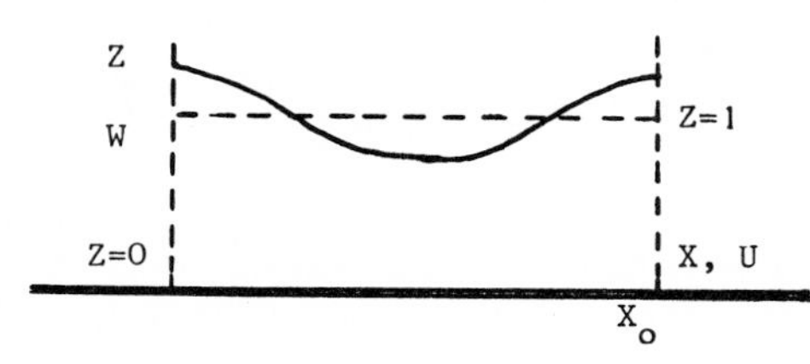

Fig.3. *GEOMETRY OF THE FLUID LAYER WITH A DEFORMABLE INTERFACE* : $Z = 1 + \eta(X,t)$. Z=1 corresponds to the level position.

ature and mass fraction of the components, respectively, $\mu\kappa / d^2$ and σ_0 for pressure and surface tension, respectively.
The open interface, $S(t)$, is located at $z = 1 + \eta(x,t)$. n denotes the outward unit normal vector to S, $n = (-\partial\eta/\partial x, 1)/N$, whereas t is the unit tangent vector $t = (1, \partial\eta/\partial x)/N$ and the curvature is $K(\eta)$,

$$K(\eta) = (\partial^2\eta/\partial x^2) / N^3 \quad \text{with} \quad N = \{ 1 + (\partial\eta / \partial x)^2 \}^{1/2} .$$

The stress balance at the deformable interface is

$$\tau_{ij}\, n_j = - (G/C)\, (\eta + A\eta^2/ 2)\, n_i + (K/C)\, \{1 - MC(\theta - \eta) - ELe(\Gamma - \eta)\}\, n_i + t_i\, (t.\nabla)\, \{ M(\theta - \eta) + ELe(\Gamma - \eta)\} \qquad i,j=1,2 \tag{2.1}$$

where θ, Γ and v account for disturbances upon the temperature, mass fraction and velocity of the initially steady rest state. The stress tensor is

$$\tau_{ij} = - p\,\delta_{ij} + \varepsilon_{ij} \quad \text{with} \quad \varepsilon_{ij} = \partial v_i/\partial x_j + \partial v_j/\partial x_i \tag{2.2}$$

δ_{ij} is the Kronecker delta and the summation convention over repeated indices is assumed.
The kinematic boundary condition at the interface is

$$\partial\eta / \partial t = N v_i\, n_i \qquad \text{on} \qquad z = 1 + \eta \tag{2.3}$$

The convention that the heat flux is prescribed at the open interface leads to the condition

$$(n.\nabla)\,\theta = (1 - N)/N \qquad \text{on} \qquad z = 1 + \eta \tag{2.4}$$

For the impurity we also prescribe its flux at the interface. This leads to a simpler analysis.
For a fluid layer bounded by a copper plate at the bottom we take there a heat conducting plate, impervious to matter transfer and mechanically rigid. Thus we have

$$v_i = \theta = \Gamma = 0 \qquad \text{on} \quad z = 0. \tag{2.5}$$

We assume that originally the fluid layer is in motionless state with steady linear distribution of temperature and solute. Thus the evolution equations for disturbances upon the motionless steady state are the Navier-Stokes, Fick, and Fourier equations for the region $0 \leq z \leq 1 + \eta(x,t)$, $-\infty < x < +\infty$

$$P^{-1}\, (\partial v_i / \partial t + v.\nabla v_i) = \partial\tau_{ij} / \partial x_j + Ra\,\theta\, k_i + LeRs\,\Gamma\, k_i \tag{2.6}$$

$$\partial\theta / \partial t + v.\nabla\theta = \nabla^2\theta + w \tag{2.7}$$

$$\partial\Gamma / \partial t + v.\nabla\Gamma = Le\,\nabla^2\Gamma + w \tag{2.8}$$

$$\nabla \cdot v = 0 \tag{2.9}$$

where $k_i = (0,1)_i$. Note that we have not included the Soret effect in Fick's mass transport equation as, for simplicity, we shall focus on the competition of the two possible gradients of temperature and solute.

3. Energy stability of arbitrarily large disturbances

We define the integral over the open and deformable interface for a quantity f as

$$\int_{S(t)} f = \int_0^{S_0(t)} f ds = \int_0^{x_0} f(z = 1 + \eta) N dx \tag{3.1}$$

where ds is an element of arc length along S(t), and $S_0(t)$ is the length of one period along x. A two-dimensional volume integral of f over a period is

$$\langle f \rangle = \int_0^{x_0} \int_0^{1+\eta} f(x,z,t) dz\, dx \tag{3.2}$$

Then we define the energy

$$E = P^{-1} \langle v^2/2 \rangle + \lambda \langle \theta^2/2 \rangle + \Lambda Le \langle \Gamma^2/2 \rangle$$
$$+ (G/C) \int_{S(t)} \tfrac{1}{2} (\eta^2 + A\eta^3/3)/N \tag{3.3}$$

where $|\eta| < 3/A$ (Boussinesquian approximation). λ and Λ are the linking parameters whose choice is dictated by the convenience in obtaining the largest parameter region of stability of the initially motionless steady state of the fluid layer. Thus a variational condition is introduced: $\delta(dE/dt) = 0$ where δ accounts for an arbitrary variation subjected, however, to the conditions earlier indicated (mass conservation, boundary conditions and all that). We have

$$\delta \left(dE/dt + \langle 2p \nabla \cdot v \rangle + \beta \int_x \eta \right) = 0 \tag{3.4}$$

The consideration of arbitrary values of the capillary or crispation number produces a formidable problem and a reasonable approach is to consider its contribution, i.e., the role of the surface deformation to a first order approximation. Thus we set

$$\eta = \eta^{(0)} + \eta^{(1)} C + O(C^2) \ ; \quad \eta^{(0)} = 0 \tag{3.5}$$

(note that first it moves and then it gets deformed which in turn affects the motion)

together with similar expansions for the remaining quantities. Then the evolution problem for disturbances upon the initial state is reduced to the following Euler-Lagrange equations

$$2\tau_{ij}^{(0)} + (Ra + \lambda)\theta^{(0)} k_i + (Rs + \Lambda) Le\, \Gamma^{(0)} k_i = 0 \tag{3.6}$$

$$(Ra + \lambda)\, w^{(0)} + 2\lambda\nabla^2 \theta^{(0)} = 0 \tag{3.7}$$

$$(Rs + \Lambda)\, w^{(0)} + 2\Lambda\, Le\, \nabla^2 \Gamma^{(0)} = 0 \tag{3.8}$$

$$\nabla \cdot v^{(0)} = 0 \tag{3.9}$$

together with the conditions on $z = 0$:

$$v_i^{(0)} = \theta^{(0)} = \Gamma^{(0)} = 0 \tag{3.10}$$

On $z = 1$ we have

$$w^{(0)} = 0 \tag{3.11}$$

$$2\tau_{ij}^{(0)} n_j^{(0)} t_i^{(0)} + M\,\partial\theta^{(0)}/\partial x + ELe\,\partial\Gamma^{(0)}/\partial x = 0 \tag{3.12}$$

$$2\lambda\partial\theta^{(0)}/\partial x - M\,\partial u^{(0)}/\partial x = 0 \quad \text{and} \tag{3.13}$$

$$2\Lambda\, Le\,\partial\Gamma^{(0)}/\partial x - E\,\partial u^{(0)}/\partial x = 0. \tag{3.14}$$

Solutions of the above posed problem can be sought in the form of exponentials $\exp(iax)$ where a denotes a Fourier decomposition mode. We can set

$$\theta^{(0)} = \{\sum_{i=1}^{6} \Omega_i \exp(q_i z) + \Omega_7 (\exp(az)-\exp(az))\} \exp(iax) \tag{3.15}$$

where the q_i are the roots of the polynomial equation

$$(q_i^2 - a^2) + (\lambda + \Lambda)a^2/4 = 0 \tag{3.16}$$

There are similar expressions for the remaining quantities.
To the lowest order approximation, necessary conditions for the onset of instability are obtained in the form of Taylor expansion in the capillary number. For instance, using the Marangoni number the motionless steady state of the fluid layer is stable provided that its value remains below

$$M^{(0)} + C\,M^{(1)}, \text{where}$$

$$M^{(1)} = -2a^{-2}(M + LeE)\int_x \tau_{ij}^{(0)} n_j^{(0)} n_i^{(0)} (\partial w/\partial z)^{(0)} \left\{\int_x \theta^{(0)} (\partial w/\partial z)^{(0)}\right\}^{-1} \tag{3.17}$$

Here the superscript (0) accounts for values at vanishing crispation number. Similar expressions have been found for the remaining control parameters (Rayleigh and Marangoni numbers).
When, however, C vanishes there is no gravity and both Marangoni numbers are positive, energy theory yields $M + E = 56.7$ for a critical wavenumber $a_c = 2.22$. Below this line the fluid layer is absolutely stable (Fig.4)

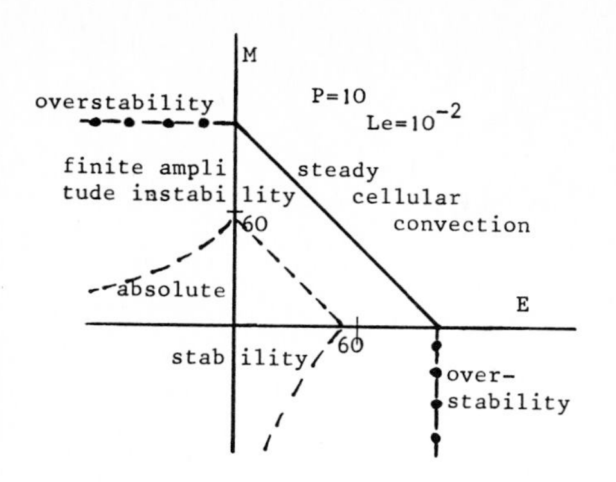

Fig.4. *ZERO-GRAVITY BENARD-MARANGONI INSTABILITY*. Solid and broken lines correspond, respectively, to linear and energy analyses. Dotted lines define regions of overstability, according to linear theory. Over these lines oscillations are expected depending on values of Prandtl and Lewis numbers.

4. Linear stability analysis

Sufficient conditions for instability of the motionless steady initial state of the layer can be obtained by means of a linear stability analysis using normal modes. As the results of this and the preceding energy analysis do not coincide there appear possibilities of subcritical modes of instability (finite amplitude steady states or oscillations) and transient oscillations. Fig.5 gives an illustration of some of the results obtained for the case of vanishing gravitational acceleration ($g = 0$). For a given value of the temperature gradient the role of the impurity is rather clear. In accordance with the sign of E (solutal Marangoni or elasticity number) there is a dramatic lowering of the threshold for thermoconvective instability or the possibility of overstable modes (oscillations). When all buoyancy phenomena in the bulk are negligible ($Ra = Rs = 0$) transition to steady convection is expected above the line

$$M+E = 8a \frac{(a^2+G)\{\beta^3 - \beta^2 (4a-1) - \beta(4a+1) - 1\} + 4a^3(\beta^2+\beta) ECS}{(a^2+G)\{\beta^3 - \beta^2 (4a^3+3) + \beta (3-4a^3) - 1\} + 32a^5 C(\beta^2+\beta)} \tag{4.1}$$

where $S = 1 - Le$, $\beta = \exp(2a)$ and a is the wavenumber. Thus in the limit of vanishing capillary number (or deformation, $C=0$) we have $M + E = 79.6$ with $a_c = 1.99$. When the Bond number is negligible, $G = 0$, the correction induced

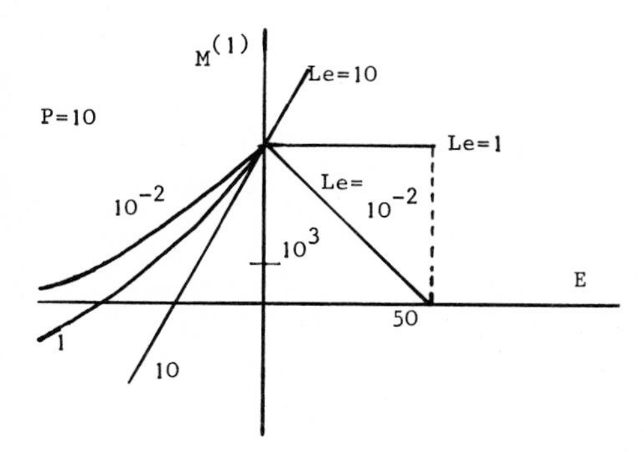

Fig.5. *BENARD-MARANGONI CONVECTION*
Corrections to Marangoni number induced by deformation of the interface in the case of steady cellular convection in the absence of gravity. The actual Marangoni number is $M=M^{(0)} + C\, M^{(1)}$, where $M^{(0)}$ is obtained from the corresponding value of E in Figure 4.

by the deformation is

$$M^{(1)} = - 1.11\ 10^3 \ \{ M^{(0)} + ELe \} \tag{4.2}$$

Thus, to a first-order approximation with positive Marangoni numbers the deformation of the interface reduces the region of finite amplitude instability. Further results from linear and nonlinear analyses can be found in references /36-39/.

Acknowledgments

The research has been sponsored in part by the Stiftung Volkswagenwerk.

References

1 H. BENARD. (a) Rév. Gén. Sci. pures appl. 11, 1261-1271(1900). (b) Ann. Chim. Phys. 23, 62-144 (1901)
2 M. BLOCK, Nature 178, 650-651 (1956)
3 E.L. KOSCHMIEDER, J. Fluid Mech. 30, 9-15 (1967)
4 J. R. A. PEARSON, J. Fluid Mech. 4, 489-500 (1958)
5 R.V. BIRIKH, J. Appl. Mech. Tech. Phys. 3,43 (1966)
6 D.A. NIELD, J. Fluid Mech. 19, 341-352 (1964)
7 C.V. STERNLING, L.E. SCRIVEN, AICHE J. 5, 514-523 (1959)
8 J.S. LANGER, Rev. Mod. Phys. 52, 1-28 (1980)
9 S. OSTRACH, In: Physico Chemical Hydrodynamics, Advance Pub., London, 1977
10 D. SCHWABE, A. SCHARMANN, J. Crystal Growth 52, 435-499 (1981)
11 C. DAUZERE, J. Physique 7, 930-934 (1908)
12 V. VOLKOVISKY, Publications Sci. Tech. Ministere de l'Air 151, Paris (1939)
13 Lord RAYLEIGH, Phil. Mag. 32, 529-546 (1916)
14 A. R. LOW , D. BRUNT, Nature 115, 299-301 (1925)
15 M.G. VELARDE, Ch. NORMAND, Sci. American 243, 78-93 (1980)
16 R. PEREZ-CORDON, M.G.VELARDE, J. Physique 36, 591-601 (1975)
17 M.G. VELARDE, R. PEREZ-CORDON , J. Physique 37, 171-182 (1976)
18 L.E. SCRIVEN, C.V. STERNLING, Nature 187, 186 (1960)
19 L.E. SCRIVEN, C.V. STERNLING , J. Fluid Mech. 19, 321-340 (1964)
20 A. ORELL, J. W. WESTWATER, Chem. Eng. Sci. 16, 127 (1961)
21 A. ORELL, J. W. WESTWATER, AICHE J. 8, 350-356 (1962)
22 C. J. F. WANNGARD, Ph.D. Diss., Royal Inst. Techn., Stockholm, 1980
23 J. PANTALONI, B. BAILLEUX, J. SALAN, M.G. VELARDE, J. Non-Equilibrium. Therm. 4, 201-218 (1979)
24 J. PANTALONI, P. CERISIER, R. BAILLEUX, C. GERBAUD, J. Physique Lett. 42, L 147- L 150 (1981)
25 A.V. HERSHEY, Phys. Rev. 56, 204 (1939)
26 K. C. D. HICKMANN, Ind. Eng. Chem. 44, 1892-1902 (1952)
27 H. JEFFREYS, Quart. J. Mech. Appl. Math. 4, 283-283 (1951)
28 K. A. SMITH, J. Fluid Mech. 24, 401-414 (1966)
29 M. BENTWICH, Int. J. Heat Mass Transfer 9, 663-670 (1966)
30 J. C. BERG, A. ACRIVOS, Chem. Eng. Sci. 20, 737-745 (1965)
31 J. C. BERG, A. ACRIVOS, M. BOUDART, Adv. Chem. Eng. 6, 61-123 (1966)
32 A. BOSE , H.J. PALMER, J. Fluid Mech. 126 , 491-506 (1983)
33 S.H. DAVIS, L. A. SEGEL, Phys. Fluids 11, 470-476 (1968)
34 V. LUDVIKSSON, E.N. LIGHTFOOT, AICHE J. 14, 620-626 (1968)
35 J.W. SCANLON, L. A. SEGEL, J. Fluid Mech. 30, 149-162 (1967)
36 S. H. DAVIS, G. M. HOMSY, J. Fluid Mech. 98, 527-553 (1980)
37 J.L. CASTILLO , M.G. VELARDE, J. Fluid Mech. 125, 463-74 (1982)
38 J.L. CASTILLO , M.G. VELARDE, J. Colloid Interface Sci., submitted for publication .
39 E. FERM , D. WOLLKIND, J. Non-Equilib. Thermodyn. 7,169-90 (1982).

Laser-Induced Autowave Processes

F.V. Bunkin, N.A. Kirichenko, and B.S. Luk'yanchuk
Institute of General Physics, Academy of Sciences of the USSR
SU-Moscow, USSR

One of remarkable features of physical, chemical and other systems, which are open for energy and/or mass exchange, is their ability to self-organization [1-3]. The processes in such systems are self-consistent and strongly dependent on initial and boundary conditions, i.e. the dynamics of these processes is an inherent property of the systems.

Various self-organization phenomena in chemically active media exposed to high-power laser radiation have recently become a subject of systematic investigations. The enhanced self-organization ability of such media is attributed to feedbacks between thermal and chemical degrees of freedom [4]: the rate of laser heating of a medium varies with changing of optical characteristics of the medium in the course of chemical reactions, the rate of reaction depending on temperature.

Investigations of this branch of laser-induced chemistry are attractive for experimentalists, since there is a vast possibility to change the sign and gain of feedback by varying the parameters of laser radiation. As a result, these systems provide a universal means of simulating various self-oscillating and autowave processes. This point is illustrated below using some examples of particular homo- and heterophase chemical processes induced by laser radiation.

1. Self-oscillating regimes of metal combustion

Self-oscillations have been observed in many experiments on laser-induced chemistry. Their nature can be explained by the example of laser combustion of tungsten in air, when high-temperature oxidation of tungsten is accompanied by evaporation of the new-formed oxide.

Let us denote the thermal effect of the oxidation reaction by W and the latent heat of oxide evaporation by L. If the target is small, the problem can be considered as a point problem and variations of temperature T and of thickness of the oxide layer x can be defined by

$$mc\,\frac{dT}{dt} = PA(x) + \rho sW\,\frac{d}{x}\,\exp\left(-\frac{T_d}{T}\right) - \rho sLv\,\exp\left(-\frac{T_v}{T}\right) - P_{loss}(T),$$

$$\frac{dx}{dt} = \frac{d}{x}\exp\left(-\frac{T_d}{T}\right) - v\,\exp\left(-\frac{T_v}{T}\right);\quad T(t=0)=T_{in},\quad x(t=0)=0, \tag{1}$$

where m, c, and s are the mass, specific heat, and area of the target surface, respectively; ρ is the oxide density; d and T_d are constants of the oxidation law; v and T_v are constants of oxide evaporation; P_{loss} is the heat loss power; P is the radiation power; A(x) is the target absorbency.

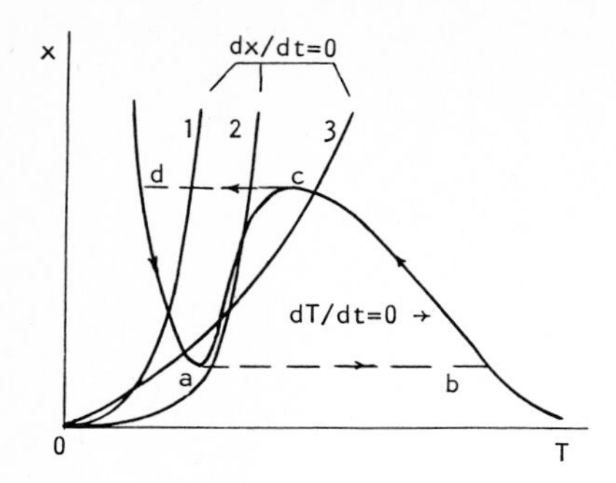

Fig. 1 Nullclines for equations (1), a-b-c-d is the phase trajectory of a limit cycle.
Curve 1: driven regime, in which the system generates a single pulse in response to a perturbation with sufficiently high amplitude.
Curve 2: self-oscillating regime, T(t) and x(t) are periodically oscillating.
Curve 3: trigger regime, a perturbation switches one stable state into another.

Now consider a simple case when A(x)=const. The set of equations (1) is analyzed in terms of the theory of nonlinear oscillations. Fig.1 shows possible intersections of the nullcline dT/dt=0, dx/dt=0 (for $T_d > T_v$ and W>L).

Different types of the system behavior correspond to different patterns of isocline intersections. The essential property of the system is that the transition from one regime to another can be produced by varying an external parameter, viz. the radiation power P.

All the above regimes have been observed in experiments on heating of tungsten and molybdenum targets by CO_2-laser radiation (P≈40 W) in air [5]. Fig.2 gives oscillograms of T(t) and dT/dt signals taken in heating of tungsten targets.

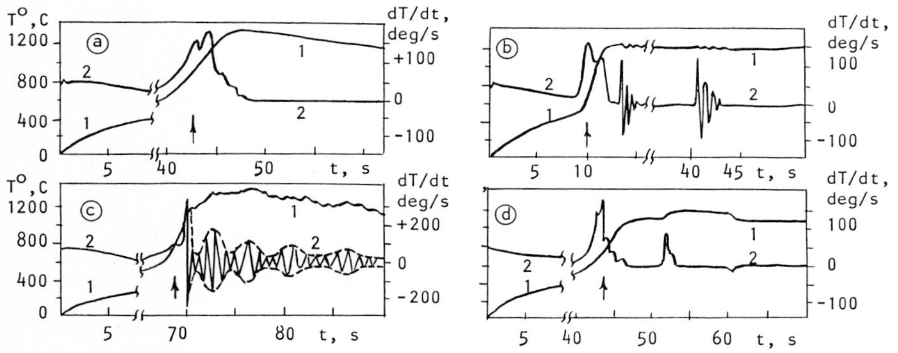

Fig. 2 Experimental oscillograms for laser heating of tungsten targets in air.
1 - T(t); 2 - dT/dt

The arrows point to the moments of target ignition. After the ignition, the following regimes of quasistationary combustion of targets are observed: a) stationary regime, b) driven regime, c) self-oscillating regime, d) trigger regime.

2. Laser heating of massive samples

Now consider the behavior of a thermal field produced by a laser beam in massive samples [4,6,7]. It is assumed that the sample occupies the semispace $z \geq 0$, the radiation being absorbed of the surface z=0. The intensity of the laser beam is distributed over its cross-section as $I(r)=I_0\exp(-r^2/r_0^2)$. If an exothermic reaction can take place on the target surface with the formation of an oxide layer, the boundary problem describing the process is

$$\frac{1}{a}\frac{\partial T}{\partial t} = \frac{1}{r}\frac{\partial}{\partial r}\left(r\frac{\partial T}{\partial r}\right) + \frac{\partial^2 T}{\partial z^2}, \quad z>0, \quad T_{(r,z)\to\infty} \to 0,$$
$$-k\left.\frac{\partial T}{\partial z}\right|_{z=0} = A(x)\cdot I(r) + \rho W\frac{\partial x}{\partial t}, \tag{2}$$

where a and k are the thermal diffusivity and conductivity, respectively. Oxide evaporation and heat exchange with the environment are neglected for simplicity.

Two limiting cases are considered here.

(a) Let the variance of absorbency be negligible: $A(x) = \text{const}$ and the oxidation law be taken in the simple form

$$\frac{\partial x}{\partial t} = d\exp(-T_d/T). \tag{3}$$

To study the stability of the thermal field, all possible steady states should be found. It can be done approximately, assuming that $T(r)\approx T_0(1-r^2/R_0^2)$ in the vicinity of r=0 at z=0 and expanding the exponent in Formula (3) into powers of r:

$$\exp(-T_d/T)\approx\exp(-T_d/T_0)\cdot\exp(-r^2/R_e^2), \text{where } R_e^2=R_0^2\cdot T_0/T_d<<R_0^2.$$

Then transforming equations (2) and (3) into an integral equation for T(r,z=0) and expanding the latter into powers of r, we derive a system of transcendental equations for T_0 and R_0. It has been shown [6] that a solution of this system exists only if I_0 and r_0 do not exceed certain critical values. In Fig.3 the boundary of the domain, where steady-state solutions of (2) and (3) exist, is shown.

In the regions lying above the boundary R(j) there are no steady states at all. In the region below the boundary there are two steady states, and inside the "triangle" there are four steady states, only one of them being stable. The "triangle" is one of the catastrophe cross-sections, namely the "swallow tail" [8]. Depending on the value of laser intensity and on the beam radius, either the system comes to a stable state or a combustion wave propagates over the target surface. In the latter case the wave parameters are determined only by the intrinsic properties of the system under rather weak constraints on the initial conditions.

(b) Now, let the reaction be negligibly exothermic, but the rate of heat input into the system increasing with absorption [7]. In the initial phase of the process [4]

$$A(x)\sim A_0 + bx^2, \quad b=4\pi^2(n^2-1)/\lambda^2; \quad bx^2<<A_0$$

(n being the refractive index of the oxide and λ being the wavelength of radiation). Assume the oxidation law to be written as

$$\frac{\partial x}{\partial t} = \frac{d_0}{x}\exp\left(-\frac{T_d}{T}\right). \tag{3'}$$

Unlike the previous case, the problem is now substantially non-stationary. Nevertheless, it can be solved in a similar way (see [7]).

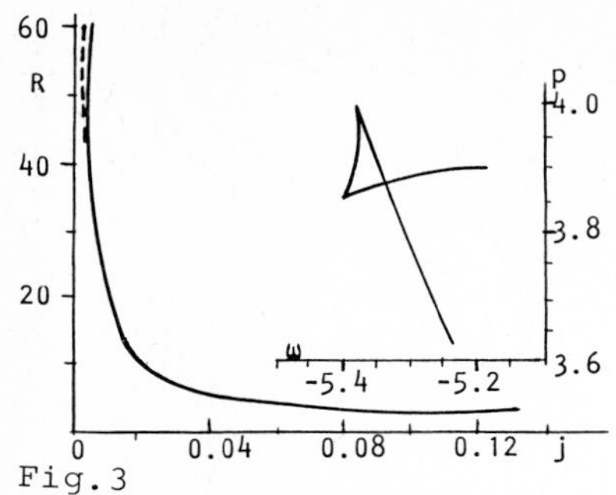

Fig.3

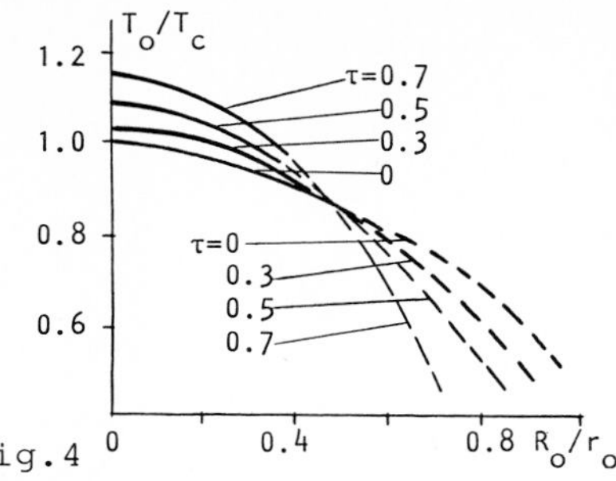

Fig.4

Fig. 3 The boundary of thermal field stability.

$R=\rho dWr_o/kT_d$, $j=AI_o/\rho dW$, $p=kT_d/\sqrt{\pi}AI_o r_o$, $\omega=\ln(\sqrt{\pi}R/j)-4\sqrt{\pi}Rj$

Fig. 4 Thermal field distribution over the surface at successive moments, $\tau = t/t_a$.

Omitting the details of calculations, we give here the approximate expressions for $T_o(t)$ and $R_o(t)$:

$$T_o(t)-T_c \simeq \frac{4}{3}\frac{T_c^2}{T_d}\ln(1/(1-\tau)), \quad \tau=t/t_a, \quad t_a \gg r_o^2/a;$$

$$R_o(t)\simeq\sqrt{2}r_o(1-\tau)^{1/3}, \quad T_c=T_{in}+\sqrt{\pi}A_oI_or_o/2k, \tag{4}$$

$$t_a \simeq \frac{\sqrt{2}}{3}\frac{T_c}{\sqrt{T_d(T_c-T_{in})}}\frac{1}{bd_o}\exp(T_d/T_c).$$

These relations are illustrated in Fig.4. As the reaction accelerates, the thermal field sharpens, thereby a localized structure appears.Recently, similar phenomena have intensively been studied for a number of problems of the theory of nonlinear thermal conductivity (see, e.g. [9] and [10]).

3. Formation of structures in homogeneous media

If the radiation is absorbed in a volume of a chemically active medium, the reverse influence of the medium upon radiation (chemical inertial nonlinearity) must be taken into account [4,12]. Let a reaction A→B run in a certain medium. The refractive indices of substances A and B are n_1 and n_2, respectively ($n_2>n_1$). The effect of radiation on the substance is described by the system of equations of geometrical optics, heat conductivity and chemical kinetics [4,12]. The analysis [4] has shown that if the beam has a Gaussian profile when entering the medium, it starts self-focusing in the course of reaction due to formation of a substance with higher optical density. Diffraction effects cause formation of necks on the beam, which are connected with the non-uniformity of the reaction along the z-axis. If the beam entering the medium has a plane front, it tends to filamentation [14]. These features are illustrated in Figs. 5 and 6.

In the case of a plane front beam at z=0, the increase of the intensity in a filament is followed by its decrease, first, because of increasing distance and, second, because of the wave of quenching due to thermal conductivity and diffusion.

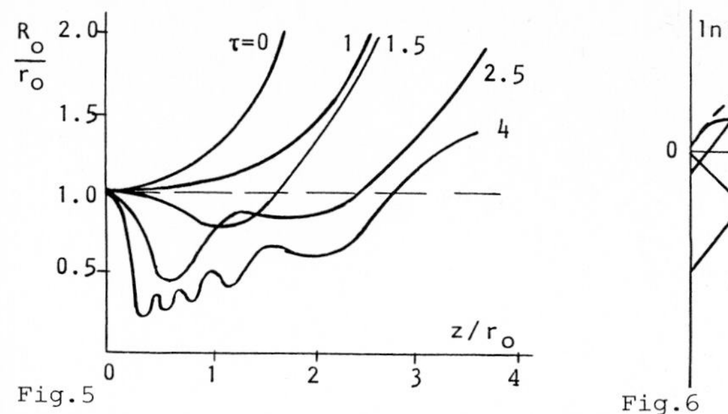

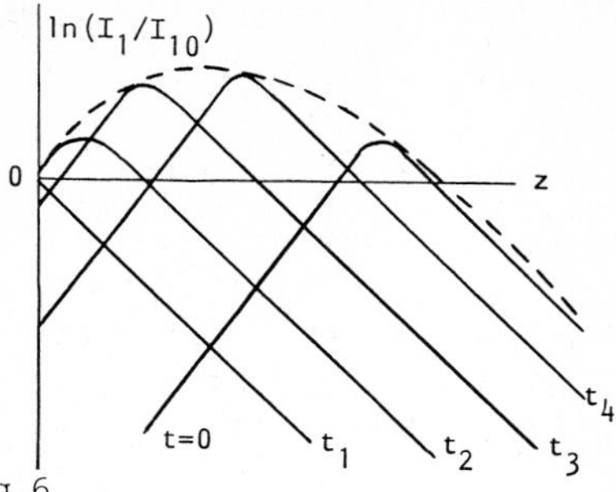

Fig. 5 Effective beam radius versus the coordinate along the beam axis in time succession (the beam having a Gaussian distribution over the radius at z=0).

Fig. 6 Radiation intensity distribution in a filament along the z-axis, the beam having a plane front at z=0.

4. Diffusion instability of gas mixtures

If thermodiffusion phenomena are substantial, i.e. fluxes of substance, proportional to the temperature gradients, are considerable, the medium may form stationary spatially non-uniform structures. Thus if a thin cell with a mixture of two non-reacting gases is irradiated with a beam, having uniformly distributed intensity, and if the radiation is absorbed in the lighter component with concentration N, then the problem can be described by the set of equations of thermal conductivity and diffusion:

$$\frac{1}{a}\frac{\partial T}{\partial t} = \frac{\partial^2 T}{\partial x^2} + \frac{I}{k}\beta N - \frac{\eta}{kh}(T-T_{in}), \quad T\Big|_{t=0} = T_{in},$$
$$\frac{1}{D}\frac{\partial N}{\partial t} = \frac{\partial^2 N}{\partial x^2} + \alpha\frac{\partial}{\partial x}\left[\frac{N(1-N)}{T}\frac{\partial T}{\partial x}\right], \quad N\Big|_{t=0} = N_o, \tag{5}$$

where α is the thermal diffusion constant ($\alpha<0$), β is the absorbency of the light component, η is the constant of heat exchange with the environment. In [15] the stability of the problem (5) was analyzed and the conditions for formation of dissipative structures were established. Fig.7 shows a periodic structure obtained when the threshold intensity is exceeded.

We believe that the above examples give the idea about simulations of various processes and about the capabilities of chemically active media to exhibit self-organization when exposed to laser irradiation.

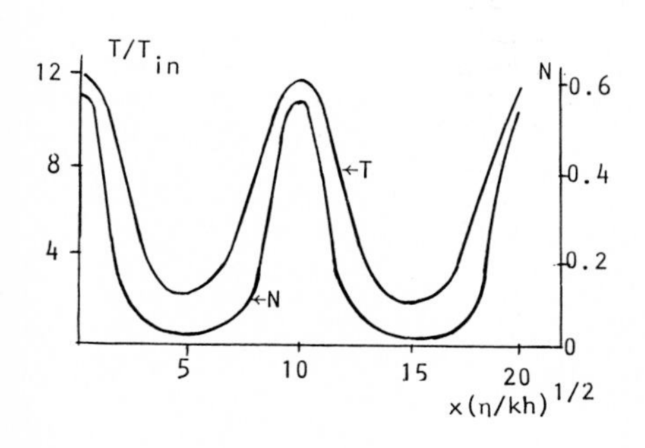

Fig. 7 Typical distributions T(x) and n(x), when a homogeneous solution is no longer stable [15]. $\alpha=-3$, $N_o=0.2$, $I\beta h N_o/\eta T_{in}=10$

References

1. G.Nicolis, I.Prigogine. Self-Organization in Nonequilibrium Systems. John Wiley, New York (1977)
2. H.Haken. Synergetics. An Introduction. Nonequilibrium Phase Transitions and Self-Organization in Physics, Chemistry and Biology. Springer, New York (1977)
3. W.Ebeling. Strukturbildung bei Irreversiblen Prozessen. Teubner, Leipzig (1976)
4. F.V.Bunkin, N.A.Kirichenko, B.S.Luk'yanchuk. Usp. Fiz. Nauk, 138, 45 (1982) (English transl. - Sov. Phys. Uspekhi, 1982, v.25)
5. V.A.Bobyrev, F.V.Bunkin, N.A.Kirichenko, B.S.Luk'yanchuk, A.V.Simakin. Kvantovaya Elektronika, 10, 793 (1983) (English transl. - Sov. J. Quantum Electronics (1983), v.13(4))
6. F.V.Bunkin, N.A.Kirichenko, B.S.Luk'yanchuk. ibid., 9, 1959 (1982)
7. N.A.Kirichenko, B.S.Luk'yanchuk. - ibid., 10, 819 (1983)
8. T.Poston, I.Stewart. Catastrophe Theory and its Application. Pitman, London (1979)
9. S.P.Kurdyumov. In: Up-to-Date Problems of Mathematical Physics and Computational Mathematics, Moscow, Nauka (1982) p.217
10. V.A.Galaktionov, S.P.Kurdyumov, A.P.Mikhailov, A.A.Samarskii. Differen. Uravneniya, v.XVII, p.1826 (1981)
11. F.V.Bunkin, N.A.Kirichenko, B.S.Luk'yanchuk. Izvestiya Acad. Sci. USSR, ser. fiz., 45, 1081 (1981)
12. F.V.Bunkin, N.A.Kirichenko, B.S.Luk'yanchuk. - ibid., 46, 1150 (1982)
13. D.T.Alimov, Sh.Atabaev, F.V.Bunkin, V.L.Zhuravskii, N.A.Kirichenko, B.S.Luk'yanchuk, A.I.Omel'chenko, P.K.Khabibulaev. Sov. J. Surface, No 8, p.12 (1982)
14. N.A.Kirichenko. Preprint No 196 (1982), P.N.Lebedev. Physical Institute of the USSR Academy of Sciences
15. F.V.Bunkin, N.A.Kirichenko, B.S.Luk'yanchuk, Yu.Yu.Morozov. Kvantovaya Elektronika, 10, No 9, (1983) (English transl. - Sov. J. Quant. Electronics (1983), v. 13 (9))

Completely Integrable Models in the Domain Walls and Interphases Theories

V.M. Eleonskii, N.E. Kulagin, L.M. Lerman, and Ja.L. Umanskii

Scientific Research Institute for Applied Mathematics and Cybernetics
Gorky State University, SU-Gorky, USSR

Many problems of nonlinear physics lead to necessity of studying completely integrable dynamical systems. For instance, the integrable models in the theory of interphases for the media with several order parameters (magnetic or segnetoelectric media) allow us to give an exhaustive classification of the interphases and to study how the number and the kind of interphases may vary due to continuous variation of structural parameters of a model [1]. The construction of completely integrable Hamiltonian systems with the potentials associated with the Landau phase transitions theory with several order parameters (such as potential, permitting a continuous transition from the potential pit with one global minimum to that with several local minima) gives us new possibilities to develop the models of phase transition (also in terms of soliton states). The existence of such classical models means that the corresponding quantum model has a coordinate system in which the Schrödinger equation variables become separated. The method of transfer matrix for quasi-one-dimensional systems is known [2] to lead it the thermodynamical approach to the eigenvalue problem for the Schrödinger type operator. Here the changes of the structure of the phase space of classical problem (for example, those connected with a change of the type and number of singular points of a Hamiltonian system) correspond to characteristic changes of the dependence of eigenvalues on the model's structural parameters. The opportunity of thorough analysis of both classical and quantum completely integrable Hamiltonian systems is of considerable interest for statistical physics of quasi-one-dimensional structures with several order parameters.

Since the mathematical model of a plane domain wall is a separatrix, doubly asymptotic to the saddle singular points, the study of the problem of domain walls corresponds to the constructions of a separatrix contour. Separatrix contour is a graph with connections between all the saddle type points. The problem is its reconstruction if the model's structural parameters vary. Then it is necessary to take into account the existence of the separatrix loops at the graph vertices, which can be considered as images of the new phase "embryos". It is very important for the theory of interphases in media with several order parameters that this problem may be analysed in detail for integrable models.

Let us consider integrable models for Hamiltonian systems with two degrees of freedom. The well-known classical integrable problems, i.e. the planar, spherical or ellipsoidal motions of a particle under the influence of a quadratic potential, may be generalized to the case of multiparameter potentials (both regular and singular), the model remaining integrable [1,3-5]. An effective approach to construct completely integrable models is based upon the hypothesis that Hamiltonian function and additional first integral are quadratic polynomials of generalized momenta. The condition of Poisson bracket vanishing yields a system of partial differential equations for the metric coefficients and the pair of conjugated potentials. The condition of solubility of this system gives a linear hyperbolic equation whose solutions define the dependence of one of conjugated potentials (for example, of that contained in the Hamiltonian) on generalized coordinates. In [1,3,4] both general integral representations of those potentials and representations as superposition of the homogeneous polynomials of two variables defining an orthogonal curvilinear coordinate system on a two-dimensional manifold (more exactly a map of the corresponding atlas) are given. In such a system the variables of Hamilton-Jacobi equations may be separated. In [1] it is shown that for the problem of planar motion of a particle two basic classes of multiparametric potentials of integrable models may be selected. The first class may be connected with such continuous transformations of the quadratic potential of a two-dimensional oscillator which lead to formation of arbitrary number of local minima and maxima if the potential tends to infinity in any direction on a plane. The second class may be related to continuous transformations of the linear potential (a particle moving in a homogeneous field) which lead to formation of local minima and maxima if the potential increases infinitely in the selected direction (i.e. when the potential barrier is preserved).

We do not know whether it is possible to construct other general and physically meaningful multiparametric integrable Hamiltonian systems with two degrees of freedom, because the structure of the set of completely integrable systems is not clear yet.

Thus, it is of special interest to develop an advanced mathematical approach to the mentioned problems. We believe that the most convenient basis is the qualitative topological theory of integrable Hamiltonian systems considering the topological structure of foliation on the system trajectories and the changes of this structure when the system parameters vary.

A Hamiltonian system with n degrees of freedom and with the Hamiltonian H is called integrable within a region D of the phase space if there exist functions H_i $(i=1,\dots,n)$ such that the Hamiltonian vector fields corresponding to them are independent in an open dense subset of D. The functions H_i are in involution and therefore flows of these vector fields commutate. Thus, the action Φ of the group R^n is correctly defined by $\Phi(a;x)=f^1_{t_I}\dots f^n_{t_n}(x)$, $a=(t_I,\dots,t_n)\subset R^n$, $x\subset D$, the orbits of Φ being the invariant manifolds for a given Hamiltonian system. Hence the problem of a Hamiltonian system orbit structure is reduced to studying foliation into action orbits.

The classical Liouville theorem about integrability describes the orbit structure in a set of points where functions H_i are independent. Namely, in a vicinity of every such point the symplectic coordinates (p,q) exist in which the action orbits are p=const (in particular, n-dimensional orbits are Lagrangian manifolds). If, in addition to that, the given common level of integrals is compact then its connected component is an n-dimensional torus T^n. Hence, of special interest is the study of a degenerate set, where the functions are dependent, as well as the description of the neighbouring structure of orbits.

For systems with two degrees of freedom this approach was given in [6]. In this case the action orbits are homomorphic (in inner topolopy) to $T^p x R^q$, $p+q\leqslant 2$. In particular, degenerate orbits may be of three types: singular points, closed curves, and lines. The examples show that orbit foliation may be very complicated. Therefore, some generic cases are to be extracted. It turned out that there exist exactly four types of singular points. The linear parts of corrsponding Hamiltonians are of the following form:

1) $H_i^o = \lambda_1^{(i)} p_1 q_1 + \lambda_2^{(i)} p_2 q_2$ - saddle-saddle singular point

2) $H_i^o = \lambda_1^{(i)} p_1 q_1 + \lambda_2^{(i)} (p_2^2 + q_2^2)/2$ - saddle-centre singular point

3) $H_i^o = \lambda_1^{(i)} (p_1 q_1 + p_2 q_2) + \lambda_2^{(i)} (p_1 q_2 - p_2 q_1)$ - focus-focus singular point

4) $H_i^o = \lambda_1^{(i)} (p_1^2 + q_1^2)/2 + \lambda_2^{(i)} (p_2^2 + q_2^2)/2$ - centre-centre singular point

Here $\delta \neq 0$ (the condition of generity), $\delta = \lambda_1^{(1)} \lambda_2^{(2)} - \lambda_1^{(2)} \lambda_2^{(1)}$.

The action orbit structure near those points is completely described by the following:
THEOREM: In some neighbourhood of a singular point with $\delta \neq 0$, action ϕ is topologically equivalent to the linear action having corresponding Hamiltonians 1)-4).

This theorem's proof, as N.K. Gavrilov noted, follows from the Rüssmann theorem [7].

The study of foliation structure near a one-dimensional orbit is reduced to a problem of the theory of singularities of differential mappings. Indeed, near a point belonging to the one-dimensional orbit there exist the symplectic coordinates in which $H_1 = x_1$, $H_2 = h(x_1, x_2, y_2)$, $h(0,0,0) = h_{x_2}(0,0,0) = h_{y_2}(0,0,0) = 0$, i.e. there exists a family of functions $h(x_1, x_2, y_2)$ depending on parameter x_1 such that at $x_1 = 0$ function $h(0, x_2, y_2)$ has a critical point at $x_2 = y_2 = 0$. In generic case this family is equivalent to one of the following families: $h_o = x_1 + x_2^2 \pm y_2^2$ or $h_o = x_2^3 + x_1 x_2 + y_2^2$ [8]. Therefore, local orbit foliation is topologically equivalent to foliation $x_1 = C_1$, $h_o(x_1, x_2, y_2) = C_2$. Consequently, one-dimensional orbits may be hyperbolic ($h_o = x_1 + x_2^2 - y_2^2$), elliptic ($h_o = x_1 + x_2^2 + y_2^2$) and parabolic ($h_o = x_1 x_2 + x_2^3 + y_2^2$). Closed one-dimensional orbits may be of all the three types; but nonclosed (bounded) orbits may be either hyperbolic or elliptic.

From the point of view of the phase transition theory the nonclosed orbits which contain in their closures singular points of the types of saddle-saddle, focus-focus and saddle-centre are of the most interest. One-dimensional orbits were shown to contain in their closures the singular points of one type, the existence of such an orbit implying a certain arithmetic relation for eigenvalues of corresponding Hamiltonian systems at those singular points [9].

Also note that a continuous family of trajectories lying at a two-dimensional action orbit which adjoins to the saddle-saddle or focus-focus singular point is destroyed when nonintegrable Hamiltonian perturbation is imposed. Instead of this family the basic loops (or separatrices) may appear. Near the basic loops there exists a countable set of loops with various numbers of winding around the basic loops [10]. This set may be called a stochastic cluster of travelling waves of the original distributed system. To find these basic loops the analytical formulas we have obtained may be used. These formulas are the generalization of the known Melnikov's formulas [11].

For the initial Hamiltonian system, one-dimensional action orbits joining the singular points are either loops or separatrices going from saddle to saddle. Stable and unstable manifolds of saddles may either transversally intersect along those loops (in corresponding energy level) or coincide. This distinction is particularly evident when we consider a perturbed non-integrable Hamiltonian system. The transversal loops hold under Hamiltonian perturbation but the behaviour of nontransversal loops is more complex. At the present time the case, when all trajectories lying on the stable manifold of a saddle are doubly asymptotic, is well known. Here on a saddle's two-dimensional stable manifold there are exactly 4 entering trajectories which are one-dimensional action orbits (separatrices). Moreover, if a loop is nontransversal then the adjacent loops in the same stable manifold are necessarily transversal. Then we have the following remarkable fact: under a perturbation, instead of two adjacent loops (transversal and nontransversal) of the integrable system a new loop appears which rounds the contour formed by two old loops, the transversal loop being preserved. The more accurate formulation of this statement requires additional notation and definitions and, therefore, is omitted. Hence, it is important to give the conditions of transversality or nontransversality of a loop or separatrix. One of these conditions may be formulated as follows: if a loop (or separatrix) lies in a region where the symplectic coordinates exist in which the Hamiltonian is separated $H(p_1,p_2,q_1,q_2) = H_1(p_1q_1) + H_2(p_2q_2)$, then the loop (separatrix) is transversal.

1. V.M.Eleonskii, N.E.Kulagin. Integrable models for the problem of a particle moving in two-dimensional potential pit, ZETF (Soviet Journal of Theoretical and Experimental Physics) (1983), v.85, No 10 (in press)
2. T.Schneider, E.Stoll. Classical Statistical Mechanics of the Sine-Gordon and Chains. Phys. Rev. B (1980) v.22, No 11, pp.5317-5338
3. V.K.Salaman, P.Kumar. Statistical Mechanics of Magnetic Chains, Phys. Rev. B, (1982) v.26, No 9, pp.5146-5152
3. V.M.Eleonskii, N.E.Kulagin. Some Novel Cases of Integrability of the Landau-Lifshits Equations. ZETF (1983) v.84, No 2, pp.616-628
4. V.N.Kolokol'tsev. Geodesic flows on two-dimensional manifolds with additional polynomial on velocities first integral. Izv.Akad. Nauk SSSR, ser.mat. (1982) v.46, No 5, pp.994-1010
5. A.Ramani, B.Dorizzi, B.Grammaticos. Painleve Conjecture Revisited. Phys. Rev. Lett. (1982) v.49, No 21, pp.1531-1541
6. L.M.Lerman, Ja.L. Umanskii. On topological structure of integrable Hamiltonian systems with two degrees of freedom. IX Int. Conf. Nonlin. Osc. (ICNO), Kiev (1981), Abstract (to appear in Proc. of Conf. at 1984, Kiev, "Naukova dumka" Publ. House)
7. H.Russmann. Über das Verhalten analytischer Hamiltonischer Differentialgleichungen in der Nähe einer Gleichgewichtlösung. Math. Ann., Band 154, Heft 4 (1964) 285-300
8. V.I.Arnold, A.N.Varchenko, S.M.Gussein-Zade. Singularities of Differentiable Mappings, Moscow, Nauka (1982)
9. L.M.Lerman, Ja.L.Umanskii. Necessary conditions for existence of heteroclinical trajectories in integrable Hamiltonian system with two degrees of freedom. Uspekhi mat. nauk. (Russian Math. Surveys), v.38, No 5, 1984.
10. L.P.Shilnikov. A contribution to the problem of the structure of an extended neighbourhood of a rough equilibrium state of saddle-focus type. Mat. Sb., v.81 (123) (1970) No 1 (Math. USSR Sbornik, v.10 (1970) No 1)
11. L.M.Lerman, Ja.L.Umanskii. On the existence of separatrix loops in four-dimensional, similar to integrable, Hamiltonian system. Prikl. Mat. i Mekh. (Appl. Math. and Mech.) v.47, No 3, (1983) pp.395-401

The Autowave Phenomena on the Surface of Crystallizing Solution

V.N. Buravtsev, A.S. Botin, and B.A. Malomed

Research Computing Center USSR Academy of Sciences, SU-142292 Pushchino, USSR
and
Research Institute for Biological Tests of Chemical Compounds, SU-Kupavna, USSR

1. Introduction

The present paper describes the phenomenon of periodic isothermal phase transitions revealed in the surface layer of aqueous solution during crystallization of some volatile substance (impurity), for example alcohol or ammonia. In space-distributed systems to which the solution surface layer belongs, the phenomenon is manifested by alternating waves of freezing and melting, which propagate over the surface layer.

In order to get the periodic regime of phase transition, the solution in a vessel was placed into a thermostatic chamber where constant temperature (below the freezing point of water) and rather small concentration of evaporating ammonia were provided by passing the nitrogen vapours (Fig.1).

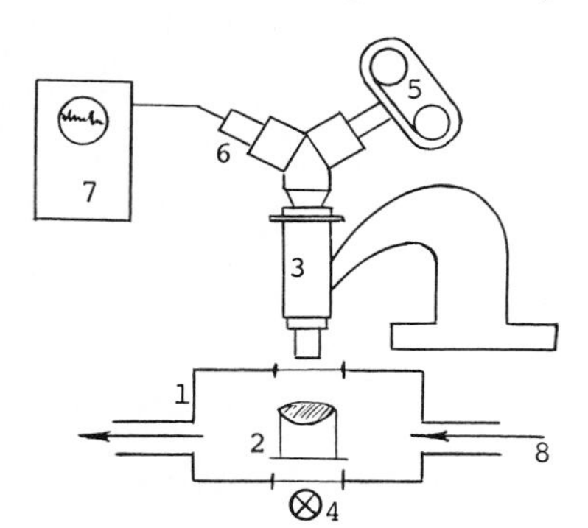

Fig. 1 Experimental set-up: 1 - chamber, 2 - vessel with ammonia solution (1 cm diameter, 1 cm height), 3 - microscope, 4 - light source, 5 - camera, 6 - photo-multiplier, 7 - oscilloscope, 8 - nitrogen vapour flow

Under such conditions the system is open because ammonia may diffuse to the surface layer from the depth of the solution and, at the same time, it may evaporate from the surface. Unlike the case with a single wave of phase transition bringing the system to an equilibrium state, in our system waves of straight and reverse phase transition may occur and propagate under constant conditions, replacing periodically each other. Thus the existence of autoregulation in this system suggests itself. Indeed, first of all, the temperature of ice melting, which corresponds to the concentration C_0 in the solution, is lower than the temperature T in the experiment. On the other hand, the solution surface cannot remain free of ice (unfrozen) for a long period of time, because the ammonia concentration in the surface layer becomes small due to evaporation so that the corresponding freezing temperature is higher than T. Therefore, neither the solid ice layer nor the ice-free surface can exist as steady states of the system. Moreover, a finite decay time of the metastable phase seems to be the delay time required for the periodic regime to appear. However, no definite conclusions can be drawn because of the lack of satisfactory theoretical description of the phase transition kinetics. Therefore, it was important to examine the possibility of autooscillation in the experiment and to give an adequate theoretical description of the phenomenon.

We hope, also, that the phenomenon considered in the present paper may serve as a simple model of the periodic isothermal structural rearrangement in more complicated systems, such as biological membranes.

A brief communication on the experimental observation of periodic phase transitions was published in [5] , and a simple version of the theory was proposed in [4].

2. Experimental Results

2.1. Evolution of the Microstructure

At the maximum melting stage the microstructure of the surface layer is formed by the randomly located solid phase "embryos" of about 10 mm in size, their shape varying from nearly spherical to very stretched (Fig.2a). Some of them contact each other and form a network on the surface, while others exist separately.

The further evolution of this structure was observed as an increase of embryos' size, their number remaining practically constant (Fig.2b,c,d). Thus, the solution surface was found to be covered with an ice cellular structure of grown-up embryos (Fig.2e). Later on the cellular structure changed into the "friable" one (Fig.2b). The latter structure then decayed into the embryos' network (refer from Fig.2f back to 2a), thus completing the evolution cycle, which was periodically repeated (Fig.2g).

2.2. The Wave Regimes

Microstructures of different parts of the surface can simultaneously be at different stages of the above described cycle, forming a surface pattern which varies with time.

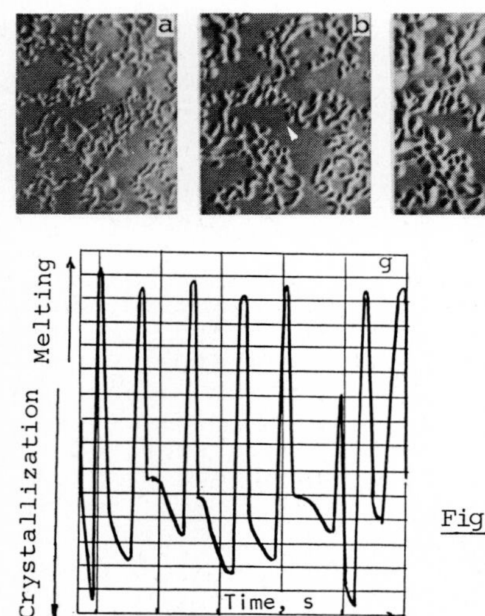

Fig. 2 a-f: Periodic evolution of the surface layer microstructure (see explanations in text); g - variations of the light flux through the solution surface during the microstructure evolution

This pattern is generated by alternating crystallization and melting waves. Two basic wave regimes can be discerned, which differ in the relative direction of motion of the alternating fronts. When the motion of the fronts are co-directed, a plane crystallization wave arises near the vessel wall and travels over the surface. It is then overtaken by a faster melting wave (Fig.3).

In the opposite case, the crystallization front also arises near the wall, but after having gone some distance it stops and goes back as a melting front. In this regime, a structure of the type of a pulsating ice-hole with periodically varying diameter (Fig.4) occurs along with the plane front.

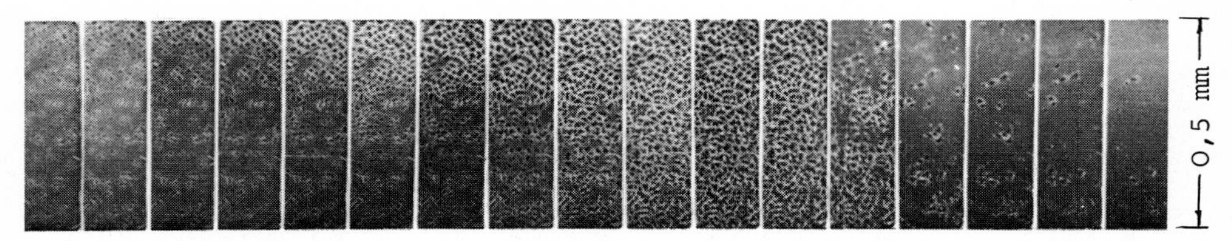

Fig. 3 Filmogram of propagation of alternating crystallization and melting fronts, both fronts moving in the same direction. 18 frames per second. Time runs from left to right.

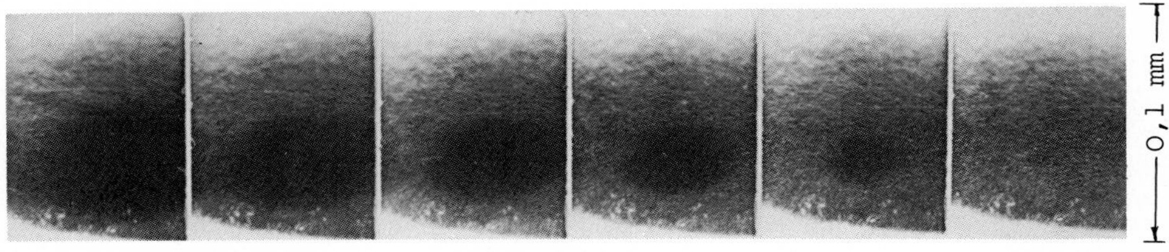

Fig. 4 Filmogram of a pulsating ice-hole in the stage of decrease of its diameter. 24 frames per second. Time runs from left to right.

3. Mathematical Model

Since the ice surface, as it appears in the above experiments, consists of separate approximately spherical ice embryos, which are distributed more or less uniformly, we shall consider the dynamics of a solitary embryo. Let it be placed into a cubic vessel of unit volume (Fig.5). Also, let an impurity diffuse into the vessel from a reservoir, where its concentration C_0 is constant, and evaporate from the upper surface layer into the space where the concentration $C_{out}=0$. The concentration $C=n/(1-\frac{4}{3}\pi r^3)$ inside the vessel is supposed to be spatially uniform (n is the number of impurity molecules in the vessel, r the radius of an embryo). Describing the diffusion of the impurity

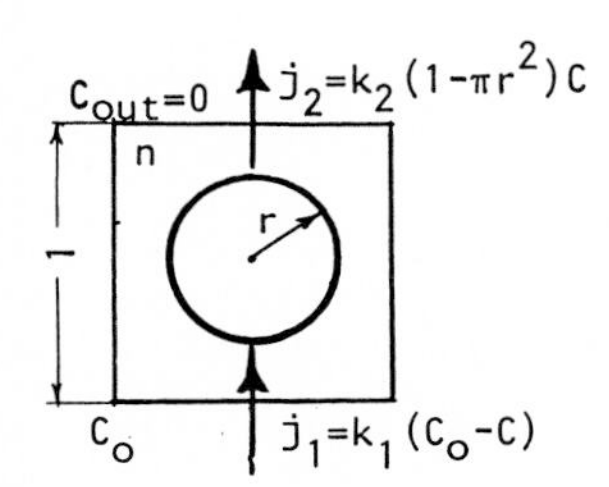

Fig. 5 A solitary embryo in the solution: r is the embryo's radius, n is the total number of ammonia molecules, j_1 and j_2 are the diffusion flux to the surface layer from the depth and the evaporation flux from the layer, respectively

to the finite-difference approximation and considering the evaporation flux to be proportional to the part of the vessel cross-section not occupied by ice, one obtains the balance equation:

$$\frac{du}{dt} = k_1(C_o - C) - k_2(1-\pi r^2)C \tag{1}$$

The evolution equation for the embryo radius is taken in the Onsager form |3|

$$\frac{dr}{dt} = -\varkappa \frac{\partial \phi}{\partial r} = -4\pi\varkappa(\Delta\mu^{(O)} r^2 - nT \frac{r^2}{1-\frac{4}{3}\pi r^3} - 2\sigma r) \tag{2}$$

where ϕ is Gibbs' potential for dilute solution [4], $\varkappa$ is the phenomenological kinetic coefficient, $\Delta\mu^{(O)}$ is the difference between the chemical potentials of pure water and ice, σ is the surface tension.

The phase plane of system (1),(2) is depicted in Fig.6a. The stationary point A is a stable node, B is a saddle, D is an unstable node. The character of C depends on the parameters. If C is located sufficiently close to B, it is an unstable focus. As C moves away from B, it first becomes a stable focus and then turns into a stable node. The calculation of the second Lyapunov's focus quantity [5] for C reveals that focus C loses stability when merging with the small-amplitude unstable limit cycle in its vicinity. With unstable focus C, model (1,2) has no limit cycle which might account for the experimentally observed oscillations. However, this model is not complete. Indeed, according to Eqs.(1),(2) the system without ice (r=0) comes to the stable point A and stays there. At the same time, this point corresponds to the supercooled solution, i.e. it is metastable. Therefore, the system will leave this state in a finite time due to fluctuations. Thus, we should supplement Eqs.(1),(2) with calculation of the probability of birth of a solid phase viable embryo in the supercooled solution. This calculation can be accomplished in the routine way (see e.g. [2]). The closed stable limit cycle thus obtained is shown in Fig.6.

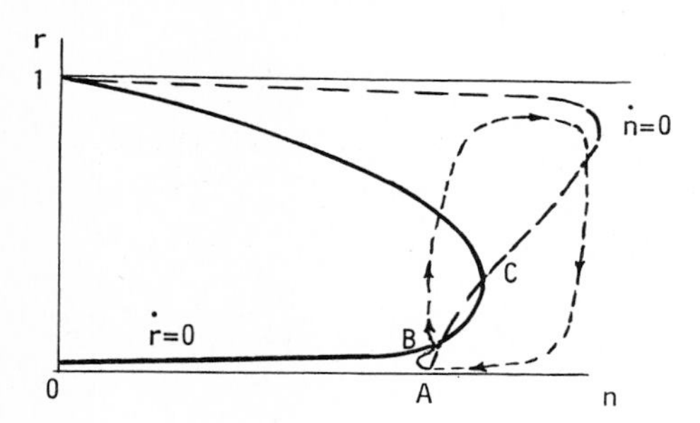

Fig. 6 The phase plane of system (4),(6), and the limit cycle (dashed and wavy lines, the latter depicts the "fluctuation" part of the cycle)

A modification of the model in which the embryo has the form of a spherical segment adjacent to the wall has also been considered (Fig.7). The phase plane of this model for sufficiently large a is shown in Fig.8a. As seen from Fig.8b, a limit cycle occurs in this case without the fluctuation stage.

It is necessary, however, that

$$a > \frac{2\sigma}{\Delta\mu^{(O)}}$$

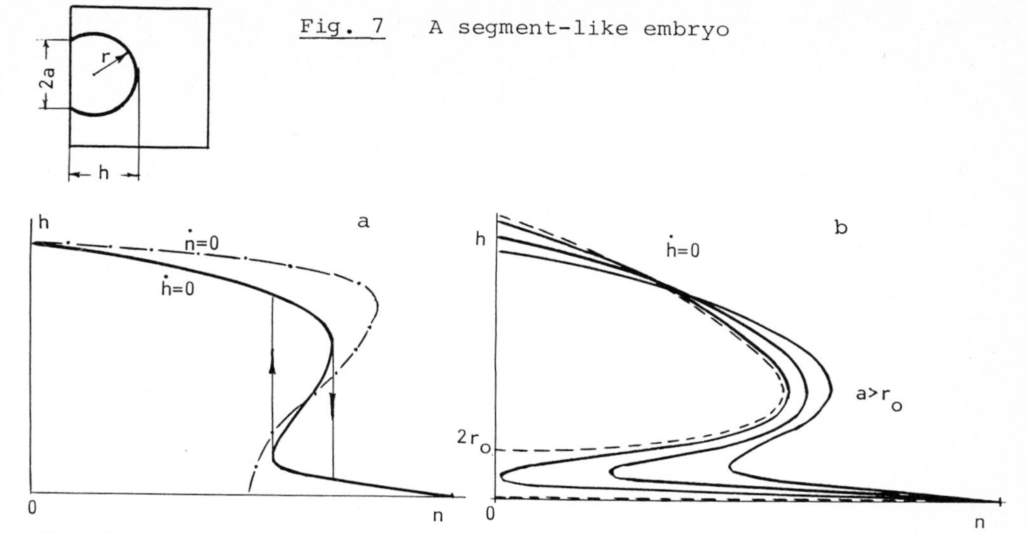

Fig. 7 A segment-like embryo

Fig. 8 a) The phase plane for a segment-like embryo. The dashed nullcline h=0 corresponds to a=0. The solid curves depict the same nullcline for $a>r_0$, displacing to the right with the growth of a. The quantity r_0 is the critical embryo size for pure water; b) the limit cycle for a segment-like embryo

i.e. that the radius of the segment base should be greater than the critical radius of the spherical ice embryo in pure water. Thus, analysis of the latter model points to the possible role of an inhomogeneity in the occurrence of the periodic regime and suggests that the embryos adjacent to the vessel wall can enter a limit cycle.

References

1. V.N.Buravtsev. Zhurn.Khim.Fiz. (Russian Journal of Physical Chemistry) 57, No 1, (1984)
2. V.N.Buravrsev, V.A.Vasilyev, B.A.Malomed, ibid. 58, No 4 (1984)
3. E.M.Lifshits, L.P.Pitaevsky. Physical Kinetics. Moscow, Nauka, (1979)
4. L.D.Landau, E.M.Lifshits. Statistical Physics, Part 1. Moscow, Nauka (1976)
5. J.E.Marsden, M.McCraken. The Hopf Bifurcation and Its Applications. Springer-Verlag, New York (1976)

Autowave Processes in Semiconductors with the Temperature-Electric Instability

Yu.V. Gulyaev, Yu.A. Rzhanov, Yu.I. Balkarei, L.L. Golik, M.I. Elinson, and V.E. Pakseev

Institute of Radioengineering and Electronics of the Academy of Sciences of the USSR
SU-Moscow, USSR

Autowave processes (AWP) can be realized in open kinetic systems which are far from the thermodynamic equilibrium. They are manifested by a variety of nonlinear waves, stationary spatial structures, domains, etc. These structures are stable to small fluctuations, independently of the initial and, often, of the boundary conditions. Many well-known phenomena observed in semiconductors with N- and S-shaped volt-ampere characteristics may be attributed to particular cases of AWP.

In the recent years AWP have been extensively investigated in biology and chemistry. Usually, they are considered in active media, where every physically small element possesses the properties of self-excitation, latent self-excitation, bi- or multistability. These elements are usually coupled with each other by transport processes of diffusive nature [1]. Every element is supposed to be independently pumped by an external source of matter and/or energy and to have a dissipative sink of matter and/or energy. Such media are exclusively rich in various AWP and we use the notion "autowave medium" just in that sense.

At the present time, the results of theoretical and experimental investigations allow one to conclude that there are a great number of ways to produce one-, two- and three-dimensional distributed autowave media on the basis of electronic, optical and thermal properties of semiconductors, insulators, ferroelectrics, and liquid crystals (see, for example, [2-9]). The excitation of these systems may be produced by electric fields, radiation of various wavelength, acoustic fields, and so on. The corresponding frequency range stretches from 1 up to 10^{11}Hz, the characteristic spatial range (wave length) being from 1 to 10^{-5}cm. Physical distributed systems can be created equavalent to a net of 10^4-10^{10} discrete active elements. In order to change smoothly the thresholds of excitation of spatial parts of the medium during the operational cycle, distributed memory can be built into autowave systems, thus essentially extending the field of possible applications of such systems.

AWP in solids can acquire some peculiarities which provide the possibility to combine equilibrium and nonequilibrium phase transitions, to exhibit interactions of autowave structures with phonons, plasmons, magnons. AWP can influence electron spectra, as well as optical, electrical and other properties of solids. Autowave media can be very useful in information processing devices. New types of memory cells, adaptive systems of perseptrone type, devices for analog simulation of complex biological processes could be created on the basis of autowave media.

This paper is devoted to the investigation of a concrete auto-wave medium based on the effect of temperature-electric instability (TEI)

observed in semiconductors more than twenty years ago [10-12]. This instability results in low-frequency oscillations of photocurrent and temperature in a semiconductor sample subjected to joule heating and illumination. To create an autowave medium a system with temperature-electric oscillations is to be built which is distributed in the direction perpendicular to the photocurrent flow. The sandwich-like "metal-semiconductor-metal" structure on a heatconducting base is shown in Fig.1a. This system may be considered as a continuously distributed net of self-excited generators, connected by diffusion of both heat and particles in the plane of the medium. In the simplest case the semiconductor layer is thin and homogeneous along the direction of the current flow.

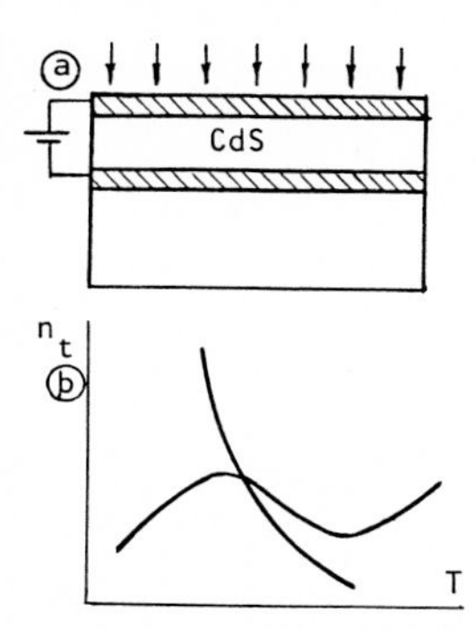

Fig. 1 a. Sandwich-like structure of a semiconductor with TEI; b. Stationary homogeneous model in coordinates (n_t, T)

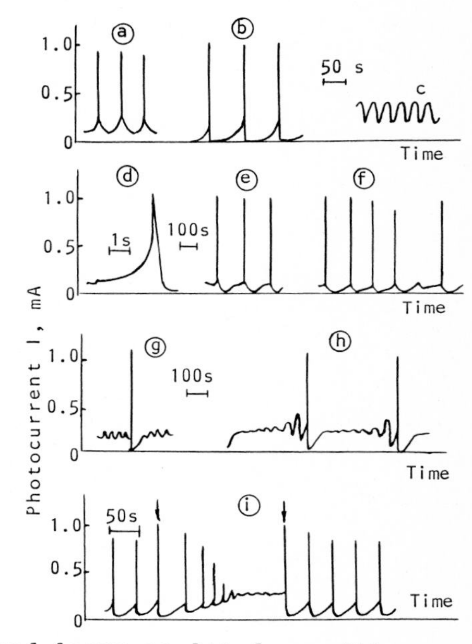

Fig. 2 Experimentally observed oscillation regimes: a,b - relaxation regimes; c - quasiharmonic regime; d - single pulse in driven regime; e,f - pulse series in driven regime with $\tau_S > \tau_R$ (e) and $\tau_S < \tau_R$ (f); g - regime with small- and large-amplitude oscillations; h - complex oscillation regime; i - switching between auto-oscillations and stationary states (arrows indicate the moments of short perturbations)

The mathematical model of the medium is described by the following equations

$$\frac{\partial n}{\partial t} = \beta k_o I - \frac{n}{\tau_R} - \frac{n}{\tau_t}\left(1 - \frac{n_t}{N_t}\right) + \frac{n_t}{\tau_B(T)} + D\nabla_{11}^2 n,$$

$$\frac{\partial n_t}{\partial t} = \frac{n}{\tau_t}\left(1 - \frac{n_t}{N_t}\right) - \frac{n_t}{\tau_B(T)} \; ; \qquad \tau_B(T) = \tau_B^o \exp\left(\frac{E_t}{T}\right) , \tag{1}$$

$$\frac{\partial T}{\partial t} = \frac{e\mu E^2}{c\rho} n - \frac{T - T_o}{\tau_c} + \varkappa \nabla_{11}^2 T,$$

where n is the free electron concentration, n_t the trapped electron concentration, T the semiconductor temperature, T_o the conducting base temperature, N_t the trap concentration, E_t the trap activation energy, k the light absorption coefficient, d the thickness of the sample, I the light flux intensity, E the electric field strength, μ the electron mobility, τ_r, τ_t, τ_B, τ_c the characteristic times of recombination, trapping, traps-band electron activation, and cooling of the sample, respectively, $\varkappa$ and D thermoconductivity and diffusion coefficients, ∇_{11} the Laplace operator in the plane coordinate system.

The described model is quite similar to the models used for some chemical and biological systems [1]. The homogeneous system (1) has only one stationary point:

$$n^* = \beta k_o I \tau_r; \quad n_t^* = n^* \left[\frac{n^*}{N_t} + \frac{\tau_t}{\tau_B(T^*)} \right]^{-1}; \qquad (2)$$

$$T^* = T_o \left\{ 1 + \frac{e\mu E^2 n^* \tau_c}{c\rho T_o} \right\} .$$

In some semiconductor samples with TEI the temperature dependence of parameter τ_r has a rather sharp maximum, where the decrease of τ_r corresponds to the effect of temperature-quenching of photoconductivity [11]. In those cases three stationary points are possible. The main control parameters of the system are: I, E, T_o, τ_c.

After suitable normalization, there are two dimensionless parameters in (1): $\varepsilon=\tau_r/\tau_B$ (T*) and $\varepsilon' = \tau_c/\tau_B$ (T*). If $\varepsilon<<1$, system (1) can be reduced to two differential equations for n_t and T. The isoclinic lines of the reduced system are shown in Fig.1b. When $\varepsilon<<\varepsilon' \leqslant 1$, two relaxational regimes are possible: the auto-oscillations and potential (driven) auto-oscillations. When $\varepsilon<<\varepsilon'\simeq 1$ the auto-oscillations become quasiharmonic. Besides, in the case when $\varepsilon\simeq\varepsilon'\simeq 1$ there appear relaxational auto-oscillations of another kind which could be called "spike" auto-oscillations. The numerical analysis of the model [1] confirms these conclusions and allows some new regimes to be predicted, provided that real temperature dependences μ(T) and τ_2 (T) are taken into account. These new regimes are: i) potential auto-oscillations superimposed on the small-amplitude quasiharmonic oscillations; ii) complex trigger regimes, when the system switches between the stationary and oscillatory states; iii) stochastic transitions from small- to large-amplitude oscillations induced by the periodic modulation of light under the conditions of regime i).

The majority of these regimes were observed in our experiments during the study of CdS monocrystals optically excited from the region of intrinsic or impurity absorption in an electric field with the strength E≃1-5 kV/cm at temperatures of the conducting base ~ 300°K. Simple auto-oscillation regimes are shown in Fig.2a-c. In the potential driven regime (Fig.2a-f), single pulses and pulse series were detected at excitation of the crystal by short stimulating light pulses. The case of Fig.2e is realized when the period of stimulating pulses τ_S is longer than the characteristic time of regeneration of the system (refractoriness time) τ_R, during which the trap population is recovered after the response pulse. In the case of Fig.2f, corresponding to the inequality $\tau_S<\tau$ the effect of arrhythmia can be observed, which is characteristic for biological active media.More complex auto-oscillation regimes consistent with the numerical simulation are shown in Fig.2g-i. In particular, Fig.2i shows the bidirectional switching between the stationary and auto-oscillation states caused by short auxiliary light pulses.

The spatially inhomogeneous auto-oscillatory regimes in system (1) were studied only numerically as yet. Fig.3 shows as an example, the generation of two travelling pulses at the boundaries of the medium and their annihilation in the collision. The considered medium in the regime of quasiharmonic auto-oscillation can be used to amplify spatially inhomogeneous signals and to extract them from intense dynamic noise. Let the light intensity I consist of three parts: the continuous light flux I_o, the inhomogeneous flux $\Delta I(x)$ and the noise $I_n(t,x)$. The additives to I weakly change the oscillation frequency but result in local phase shifts. The weak time-independent signal ΔI is gradually accumulated, while the noise signal I_n is being averaged and practically vanishes. As a result, the signal ΔI can be extracted even from intense noise. The results of numerical simulation are shown in Fig.4. The temperature distribution T(x) is recorded during each oscillation period in the moments when the action of ΔI reaches its maximum. One can see the gradual accumulation of the phase shift between the signal and fluctuating noise. Long exposure to ΔI results in spreading of the phase due to diffusion, the recognition of the region $\Delta I(x)$ being thereby deteriorated. To conclude, we should emphasize that the simple semiconductor medium considered above exhibits a quite nontrivial behaviour.

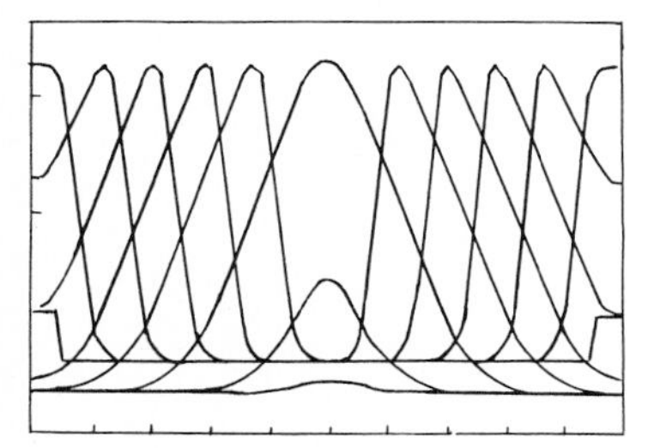

Fig.4

Fig.3

Fig. 3 Collision and annihilation of travelling pulses

Fig. 4 The extraction of the duty signal

References

1. V.A.Vasiliev, Yu.M.Romanovskii, V.G.Yahno. Usp. Fiz. Nauk, 128, 625 (1979)
2. Yu.I.Balkarei, M.G.Nikulin. Fiz. Tech. Polupr., 10, 1455 (1976)
3. B.S.Kerner, V.V.Osipov. JETP, 71, 1542 (1976)
4. B.S.Kerner, V.V.Osipov. Fiz. Tech. Polupr., 13, 721 (1979)
5. B.S.Kerner, V.V.Osipov. Fiz. Tverd. Tela, 8, 2342 (1979)
6. B.S.Kerner, V.F.Senkevich. Pisma v JETP, 36, 359 (1982)
7. Yu.I.Balkarei, M.G.Nikulin, M.I.Elinson. Solid state self-wave media. In: "Autowave processes in Systems with Diffusion", Gorkii, p.117 (1981)
8. L.L.Golik, V.N.Nemenuschi, M.I.Elinson, Yu.I.Balkarei. Pisma v JTP, 7, 813 (1981)
9. L.L.Golik, A.V.Grigoriants, M.I.Elinson, Yu.I.Balkarei. Optics Communications, 46, 51 (1983)
10. V.L.Vinetskii. Fiz. Tverd. Tela, 11, 1402 (1969)
11. S.G.Kalashnikov, V.I.Pustovoit, G.S.Pado. Fiz. Tech. Polupr., 4, 1255 (1970)
12. Y.Seki, F.Endo, T.Irie. Jap. J. Appl. Phys., 19, 1667 (1980)

Thermal Wave Propagation in a Superconducting System as an Autowave Process

Yu.M. Lvovsky

Institute of Thermophysics, Academy of Sciences of the USSR
SU-Novosibirsk, USSR

A superconducting system (SCS) may be treated as an object where the autowave processes can be realized, which essentially affect its operational characteristics and determine the uppermost achievable parameters and stability regions. There are several lines of argument for this. The thermal state of a SCS is described by a quasilinear parabolic equation (in the general case, by a system of equations). The heat released by the current in an overheated superconductor serves as a distributed energy source. Strong nonlinearities in the temperature dependences of heat release and of equation coefficients may generate various autowave regimes (such as development of local overheated regions or propagation of thermal waves, etc.), which are dangerous for stable SCS operation. When a local overheating region with the temperature above a critical value T_c (so-called normal zone) arises in a current-carrying superconductor, it can then propagate along with a constant velocity bringing the system into the emergency (normal) state. The dependence of propagation velocity on the current is one of the most important characteristics of SCS. Earlier the velocity was evaluated in the framework of the simplified model [1] with constant coefficients or by numerical simulation of a given system. Meanwhile, the autowave approach and analogy with combustion theory and other fields will enable one not only to elaborate a general view on the superconductor stability problems but also to obtain new concrete results.

The present paper applies the theory by Kolmogorov, Petrovsky, Piskunov (KPP) [2], which is widely used in autowave problems [3,4], to the analysis of thermal wave propagation in SCS. It allows the velocity/current dependence to be derived with good accuracy for the general nonlinear case without recourse to numerical calculations.

Equation of normal zone propagation. Limiting velocity.

The thermal state of a helium-cooled superconductor of perimeter P and cross-section S with current I is described by the one-dimensional equation of heat conduction

$$C(T)\frac{\partial T}{\partial t} = \frac{\partial}{\partial x}\left(\lambda(T)\frac{\partial T}{\partial x}\right) + \Phi(T)/S \qquad (1)$$

with the source $\Phi = W - Q$. The heat release W in a composite superconductor is [1]

$$W = \begin{cases} 0 & \text{at } T_b < T < T_r \text{ - superconducting state} \\ \dfrac{T-T_r}{T_c-T_r}\, I^2\rho/S & \text{at } T_r < T < T_c \text{ - resistive state} \\ 1 & \text{at } T > T_c \text{ - normal state} \end{cases} \qquad (2)$$

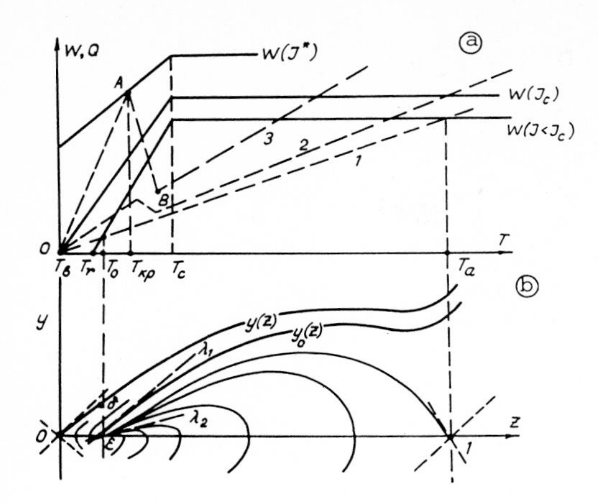

Fig. 1 a - Heat release and heat transfer curves for a superconductor with current; b - Phase portrait corresponding to system (3)

The heat flowing into a cooling bath is $Q=HP\ (T-T_b)$ (curve 1 of Fig.1). Here T_b is the bath temperature, $T_r=T_b+(T_c-T_b)(1-I/I_c)$ is the temperature of transition to the resistive state. At $I\to I_c, T_r\to T_b$. The system is bistable and the stable equilibrium states $T=T_b$ and $T=T_a$ related to the boundary conditions $(\partial T/\partial x)_{x\to\mp\infty}=0$ are realized at the left and right boundaries, respectively. For a travelling wave $T(x,t) = T(\xi=x+vt)$ the substitution [5]

$$\left(\lambda(T)/\int_{T_b}^{T_a} CdT\right)\ dT/d\xi = y\left(z = \int_{T_b}^{T} CdT/\int_{T_b}^{T_a} CdT\right)$$

with arbitrary $\lambda(T)$, $C(T)$ reduces eq.(1) to the problem

$$ydy/dz = Vy - \varphi(z) \tag{3}$$

with the boundary conditions

$$y(0) = y(1) = 0. \tag{4}$$

The source

$$\varphi(z) = \lambda(W-Q)/\left[SC\int_{T_b}^{T_a} CdT\right] \tag{5}$$

is negative at $0<z<\varepsilon$ $(T_b<T<T_o)$ and positive at $\varepsilon<z<1$ $(T_o<T<T_a)$ (see Fig.1). An analytical expression for V cannot be found in such a general formulation [3,4]. Nevertheless, the dependence V(I) can be defined for two limiting cases.

The first case is a so-called minimal propagating current I_m, corresponding to V=0 [6]. From (3) and (4) it follows that

$$V = \int_0^1 \varphi(z)dz/\int_0^1 y(z)dz. \tag{6}$$

At V=0 equation (3) yields $y=\sqrt{-2\int_0^z \varphi dz}$ and from (6) we obtain

$$\left(\frac{dV}{dI}\right)_{I_m} = \frac{\int_0^1 |\partial\varphi(z,I)/\partial I|\,dz}{\int_0^1 dz\left[2\int_0^z - \varphi(u)\,du\right]^{1/2}} = \frac{\int_{T_r}^{T_a} \lambda(\partial W/\partial I)_{I_m}\,dT}{\int_{T_b}^{T_a} dT\left[2sc^2\int_{T_b}^{T}\lambda(Q-W)\,dT\right]^{1/2}} . \tag{7}$$

In the opposite limit, at a critical current I_c, the velocity achieves its maximum value V^*. At $I \to I_c$, $T_o \to T_r \to T_b$ $(\varepsilon \to 0)$ and the source $\varphi(z)$ tends to some positive KPP-function [2]. Here we have

$$V^* = 2\sqrt{\frac{\lambda(T_b)}{sc^2(T_b)}\,\frac{d}{dt}(W-Q)} \qquad \text{and} \tag{8}$$

$$(dV/dI)_{I_c} \to \infty. \tag{9}$$

Validity of the transition to the limit $\underset{I \to I_c}{V \to V^*}$ and of formula (9) are discussed below.

<u>Transition to the positive source limit in KPP problem.</u>

In the KPP theory [2], a positive source $\varphi(z)$ was supposed to satisfy the convexity condition:

$$0 \leqslant \varphi(z) \leqslant \varphi'(0)\cdot z, \qquad 0<z<1. \tag{10}$$

Here the stable wave velocity is given by the bifurcational value $2\sqrt{\varphi'(0)}$. The case when the positive source does not satisfy the condition (10) is considered in [7]. The phase portrait of a positive source problem corresponds to the band $\varepsilon \leqslant z \leqslant 1$ in Fig.1, provided that $z=\varepsilon$ is replaced by $z=0$. The necessary and sufficient condition for a solution to exist is cutting the line $z=1$ by the separatrix $y_o(z)$ (the curve touches the line $y=\lambda_1 z$ at the node, $\lambda_{1,2} = V/2 \pm \sqrt{V^2/4-\varphi'(0)}$) at a certain point $y_o(1)>0$ [7]. From (3) we have $\partial y_o(1)/\partial V > 1$ [7]. Therefore the velocity spectrum is continuous and bounded below, the lower boundary V^* being determined by one of the two different mechanisms cancelling the solution. The first is the node bifurcation (here $V^*=2\sqrt{\varphi'(0)}$) and the second is the node separatrix encountering the saddle (1,0), here $V^*>2\sqrt{\varphi'(0)}$. For convex sources, it is the bifurcation that determines the boundary of the spectrum since at $V^*=2\sqrt{\varphi'(0)}$ it follows from (3) and (10) that $y_o(1) \geqslant \sqrt{\varphi'(0)}$. On the contrary, for strongly nonconvex sources the separatrix $y_o(z)$ can fall into the saddle (1.0) before the bifurcation develops [7].

It should be noted, however, that the above positive source problem is to a great extent an idealization, since it is impossible to maintain an infinite medium at the unstable point T_o because of perturbations. In fact, the unstable equilibrium point is preceded usually by a small region ("threshold" with $\varphi \leqslant 0$) which prevents spontaneous growth of instabilities. Therefore, the feasibility to perform the limit transition from the problem with a threshold to the problem with a positive source is of crucial importance in various applications of KPP-theory.

The solution of the problem with a threshold ε is given by the trajectory going from saddle (0,0) to saddle (1,0). If the limiting (at $\varepsilon\to 0$) source is convex the node separatrix y_0 passes above the saddle (1,0) at $V \geq 2\sqrt{\varphi'(\varepsilon)}$ and separates the required trajectory $y(z)$ from it (see Fig.1). Hence, here a solution can exist only at $V<2\sqrt{\varphi'(\varepsilon)}$, the point $z=\varepsilon$ being a focus. Putting $-kz \leq \varphi\ (z<\varepsilon) \leq 0,\ k>0$ we obtain from (3) $V\varepsilon \leq y(\varepsilon) = \delta \leq (V/2+\sqrt{V^2/4+k})\varepsilon$, hence, $\delta \approx \varepsilon$. Nearby the focus, the trajectories are given by

$$\ln\left[y^2-Vyz_1+z_1^2\varphi'(\varepsilon)\right]+\frac{2V}{\sqrt{4\varphi'(\varepsilon)-V^2}}\ \text{arc tg}\frac{2y-Vz_1}{z_1\sqrt{4\varphi'(\varepsilon)-V^2}} = \text{const}, \tag{11}$$

where $z_1=z-\varepsilon$. For the solution sought, $y(z_1=0)=\delta\simeq\varepsilon$, and $z_1(y=0)\simeq 1$ (hitting the saddle (1,0)), and from (11) it follows

$$2\sqrt{\varphi'(\varepsilon)} - V = \frac{\pi^2\sqrt{\varphi'(\varepsilon)}}{(\ln\varepsilon)^2} + O\left(\frac{1}{\ln^2\varepsilon}\right). \tag{12}$$

Thus, V tends to $V^*=\lim_{\varepsilon\to 0} 2\sqrt{\varphi^1(\varepsilon)}$ from below having $(dV/d\varepsilon)_{\varepsilon\to 0}\to\infty$.

If the source is nonconvex (curve 2 of Fig.1) and the limiting solution is given by the "separatrix mechanism", the node remains at the point $z=\varepsilon$. Here, instead of (12) the estimate $V^*-V \approx \varepsilon^{1-\lambda_2/\lambda_1}$ can be obtained. Hence, V also tends to V^* from below but more quickly and $(dV/d\varepsilon)_{\varepsilon\to 0}\to\infty$, although now V^* is determined not by the local value υ', but by the whole of the limiting function $\varphi(z)$.

Note that for the thermal wave in the SCS under consideration $\varepsilon\simeq 1-I/I_C$.

Dependence of propagation velocity on the current.

For real conductors at helium temperatures, the relationships $C\sim T^3$, $\lambda\sim T$ can be taken as model ones. In dimensionless parameters

$$i = I/I_C,\quad v = V\left[hP\lambda/SC^2\right]_{T_C}^{-1/2},\quad z = \left[(T/T_b)^4-1\right]/\gamma,$$
$$\gamma = \left[1+\alpha i^2(\mu-1)^4\right]-1, \tag{13}$$

the limiting velocity expression becomes

$$v^* = 2\mu^{5/2}\sqrt{\alpha-1} \tag{8a}$$

where $\alpha = I_C^2\rho/hPS(T_C-T_b)$ is the Stekly parameter [1] and $\mu = T_C/T_b$ is the parameter of nonlinearity.

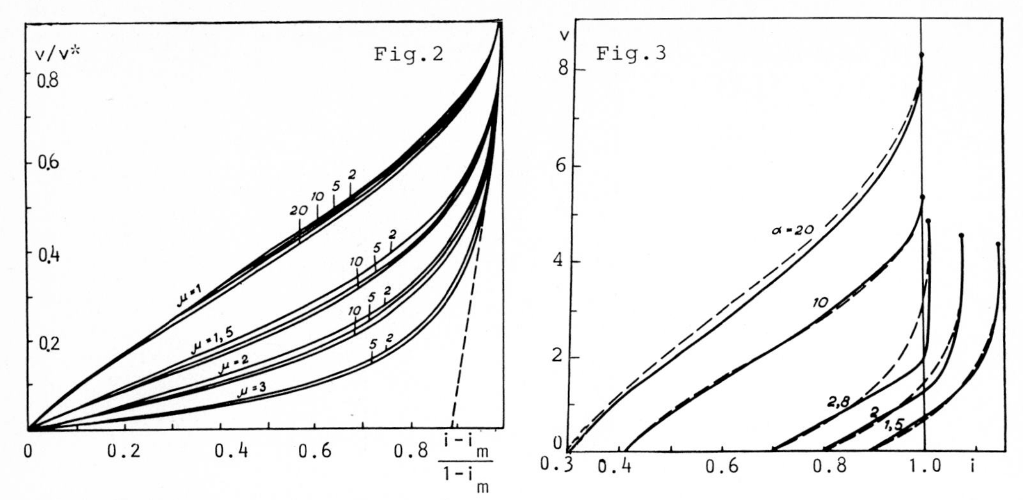

Fig. 2 Normalized dependences of wave velocity on current. Figures at curves denote α values

Fig. 3 Dependence $v(i)$ with boiling crisis taken into account. Solid lines - numerical calculation, dotted - formula (14)

The obtained values of velocity and its derivative at the two points ($v=v^*$ and $dv/di\to\infty$ at $i=1$; $v=0$ and formula (7) at $i=i_m$) allow the dependence $v(i)$ to be approximated with a good accuracy. To compare the approximating values with the exact ones, the problem (3)-(5) was solved numerically. Fig.2 represents numerical dependences in normalized coordinates. It is clearly seen that all curves at $i\to 1$ tend to (12) (dotted line). The normalized derivative $v_1=(dv/di)i_m(1-i_m)/v^*$ is the only characterization of the curves. The dependence

$$v/v^* = 1-\left[1-(v_1/2)\ln(1-(i-i_m)/(1-i_m)\right|^{-2}-$$

$$- v_1(1-0.4v_1)\left[(i-i_m)/(1-i_m)\right]^2(1-i)/(1-i_m), \tag{14}$$

approximates them with discrepancy about 10% and hence is practically applicable.

With due regard to the boiling crisis effect (curve 2, Fig.1), formula (8a) should be replaced by

$$v^* = 2\mu^{5/2}\sqrt{\alpha-\chi}, \quad \chi = h_{nucl}/h_{film}. \tag{8b}$$

The source becomes highly nonconvex with decreasing α and at a certain α_o the expression (8b) becomes invalid. Nevertheless, the numerical calculation showed that for $\mu=2$ or greater the value of α_o insignificantly exceeds χ.

For weakly stabilized superconductors with $\alpha<\chi$, the wave exists even at $I>I_c$, the main part of SCS being in the resistive state. At the value I^* (Fig.1) the maximum velocity is achieved. Here, the relationship (14) can be applied to the whole interval I_m-I^*. As seen from

Fig.3, formula (14) approximates the curves adequately for all values of α which are not close to χ.

References

1. V.A.Altov, V.B.Zenkevitch, M.G.Kremlev, V.V.Sytchev. Stabilization of Superconducting Magnet Systems. Plenum Press (1977)
2. A.N.Kolmogorov, I.G.Petrovsky, N.S.Piskunov. Bull. Moscow University, Sect.A, (1937) No 6
3. V.I.Krinsky, A.M.Zhabotinsky. Autowave Processes in Systems with Diffusion , Gorky (1981) p.6-32
4. V.A.Vasiljev, Yu.M.Romanovsky, V.G.Yakhno. Uspekhi fiz.nauk (1979) v.128, p.625-666
5. Ya.B.Zeldovich, G.I.Barenblatt, V.B.Librovich, G.M.Makhviladze. Mathematical Theory of Combustion and Explosion. Moscow, Nauka, (1980)
6. B.J.Maddock, G.B.James, W.T.Norris, Cryogenics (1969) v.19, p.261-273
7. V.Ye.Gluzberg, Yu.M.Lvovsky. J. Chem. Phys. (Russ.) (1982) v.1, p.1546-1550

Theory of Development of Large-Scale Structures in Hydrodynamical Turbulence

R.Z. Sagdeev, S.S. Moiseev, A.V. Tur, G.A. Khomenko, and V.V. Yanovsky
Institute for Space Research, Academy of Sciences of the USSR, SU-Moscow, USSR

This paper is aimed to propose a theory of self-organization of large vortex motions in a turbulent medium. The hydrodynamical turbulence is an example of nonequilibrium systems commonly met in nature. Turbulence is usually understood as an unordered motion of a fluid such that the energy of a large-scale perturbation is transferred to random chaotic motions of smaller and smaller scales and eventually dissipates. This concept of turbulence seems to be inconsistent with coherent structures existing in a turbulent medium. In the present paper we show, however, that large-scale vortex motions arise in a turbulent medium, provided that this medium is homogeneous, isotropic, and gyrotropic. The scale of the generated vortices, L, is much greater than the correlation length of the turbulence, λ, and is determined by the internal characteristics of the turbulence as $L \sim E/I$, where $E=\int\langle\vec{V}\vec{V}\rangle d\vec{r}$ is the turbulent energy and $I=\int\langle\vec{V}\cdot\nabla\times\vec{V}\rangle\, d\vec{r}$ is the topological invariant. This result is obtained by direct averaging of the Navier-Stokes equation without use of phenomenological models. Some solutions of the averaged equations exhibit large-scale instabilities with the maximum of increment on the scale L. We should emphasize that the generated structures are not the remnants of the central flow (of the type of Karman vortices) but are produced just by the gyrotropic turbulence. The proposed mechanism of generation of vortices is not valid for two-dimensional turbulence, where $\vec{V}\cdot\nabla\times\vec{V}=0$, since in contrast with the latter case there is a specified scale of motions.

The basic equations for the turbulent velocity field, V^t, and density, ρ^t, are

$$\frac{\partial V_i^t}{\partial t} + V_k^t \frac{\partial V_i^t}{\partial x_k} = \nu_o \Delta V_i^t - \frac{C_o^2}{\rho_o} \frac{\partial \rho^t}{\partial x_i}\,, \qquad \frac{\partial \rho^t}{\partial t} + \frac{\partial}{\partial x_k}(\rho^t V_k^t) = 0. \tag{1}$$

Here ν_o is the kinematic viscosity, C_o is the sound velocity (we consider a polytropic medium with $\gamma=2$), $\langle V^t\rangle=0$, $\langle\rho^t\rangle=\rho_o$. The angular brackets denote the ensemble of averaged quantities.

The turbulence is considered to be homogeneous, isotropic, stationary and gyrotropic. Since we are interested in the organization of large-scale ($L\gg\lambda$) and slow ($T\gg\tau$, τ is the turbulent correlation time) motions, we consider the turbulent velocity field as δ-correlated in time. Correlation tensors of the random fields V^t and ρ^t can be written as

$$\langle V_i^t(\vec{x}_1 t_1)\, V_j^t(\vec{x}_2 t_2)\rangle = \left[C(r)\delta_{ij} + B(r)r_i r_j + G(r)\varepsilon_{ijl}r_l\right]\delta(t_1-t_2)\,,$$
$$\langle\rho^t(\vec{x}_1,t_1)\, V_i^t(\vec{x}_2,t_2)\rangle = D(r)r_i\delta(t_1-t_2)\,; \quad \vec{r}=\vec{x}_1-\vec{x}_2. \tag{2}$$

Here the functions C, B and D are scalars and G is a pseudoscalar (note that $\langle \vec{V}^t(x) \cdot \nabla \times \vec{V}^t(x) \rangle \sim G(0)$). For the sake of simplicity, the fields V^t and ρ^t are assumed to be Gaussian.

Let us introduce at t=0 a weak large-scale perturbation $\langle V_o \rangle \ll V^t$. For t>0, the interaction of the perturbation $\langle V_o \rangle$ with the turbulent velocity field results in the development of the mean flow $\langle V \rangle$, while the random velocity field is modified by an additional component $\tilde{V}$, $\tilde{V} \ll V^t$. Note that the statistical properties of the background velocity field V^t are fixed and independent of the mean flow. Thus, we represent the total velocity as

$$\vec{V}_i = \langle \vec{V}_i \rangle + \tilde{V}_i + V_i^t, \qquad \langle \tilde{V} \rangle = 0,$$

and, similarly,

$$\rho = \langle \rho \rangle + \tilde{\rho} + \rho^t, \qquad \langle \tilde{\rho} \rangle = 0.$$

We substitute these expressions into (1), average over an ensemble, and linearize in $\langle V \rangle$ to obtain the following equations

$$\frac{\partial \langle V_i \rangle}{\partial t} + \langle \tilde{V}_k \frac{\partial V_i^t}{\partial x_k} \rangle + \langle V_k^t \frac{\partial \tilde{V}_i^t}{\partial x_k} \rangle = \nu_o \, \Delta \langle V_i \rangle - \frac{c_o^2}{\rho} \frac{\partial \langle \rho \rangle}{\partial x_i},$$
$$\frac{\partial \langle \rho \rangle}{\partial t} + \frac{\partial}{\partial x_k} \left[\langle \rho^t \tilde{V}_k \rangle + \langle \tilde{\rho} V_k^t \rangle + \langle \rho \rangle \langle V_k \rangle \right] = 0. \qquad (3)$$

To close the set of equations (3), we calculate the mean values $\langle \tilde{V}_k V_i^t \rangle$, $\langle \tilde{\rho} V_i^t \rangle$ and $\langle \rho^t V_k \rangle$ using the approach developed in references [1] and [2], which is based upon the Novikov-Furutsu formula [3,4]. In the framework of this approach, $\tilde{V}$ and $\tilde{\rho}$ are interpreted as functionals of V^t and ρ^t. To obtain the equation that determines the functional $V|V^t|$, we substitute the expressions for the total velocity, V, and density, ρ, into (3) and subtract the equations (1) and (3) for $\langle V \rangle$ and V^t respectively from the result. The averaged quantities which appear in (3), for instance $\langle \tilde{V} V^t \rangle$, are then expressed in terms of integrands of the correlation tensors (2) and the corresponding variational derivatives, for instance, $\delta\tilde{V}/\delta V^t$. As a result, we obtain the coupled equations for the mean fields

$$\frac{\partial V_i}{\partial t} - \frac{G(0)}{2} \left[\nabla \times \langle \vec{V} \rangle \right]_i = \partial \Delta \langle V_i \rangle - \frac{c_o^2}{\rho_o} \frac{\partial \langle \rho \rangle}{\partial x_i}, \qquad (4)$$

$$\frac{\partial \langle \rho \rangle}{\partial t} + \nabla \cdot (\langle \rho \rangle \langle \vec{V} \rangle) + \frac{D(0)}{2} \nabla \cdot \langle \vec{V} \rangle = \frac{C(0)}{4} \nabla \langle \rho \rangle, \qquad (5)$$

where $\nu = \nu_o + C(0)/4$.

In equation (4) the solenoidal part of the velocity field can be extracted as follows. Taking the curl of (4), we obtain finally the equation |6|

$$\frac{\partial \Omega}{\partial t} - \frac{G(0)}{2} \operatorname{rot} \Omega = \nu \Delta \Omega, \qquad (6)$$

where $\Omega = \operatorname{rot} \langle V \rangle$.

Equation (6) has unstable solutions. The increment γ is maximal, $\gamma_{max} = G^2(0)/16\nu$ for the mode of the length scale $L = 4\nu/G(0)$. Since $G \sim \langle V \text{ rot } V\rangle$ and $E = \int \langle V^2\rangle \, dr$ we can estimate the scale of the fastest growing mode as

$$L = \frac{E_T}{\langle V \text{ rot } V\rangle} \sim \frac{\int \langle V^2\rangle \, dr}{\int \langle V \text{ rot } V\rangle \, dr} . \tag{7}$$

Expression (7) is noteworthy due to the fact that it includes only integrals of motion: the energy invariant and the topological one. The latter, $I = \int \langle V \text{ rot } V\rangle \, dr$, characterizes the linkages of vortex lines with stream lines.

We should note that the helical magnetohydrodynamical turbulence can intensify the initially weak magnetic field. The equation that describes this effect (the dynamo equation) has the same structure as (6). The increments of both instabilities are of the same order of magnitude [5].

The results obtained establish theoretically the existence of large-scale instability of the mean velocity field in the helical turbulence. As follows from equation (6), the structures which arise from this instability are also helical (gyrotropic), i.e. $\langle V\rangle\langle \text{rot } V\rangle \not\equiv 0$. We shall call them "Gyrostructures" (for Greak "gyros"), or, to abbreviate, "G-structures". Unfortunately, we are not aware of any experimental results which confirm directly the existence of G-structures. However, the following arguments can serve as an indirect confirmation of the proposed theory. It is well known that convection in a rotating system (for example, in the Earth's atmosphere) becomes helical under the influence of the Coriolis force. If one considers a motion with the time and length scales much greater than the corresponding scales of the convective cell, the convective motions may be viewed as random and can be averaged out. Hence, the G-structure can be generated, provided the suitable initial disturbance is available. Tropical cyclones, which originate in regions of enhanced atmospheric convection, can be associated with this effect.

In this paper we discussed a model where turbulent fluctuations are considered as a random process δ-correlated in time. When the correlation time is small but finite, $g(0)$ in (6) should be replaced by $Mg(0)$, where $M = \sqrt{\langle (V^t)^2\rangle}/c$ is the Mach number. As a result, the increment γ acquires a factor M^2.

References

1. S.S.Moiseev, A.V.Tur, V.V.Yanovsky. Izv. Vyssh. Uchebn. Zaved. Radiofiz., 20, 1033 (1977)
2. S.S.Moiseev, R.Z.Sagdeev R.Z., A.V.Tur, V.V.Yanovsky. In Nonlinear Waves, Nauka, Moscow (1979) pp.105-115
3. B.A.Novikov. JETP, 47, 1919 (1964)
4. K.Furutsu. JOSA, 62, 240 (1972)
5. M.Steenbeck, F.Krause, K.-H.Rädler. Z.Naturforsch., 21A, 369 (1966)
6. G.M.Zaslavsky, S.S.Moiseev, R.Z.Sagdeev, A.V.Tur, G.A.Khomenko, V.V.Yanovsky. Fiz. Plazmy, 9, 62 (1983)

The Role of Fluctuations for Self-Organization in Physical Systems (an Exemplary Case of Transition from a Laminar to Turbulent Flow)

Yu. L. Klimontovich

Physics Department, Moscow State University, SU-117192 Moscow, USSR

The transition from a laminar flow to a turbulent one is considered as an example of nonequilibrium phase transition, converting the system into a state of higher order. The degree of order is characterized by entropy S and entropy production σ. The values of S and σ for the turbulent state are compared with those for an imaginary (unstable) laminar flow with the same Reynolds number (in the calculation of S) or with the same tension at the wall (with equal R_* values) in the calculation of entropy production.

The treatment is based on the semi-empiric theory [1-3]. For the flow in a plane channel (x-axis is directed along the channel; y-axis is normal to the plane) the order parameter is represented by the tangential Reynolds stress. To close the equation for the mean velocity the Prandtl-Karman relation

$$\langle \delta U_x \delta U_y \rangle = -(\nu_T - \nu)\frac{dU}{dy}, \tag{1}$$

is used which defines the turbulent viscosity. By definition, for the laminar flow $\nu_T = \nu$.

The turbulent viscosity can also be defined via the entropy production by the relation [5]

$$\sigma = \frac{\rho\nu}{2T}\left[U_{ij}^2 + \langle(\delta U_{ij})^2\rangle\right] = \frac{\rho}{2T}\nu_T U_{ij}^2, \tag{2}$$

which for the Couette and Poiseuille flows can be converted into [3]

$$\sigma = \frac{\rho v_*^4}{T}\frac{1}{v_*(y)}; \qquad \frac{\rho v_*^4}{T}\frac{1}{\nu_T(y)}\frac{y^2}{h^2}, \qquad -h \leq y \leq h, \tag{3}$$

Here and in what follows the expression before the semicolon pertains to the Couette flow, that after the semicolon to the Poiseuille flow; ρ is the density, T the temperature, v_* the dynamic velocity. Further on two Reynolds numbers will be employed:

$$R = \frac{U2h}{\nu}, \qquad R_* = \frac{v_* h}{\nu}. \tag{4}$$

The dynamic velocity is defined by

$$v_*^2 = \nu\left|\frac{dU}{dy}\right|_{y=h}; \qquad \frac{\Delta P h}{\rho l}. \tag{5}$$

The relationship between Reynolds numbers expresses the drag law, which for the laminar and the well-developed turbulent Couette flows acquires the form

$$R = 2R_*^2, \qquad R = 2\, R_c^o R_* \ln\left(\frac{2R_*}{R_c^o}\right) + 2\delta R_*, \tag{6}$$

where $R_c^o = \varkappa^{-1} = 2.5$ is the critical Reynolds number for the smallest turbulent scale; $\varkappa$ is the Karman constant; $\delta=7.8$ is the constant characterizing the thickness of the laminar sublayer [1,3]. We fix v_*, i.e. the tension at the wall or the differential pressure (for the Poiseuille flow). Let us compare entropy production in the turbulent flow with that in an imaginary (unstable) laminar flow with the same v_*, at $R>R_c$ and viscocity ν. Averaging over the cross-section we obtain [3]:

$$\frac{\sigma_T}{\sigma_L} = \frac{R}{2R_*^2}; \qquad \frac{2R}{R_*^2} \leqslant 1. \tag{7}$$

The equality sign corresponds to the laminar flow (at $R<R_c$). In particular, in the transition region and for the well-developed turbulence,

$$1 - \frac{\sigma_T}{\sigma_L} \sim \frac{R-R_c}{R_c} << 1, \qquad \frac{\sigma_T}{\sigma_L} \sim \frac{R_c^o}{R_*} \ln R_* < 1. \tag{8}$$

Thus we see that the entropy production in the turbulent state is lower than the entropy production in the corresponding (fixed v_*) imaginary (unstable) laminar flow.

Let us carry out the appropriate calculation of entropy. To do this, we use the expression (8.8) from Chap.13 in Ref.[2] which defines the entropy difference in local equilibrium and nonequilibrium states. For an incompressible fluid, this expression brings to

$$T(S-S_o) = -\frac{\rho V}{2} \langle (\delta U)^2 \rangle, \tag{9}$$

where $\langle(\delta U)^2\rangle$ is the cross-section-average dispersion of velocity pulsations; $\langle(\delta U)^2\rangle \simeq v_*^2$ for the well-developed turbulence. The quantity v_* can also be used for expressing the dispersion of nonequilibrium molecular fluctuations in the laminar flow, as well as the nonequilibrium contribution to the intensity of the relevant Langevin source. In the local equilibrium approximation, the latter is R times smaller than the main part and thus can be neglected. As a result, we can finally obtain the expression for the entropy difference in the laminar and turbulent flows with the same Reynolds number $R>R_c$

$$\frac{T}{V}(S_L - S_T) = \frac{\rho}{2} \langle(\delta U)^2\rangle 0. \tag{10}$$

In the transition region the difference is given by

$$\frac{T}{V}(S_L - S_T) \sim \frac{R-R_c}{R_c} \geqslant 0, \quad R \geqslant R_c. \tag{11}$$

Here we have disregarded the molecular fluctuations; their action smoothes out the critical reduction of entropy. The different choice of conditions of the imaginary laminar flow in calculations of σ and S is due to the different structures of these functions.

In the semi-empiric Prandtl-Karman approximation, the order parameter is represented by the Reynolds shear tension (1) which behaves critically in the transition region. The diagonal elements of Reynolds tensor are nonzero for the laminar flow as well. For both turbulent and laminar flows we obtain that $\langle(\delta U)^2\rangle \simeq v_*^2$. The peculiar property of Reynolds stress tensor is responsible for strong dissimilarity of turbulization of the plane Couette and Poiseuille flows. Let us illustrate this point with the most simple model available.

We use a quasi-linear approximation, this implies that the Orr-Sommerfeld equation is considered together with the exact (though non-closed) equation of averaged flow. For the sake of simplicity the Orr-Sommerfeld equation is replaced by the second-order equation with effective dissipation $\gamma=U/R^{1/3}$. For the Couette flow this set of equations results in two integral relations [4]:

$$\int_{-h}^{+h} \frac{\gamma}{(U(y)-C)^2+\gamma^2} \frac{d^2U}{dy^2} dy = 0 \tag{12}$$

$$\langle \delta U_x \delta U_y \rangle_{y=0} = \frac{1}{\nu k} \int_0^h \frac{\gamma}{(U(y)-C)^2+\gamma^2} \frac{d}{dy} \langle \delta U_x \delta U_y \rangle dy \langle (\delta U_y)^2 \rangle . \tag{13}$$

The first of them indicates the necessity of existence of an inflection point in the averaged flow profile (Rayleigh criterion). The second equation determines the condition of steady-state turbulence. The transition here is possible for any small excess over the threshold; in this sense the turbulence develops "softly". For static perturbations (c=0) with the scale $2hk \approx 1$ the critical Reynolds number is defined by

$$R_c^{1/3} = \frac{4}{3} \ln R_c . \tag{14}$$

The corresponding integral relation for the Poiseuille flow is of quite different structure [4]

$$\int_0^h \frac{1}{(U(y)-C)^2+\gamma^2} \frac{d}{dy} \langle \delta U_x \delta U_y \rangle dy = \int_0^h \frac{dy}{(U(y)-C)^2+\gamma^2} \frac{\Delta p}{\rho l} . \tag{15}$$

This condition can be satisfied only with finite increments over R_c, when the maximum of the function $\langle \delta U_x \delta U_y \rangle$ lies close to the wall (on the border of the laminar sublayer). Then the formula [4] can be rewritten as follows

$$\frac{d}{dy} \langle \delta U_x \delta U_y \rangle = \frac{v_*^2}{h} \quad \text{and from (15)} \quad v_*^2 = \frac{\Delta p h}{\rho l} . \tag{16}$$

The latter equation is satisfied by virtue of definition of the dynamic velocity for the Poiseuille flow (cf.(15)). In this sence the onset of steady-state turbulence is "hard".

Thus the transition from a laminar to turbulent flow is considered as a nonequilibrium phase transition to a more ordered state. Since these transitions are of the second kind, two types of fluctuations could be expected: natural (molecular) and coherent ones. In this particular case the order parameter is represented by the Reynolds shear tension $\langle \delta U_x \delta U_y \rangle$ and the degree of order is characterized by entropy and entropy production.

References

1. H.Schlichting. Grenzschicht-Theorie. Verlag G.Braun, Karlsruhe, (1965).
2. Yu.L.Klimontovich. Statistical Physics. Moscow, Nauka,(1982).
3. Yu.L.Klimontovich, H.Engel-Herbert. Averaged Steady-State Couette and Poiseuille Flows in an Incompressible Fluid. Zh. Tekh. Fiz. (1983).
4. Yu.L.Klimontovich.The Arising of Turbulence in Couette and Poiseuille Flows. Pis'ma v Zh. Tekh. Fiz., 9, No. 18 (1983).
5. Yu.L.Klimontovich. The Decrease of the Entropy in the Process of Self-Organization. S-Theorem. Pis'ma v Zh. Tekh. Fiz.,9, (1983).

Stochastization of Nonstationary Structures in a Distributed Oscillator with Delay

V.A. Kats and D.I. Trubetskov
Saratov State University, SU-Saratov, USSR

1. Introduction

Self-organization of stochastic nonstationary structures ("order-chaos" transitions) and evolution of their properties and characteristics ("chaos-chaos" transitions) are, nowadays, the most attractive phenomena of nonlinear physics [1].

We report here the results of experimental observations of different "order-chaos" and "chaos-chaos" transitions in a distributed oscillator with delayed feedback. The experimental oscillator is a closed circuit consisting of a nonlinear amplifier, a cutoff filter (a reentrant resonator) and an ultrasonic delay line. The nonequilibrium nonlinear medium, "electron beam - travelling EM wave" has been used as an active oscillator. The feedback parameter $\lambda=10\ \lg\ (P_1/P_2)$, where P_1 and P_2 are the amplifier input and output wave powers, characterizes the degree of equilibrium. Fundamental modes of the delayed system form an infinite quasi-equidistant spectrum with intermode distance 1/T, where T is the delay time. Such an oscillator with delayed feedback is a distributed dynamic system with infinite number of degrees of freedom [2,3]. Thus, there is no contradiction in stating that many various attractors, corresponding to motion on different principal modes, occur in the given multimode system and several scenarios of "order-chaos" transition may evolve concurrently in different regions of the phase space. Therefore our physical system can exhibit many routes to the chaos onset, depending on the initial state [4]. The initial state of the system involved is determined by the frequency of the principal mode and by the central frequency of the filter.

2. Modulation Instability. Transitions to Chaos

With increasing λ, the oscillations on the principal mode arise with the maximal linear increment. This stationary regime loses stability with further growth of λ and a regularly pulsing structure appears as the amplitude self-modulation with the period 2T. In the phase space plotted in Fig.1 the stable limit point is changed to a stable limit cycle (Hopf bifurcation). (Phase portraits of the attractors shown in Fig.1 have been obtained experimentally).

With the growth of nonequilibrium a stochastic nonstationary structure is being formed in the system as a result of modulation instability, and a simple one-mode chaotic regime arises. In the phase space this structure is a limited stochastic set (a strange attractor) with low dimension near the critical point ($D \geq 2$ [3]). Depending on the initial state, the distributed system revealed different routes of transition to chaos as a result of successive instabilities (bifurca-

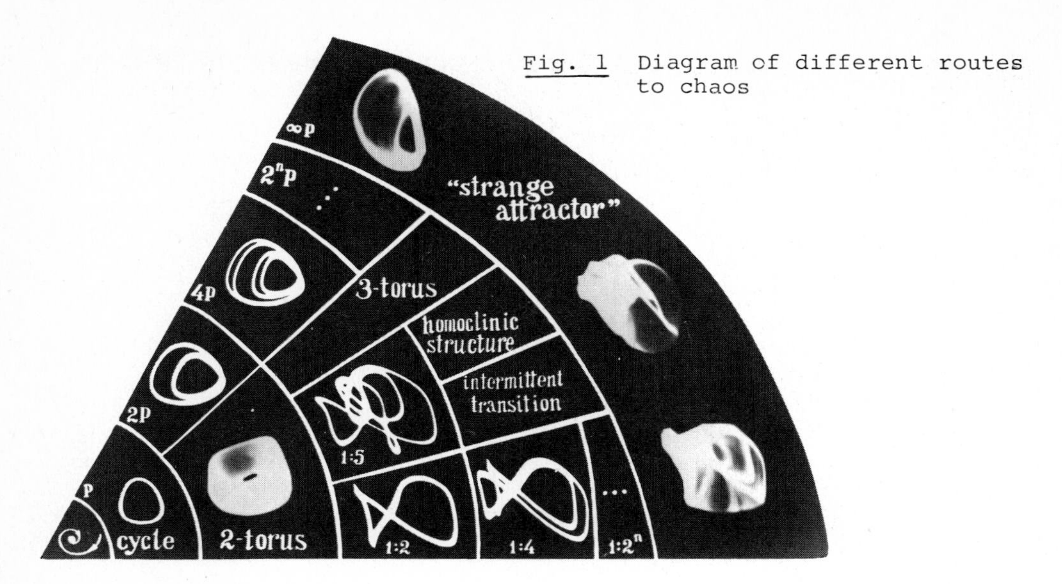

Fig. 1 Diagram of different routes to chaos

tions), which are characteristic of simple dynamic systems (see Fig.1).

Route 1 is a sequence of period doubling bifurcations of stable periodic motion, characterizing a regular structure (the Feigenbaum scenario [5]). In the phase space of the system, the initial stable cycle P loses stability and initiates a stable limit cycle of doubled period 2P, which in turn loses stability with generation of a new stable cycle of period 4P and so on. (The values of the Feigenbaum constants have been determined experimentally: $\delta=4.232\pm0.1$, $\alpha=2.51\pm0.1$).

Route 2 is the destruction of an ergodic winding of a three-dimensional torus, which is the image of quasiperiodic motion with three incommensurable frequencies. Such a regular structure is a result of three successive Hopf bifurcations (the Ruelle-Tachens scenario [6]).

Route 3 is the destruction of a quasiperiodic motion with two incommensurable frequencies as a result of synchronization of frequencies (phase locking) and the onset of regular periodic structure, which can be presented mathematically as a stable limit cycle on a two-dimensional torus (the case of resonance). With the growth of λ, the structure is subjected to further stochastization as a result of either disappearance (resonance 1:5) or a soft loss of stability through the period doubling sequence (resonance 1:2) of the cycle on the torus. In the first case either a hard onset of chaos, which is probably associated with the appearance of homoclinic structures, or intermittent chaos (Fig.2), arising from the confluence of stable and unstable periodic motions, has been experimentally observed [7,8].

Route 4 is a period doubling bifurcation sequence of the torus. The stable ergodic winding of the two-dimensional torus appears near the stable aperiodic orbit, which loses stability via Hopf bifurcation. As λ is increased, the initial periodic motion with frequency f_1 doubles the period in the presence of a weak modulation of the incommensurable frequency f_2 and the ergodic winding of the two-dimensional torus is now located near the doubled periodic orbit (see Fig.3). Such a transition is a characteristic feature of non-autonomous systems, where f_2 is the frequency of an external signal.

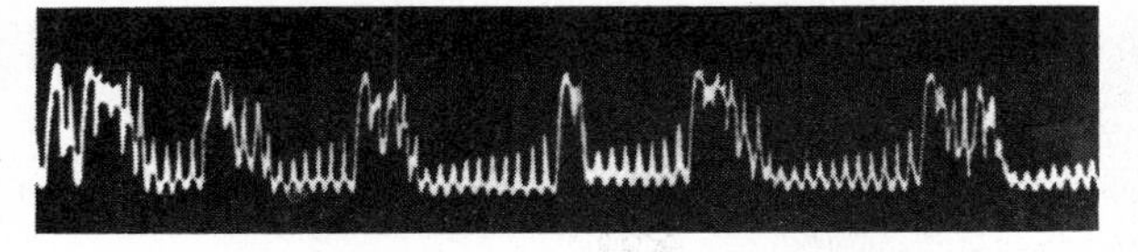

Fig. 2 Experimental diagram of intermittent chaos

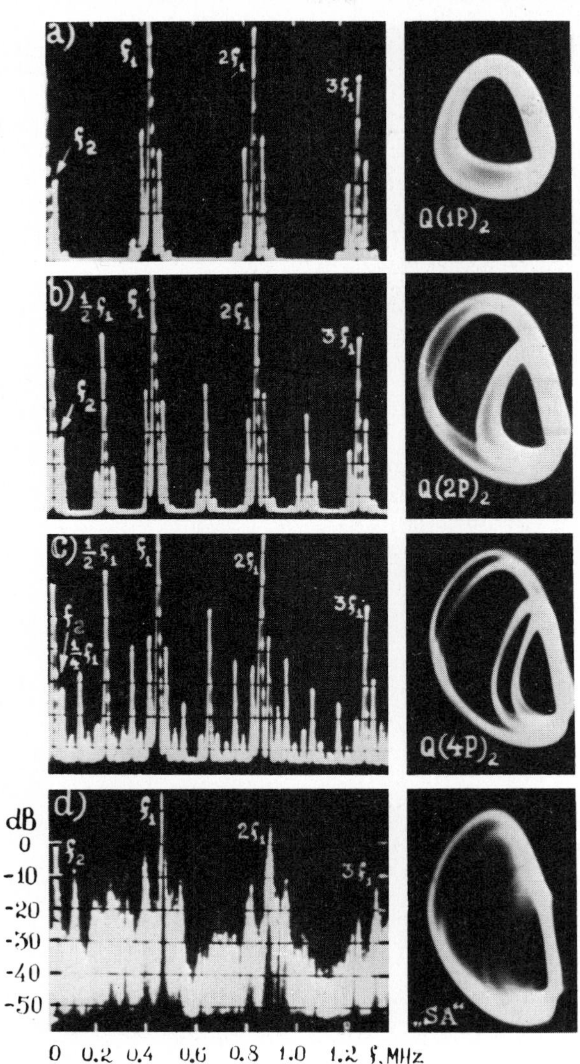

Fig. 3 Evolution of spectral density and of phase portrait of the non-stationary structure when transiting to chaos via the sequence of doublings of the torus

3. Transitions in Chaos

We should draw attention to the transitions in chaos which are accompanied by a change of all principal characteristics of stochastic non-stationary structures when the degree of non-equilibrium is increased beyond the critical point. The character of these changes depends on the evolutionary prehistory of a dynamic system. Transitions of four main types have been experimentally observed.

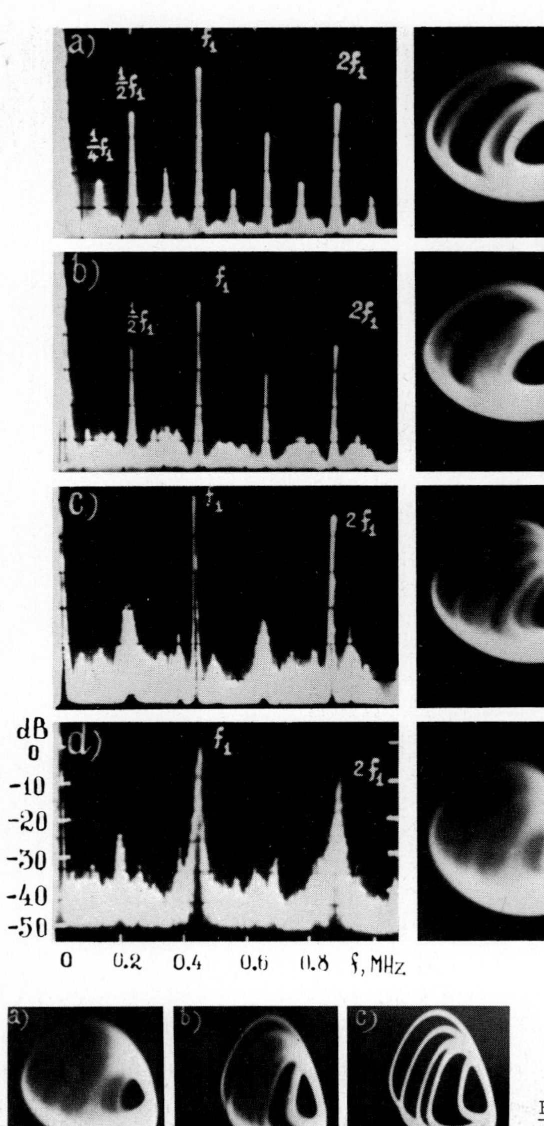

Fig. 4 Inverse cascade of doublings of the noisy period of the Feigenbaum attractor

Fig. 5 The birth of stable periodic motion from trajectory bunching of the Feigenbaum attractor

1. With the growth of λ the number of turns of the Feigenbaum attractor tape reduces by half, leading to a cascade of inverse doublings of the noisy period (Fig.4).

2. The strange attractor suddenly disappears when stable and unstable periodic motions arise from trajectory bunching, i.e. a coherent structure in chaos (a "noise-free window") appears (Fig.5). The stochastic structure arises again through either the sequence of the period doublings of stable periodic motion, or the confluence of a stable and unstable periodic orbits (intermittent chaos).

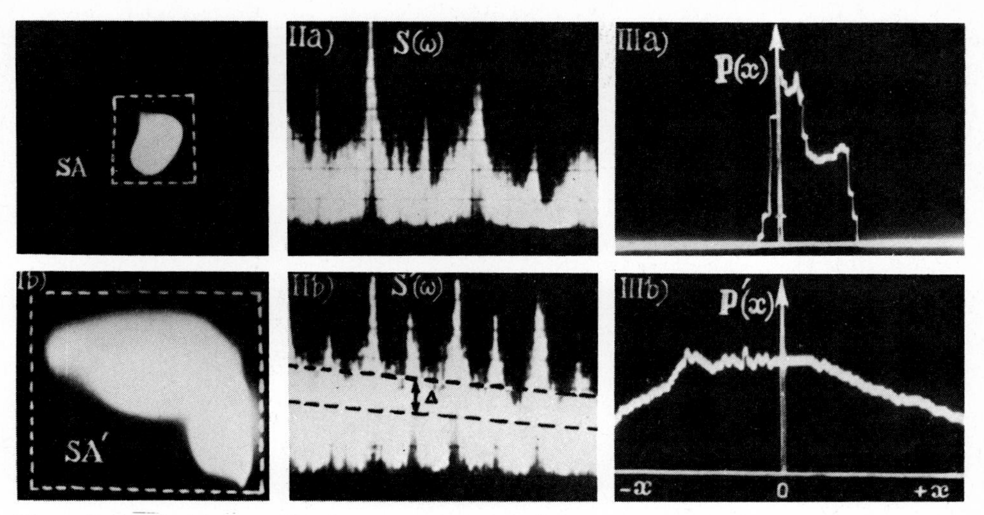

Fig. 6 Evolutions of phase portrait (I), spectral density (II), and probability distribution (III) of stochastic structure in "chaos-chaos" transitions; a) before transition, b) after transition

3. The strange attractor with high dimension disappears in a leap, when its boundaries cross the basin of the regular attractor (a fixed point).

4. The area of localization of the strange attractor in the phase space increases with λ (see Fig.6÷I), this is probably due to absorption of adjacent non-attracting stochastic sets. From the physical viewpoint this transition is connected with inclusion of new principal modes of the system into the process of stochastic motion. Thereby the intensity and the degree of disorder increase, the spectral density S(w) of the motion changes (Fig.6-II), and the probability distribution tends to a Gaussian one (Fig.6-III). The experimental oscillogram of such a motion includes fragments of stochastic oscillations of low intensity interrupted by stochastic bursts of high intensity, i.e. "chaos - chaos" intermittence [1] (see Fig.7).

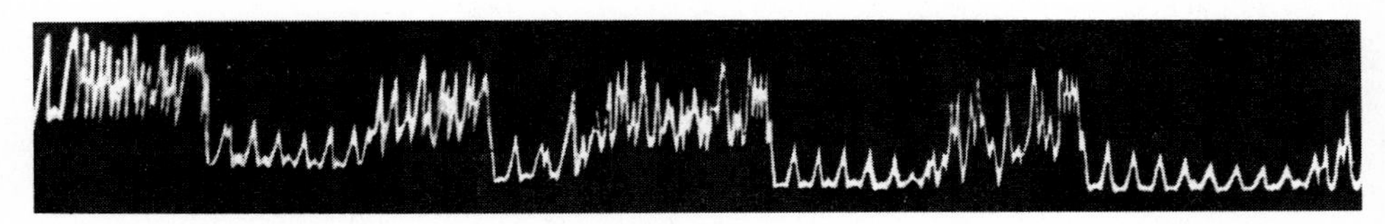

Fig. 7 Experimental oscillograms of "chaos-chaos" intermittence

We can suggest that the gas of strange attractors with initially small dimensions is the image of developed stochasticity in the phase space of a strongly non-equilibrium system.

The authors' especial thanks are due to M.I Rabinovich for valuable comments.

References

1. A.V.Gaponov-Grekhov, M.I.Rabinovich, I.M.Starobinets (the article in this book)
2. A.A.Vitt. ZhTF, 6, 1459 (1936)
3. J.D.Farmer. Physica, D4, 366 (1982)
4. J.-P.Eckmann. Rev. Mod. Phys., 53, 643 (1981)
5. M.J.Feigenbaum. J. Stat. Phys., 19, 25 (1978)
6. D.Ruelle, F.Tachens. Commun. Math. Phys., 20, 167 (1971)
7. V.S.Afraimovitch, L.P.Shilnikov. Doklady Akad. Nauk SSSR, 219, 1281 (1974)
8. P.Manneville, Y.Pomeau. Physica, D1, 219 (1980)

Experimental Study of Rossby Solitons and Dissipative Structures in Geostrophical Streams

S.V. Antipov, M.V. Nezlin, V.K. Rodionov, E.N. Snezhkin, and A.S. Trubnikov
I.V. Kurchatov Institute of Atomic Energy, SU-Moscow, USSR

1. Rossby Soliton

First experimental observations and studies of Rossby solitons have been described in [1,2]. Rossby solitons correspond to the lowest frequency limit of generation of gravitational waves in a rotating layer of shallow water. Theory [3] predicts the following properties of this phenomenon:

1) The characteristic soliton diameter (Fig.1 and 2a) should essentially exceed (at least 3 times) the so-called Rossby-Obukhov radius

$$r_R = (g^* H_0)^{1/2}/f_0, \tag{1}$$

where g^* is the free fall acceleration (the centrifugal force is taken into account), H_0 is the depth of a shallow water layer (or equivalent height of the atmosphere), $f_0 = 2\Omega_0 \cos\alpha$ the Coriolis parameter; here Ω_0 is the angular rotation velocity, $\alpha = (\pi/2)-\varphi$, where φ is a geographical latitude.

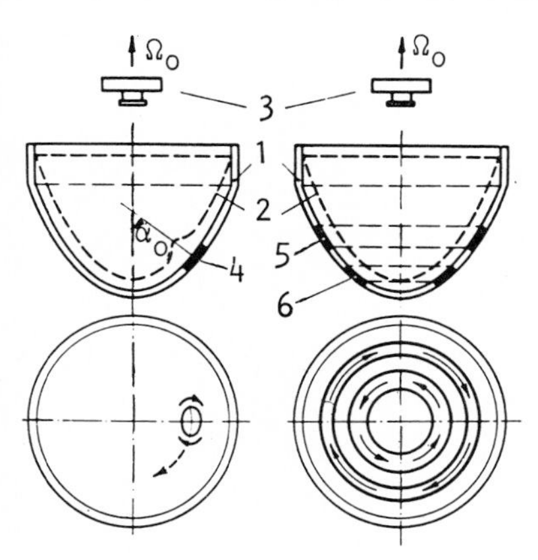

Fig. 1 Experimental devices [1,2,4] for a) observation of Rossby solitons in shallow water (2), rotating along with the paraboloid (1); b) generation of steady-state vortex structures by counterstreaming flows. Rings (5,6) rotate independently of the vessel. 3 - camera rotating with the vessel anticlockwise

2) A soliton behaves like an anticyclone, i.e. it forms a hill of the liquid rotating around the vertical axis in the direction opposite to the global rotation of the system with the angular velocity $\omega < \Omega_0$.

3) A soliton drifts against the direction of global rotation with the velocity close to the Rossby velocity, which under the given experimental conditions is expressed as

$$V_R = H_0 \Omega_0 \sin\alpha. \tag{2}$$

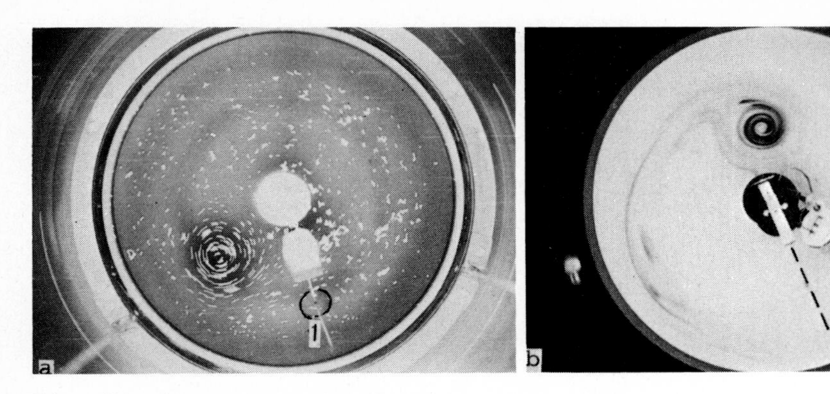

Fig. 2 a) Rossby soliton - an anticyclone drifting clockwise and visualized by white fragments; b) The same vortex produced in position 1 and coloured by dye particles in position 2, 20 sec after the onset of the drift. During its lifetime (limited by viscosity) the vortex propagates without dispersion spread or nonlinear deformation exhibiting the property of a solitary wave

Typical examples of Rossby solitons (Fig.2a,b) were obtained in a shallow water layer rotating around the vertical axis together with the vessel whose bottom was parabolic in shape (Fig.1a). It has also been shown that according to the theory [3] the cyclone type solitons with the size significantly larger than the radius (1) cannot retain their shape and decay rapidly.

However, some new important properties of Rossby solitons were found [1,2] which did not fit the theory [3], namely: a) stability Fig.1b,2a (the lifetime of a soliton (~20s) is close to the viscosity time; during this time the soliton covers a distance equal to at least 10 of its own diameters); b) "attractiveness" (any initial disturbance brings eventually to a number of Rossby solitons (Fig.1b); c) inelastic nature of collisions of solitons (they either merge in fast collision or destroy each other in slow collision to produce flows (Fig.2b); d) high efficiency of the transport of particles in the medium (Fig.1b) (the characteristic rotation velocity of particles is greater than the velocity of the drift of a soliton).

2. Generation of Steady-State Dissipative Anticyclonic Structures by Counterstreaming Flows with the Characteristic Dimension of Transverse Velocity Gradient Exceeding the Rossby-Obukhov Radius

To produce the counterstreaming flows in rotating shallow water, the separate parts of the bottom of paraboloid (rings 5,6 in Fig.1b) were moved towards each other and drove adjacent layers of the liquid [4,2b]. In the given series of experiments, the characteristic size of transverse velocity profile significantly exceeded the radius r_R. It was observed that when the velocity exceeded some threshold and vorticity of the flow was anticyclonic, then the flows underwent the large-scale Kelvin-Helmholtz instability and generated steady-state dissipative structures in the form of chains of solitary anticyclons. The size of these anticyclons along the liquid surface is much greater than the depth H_o and essentially exceeds the radius r_R. The number of vortices on the perimeter of the system can vary from eight (at small excess over the critical velocity) to two (at a greater excess over the critical velocity). These vortices [4,2b](Fig.3) are similar to Rossby solitons. They have relative elevation (in respect to the initial depth of the liquid) of 0.5-1 and the size of several Rossby-Obukhov radii r_R. The vortices drift against the direction of the gross liquid rotation with the velocity close to that determined by formula (2). When the vorticity of flows is cyclonic, the large vortices with the size exceeding the radius r_R are not formed. This set of experimental data reminds of the cyclon-anticyclon asymmetry in Jupiter's atmosphere. There, almost all long-lasting vortices are anticyclons and the largest of them, "Great Red Spot", exists and propagates west-ward just in that range of latitude where the vorticity is anticyclonic. At the same time, no large vortex is in the northern region where the amplitude of flows is even larger but the vorticity is cyclonic [5] . Thus, the above experiments confirm the contemporary theory of Jupiter's Great Red Spot based upon the soliton approach.

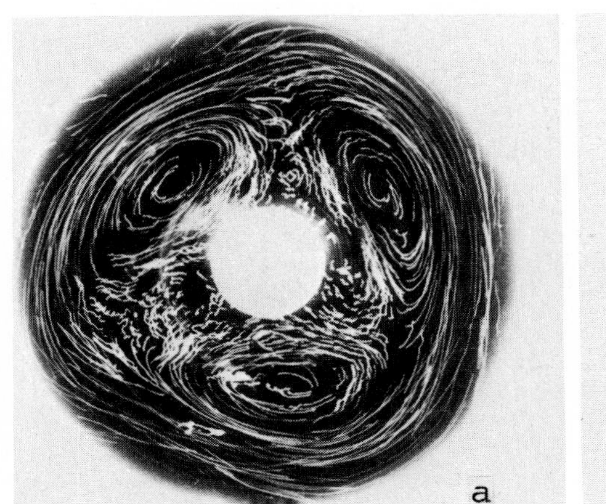

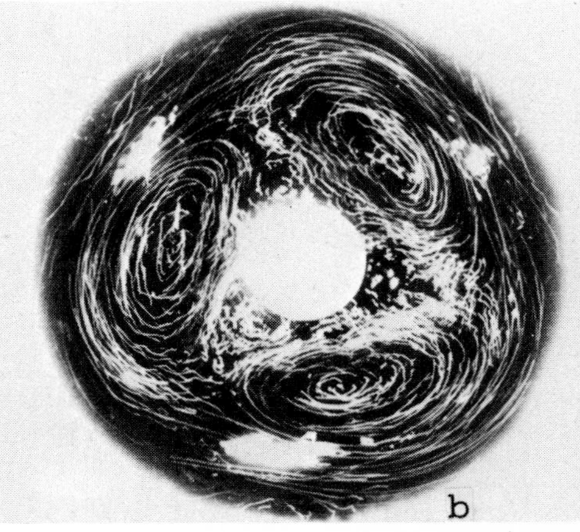

Fig. 3 Generation of steady-state dissipative anticyclonic structures drifting opposite to the vessel rotation at anticyclonic vorticity of counterstreaming flows

3. Generation of Steady-State Dissipative Cyclonic Structures by Counterstreaming Flows with the Characteristic Dimension of the Transverse Velocity Gradient Less than Rossby-Obukhov Radius

In this series of experiments the moving parts of parabolic bottom (rings 5,6 in Fig.1b) were placed without spacings (Fig.4). Depending on the direction of vorticity of counterstreaming flows, large vortices with sizes larger than r_R of both anticyclonic and cyclonic polarity are generated in this case. An example of cyclone chains is presented in Fig.5 (Ref. [2c]). The existence of cyclone-anticyclone asymmetry of vortices under conditions described in Sec.2 and the absence of this asymmetry under conditions described in Sec.3 can be explained in the following way. As shown above, the solitary anticyclonic vortices (Rossby solitons) are stable, while cyclones rapidly decay. Therefore, the experimental conditions indicated in Sec.2, under which the generation of the vortex occurs only at the periphery of the vessel (Fig.4a), turn out to be sufficient to maintain the stable anticyclone and insufficient to maintain the unstable cyclone. To generate a cyclone the geometrical arrangement of the type shown in Fig.4b must be used.

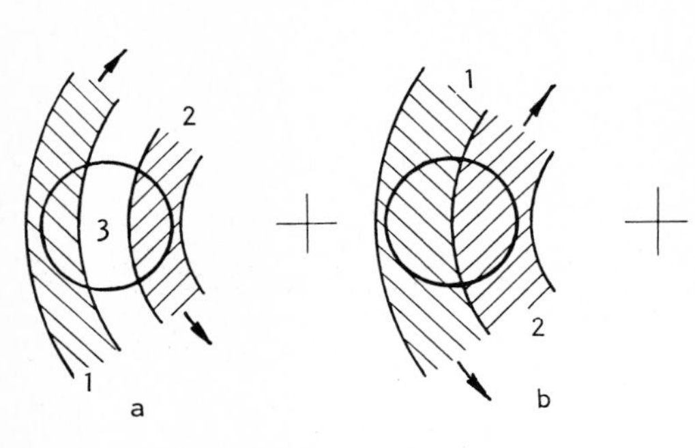

Fig. 4 Excitation of vortex 3 by rings 1,2 moving towards each other. The vessel axis is shown by a cross. Direction of rotation for rings at anticyclonic (a) and cyclonic (b) vorticity of counter-streaming flows is shown by arrows

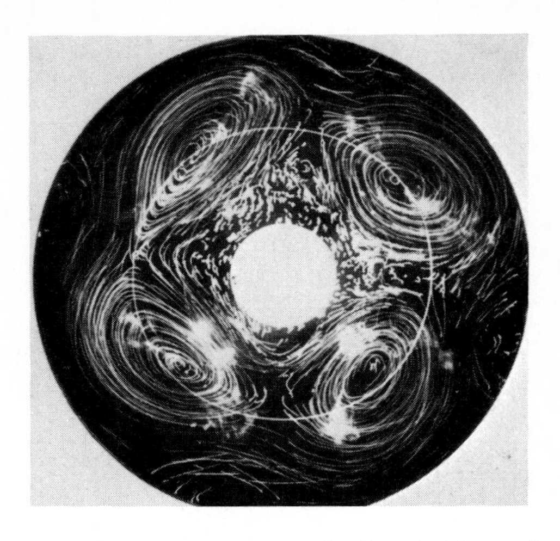

Fig. 5 Excitation of the 4-th mode of Kelwin-Helmholtz instability. The line of velocity "jump" passes through the centres of cyclonic vortices

The conditions of vortex generation described above could be plausibly compared to the conditions of cyclogenesis existing on the Earth [6] where the characteristic dimension of the transverse velocity gradient of flows is less than the relevant Rossby-Obukhov radius r_R=3000 km.

To conclude, we remind once more that the generation of dissipative vortex structures by counterstreaming liquid flows is induced by Kelvin-Helmholtz instability. The latter can be excited within the limited velocity range when the amplitude of a velocity "jump" lies between the lower and upper thresholds of instability. The existence of the upper threshold is due to the effect of the "supersonic" stabilization of instability of the tangential break predicted by Landay and observed experimentally in [2c].

References

1. S.V.Antipov, M.V.Nezlin, E.N.Snezhkin, A.S.Trubnikov. a) Pis'ma Zh. Exp. Teor. Fiz.(1981),v.33, p.368; b) Zh. Exp. Teor. Fiz. (1982) v.82, p.145
2. S.V.Antipov, M.V.Nezlin, V.K.Rodionov, E.N.Snezhkin, A.S.Trubnikov. a) Pis'ma Zh. Exp. Teor. Fiz. (1982), v.35, p.521; b) Zh. Exp. Teor. Fiz. (1983) v.84, p.1357; c) Pis'ma Zh. Exp. Teor. Fiz. (1983) v.37, p.319
3. V.I.Petviashvill. Pis'ma Zh. Exp. Teor. Fiz. (1980) v.32, p.632
4. Nezlin M.V., E.N.Snezhkin, A.S.Trubnikov. Pis'ma Zh. Exp. Teor. Fiz. (1982) v.36, p.190
5. B.Smith, J.Hunt. Jupiter (University of Arizona Press, 1976); A.R.Ingersoll. Scientific American (1981) v.245, p.66
6. F.V.Dolzhanskij. Izv. Akad. Nauk SSSR, Fizika atmosfery i okeana (1981) v.17, p.563; Chernous'ko Yu.L. Izv. Akad. Nauk SSSR, Fizika atmosfery i okeana (1980) v.16, p.423

Part III

Mathematical Backgrounds of Autowaves

A Theory of Spiral Waves in Active Media

A.S. Mikhailov
Moscow State University, SU-Moscow, USSR

1. Introduction

Rotating spiral waves (SW) represent one of the most fascinating examples of the self-organization phenomena. Along with the leading centers and vortices they represent the basic types of autonomous wave generators in active media. Experimental data indicate [1,2] that all SW have the same rotation frequency which is determined only by the properties of the medium and not by the initial conditions. Hence the analytical estimation of the SW frequency is an important task of any theoretical treatment.

First theoretical studies of SW were performed in the framework of an axiomatic model, where the active medium was approximated by a net of discrete automata with the three possible states of rest, excitation and refractoriness. Using this model, N.Wiener and A.Rosenblueth [3] have investigated rotation of SW around a hole. Later, I.S.Balakhovskii [4] has found SW solutions for a homogeneous active medium. According to [4], the period of SW is equal to the recovery time, i.e. to the sum of durations of the excitation and the refractoriness stages. Further developments within the axiomatic model were reviewed in [5].

Undoubtedly, the axiomatic model is oversimplified and can make only qualitative predictions. The higher level of description is provided by the reaction-diffusion-equations. Many results (see [6-9]) have been obtained for a particular class of the reaction-diffusion systems, i.e. for the quasi-harmonical or λ-ω models. The problem of finding the rotation frequency of SW in quasi-harmonical media was solved in [9].

It should be emphasized that the most important applications, such as the BZ reaction or heart tissue fibrillation are not described by the quasi-harmonical models. Here the waves and oscillations are of relaxation type which means that they are characterized by two very different time-scales.

The aim of the present paper is to outline the analytical procedure that permits to evaluate the rotation frequency of SW in relaxation models of active media. The detailed proof and calculations, as well as the application to particular models, can be found in [10-12].

2. Periodic wave propagation

We consider two-component reaction-diffusion systems

$$\dot{u} = f(u,v) + \Delta u, \qquad \varepsilon^{-1}\dot{v} = g(u,v), \tag{1}$$

with a single diffusive component u. The parameter ε is assumed to be small ($\varepsilon<<1$) so that the model contains a relaxation. The null-cline f=0 is S-shaped (Fig.1), while the null-cline g=0 is monotonous. The null-clines intersect one another in such a way that there is only one stationary point P which corresponds to the rest state of the medium.

The medium described by Eqs.(1) can support propagation of solitary and periodic waves (see [13]). One-dimensional periodic wave solutions have the form

$$u=u(\xi), \qquad v=v(\xi), \qquad \xi=kx-\omega t, \tag{2}$$

where $u(\xi)$ and $v(\xi)$ are the 2π-periodic solutions of the equations

$$-\omega u'=f(u,v)+k^2u'', \qquad -\varepsilon^{-1}\omega v'=g(u,v). \tag{3}$$

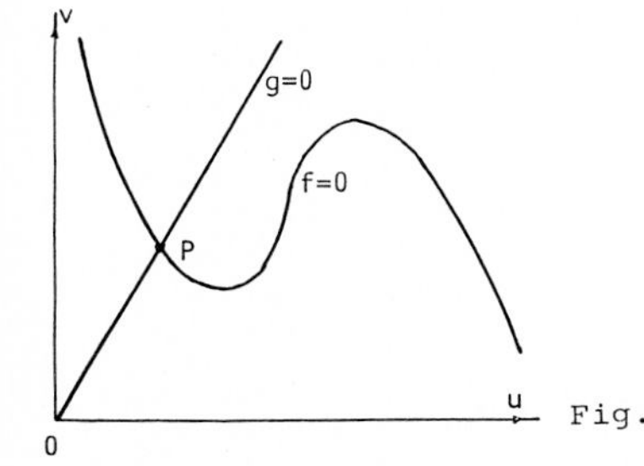

Fig. 1

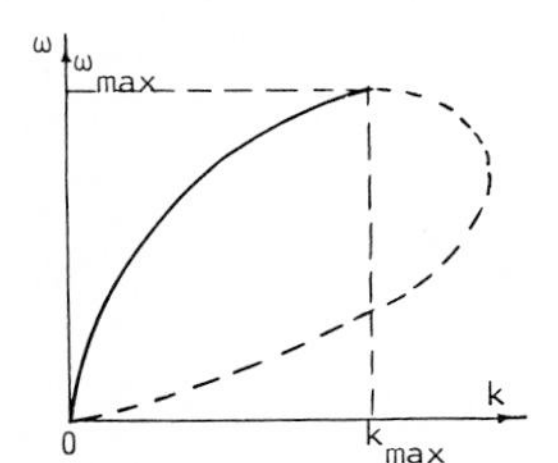

Fig. 2

It is well-known ([13,11]) that in active media the spatial period of a periodic pulse train completely determines the amplitude, the profile and the propagation speed of the pulses. It means that, although such waves are nonlinear, we can define the dispersion relation $\omega=\Omega(k)$ between the wave number k and the wave frequency ω. The typical form of the dispersion law for the relaxational active media (1) is shown in Fig.2. The dashed line indicates the unstable solution. We note that there are the maximal frequency ω_{max} and the minimal time period $T_{min}=2\pi/\omega_{max}$ for periodic wave propagation that correspond to the minimal spatial period of stable propagation $L_{min}=2\pi/k_{max}$. Note also that (2) describes the circumferential propagation of the pulse with circumference radius $r\to k^{-1}$. The limit $k\to 0$ corresponds to a solitary pulse.

In a two-dimensional medium the fronts of the propagating waves may be curvilinear. The speed of a wave depends on its local curvature K. There is a critical curvature K_{cr} and the waves with $K>K_{cr}$ are not able to propagate. It is important that for $K=K_{cr}$ the propagation speed is not zero [14].

3. Spiral Waves

Now we consider the spiral waves in an active two-dimensional medium and intend to show how the SW frequency ω_{sw} can be determined by using the dispersion relation $\omega=\Omega(k)$.

Suppose that the instantaneous shape of a spiral is given by some function $\varphi=\chi(r,t)$ in the polar coordinates r and φ. Then, by substituting

$$u=u(\varphi-\chi(r,t),r), \qquad v=v(\varphi-\chi(r,t),r) \tag{4}$$

into (1), we obtain

$$-\dot{\chi}u'=f(u,v)+Qu'+k^2u''+J, \qquad -\varepsilon^{-1}\dot{\chi}v'=g(u,v), \quad \text{where} \tag{5}$$

$$Q = \frac{1}{r}\frac{\partial}{\partial r}(r\chi_r), \qquad k^2=\chi_r^2 + \frac{1}{r^2}, \tag{6}$$

$$J = \frac{1}{r}\frac{\partial u}{\partial r} + \frac{\partial^2 u}{\partial r^2} + 2\chi_r \frac{\partial^2 u}{\partial r \partial \varphi}, \quad \text{and}$$

$$u' = \partial u/\partial \varphi, \qquad \dot{\chi} = \partial \chi/\partial t, \qquad \chi_r = \partial \chi/\partial r.$$

The major advance is omitting the term J in Eqs.(5). If we take this approximation, the validity of which and related limitations are discussed below, then at every value of r Eqs.(5) are reduced to the equations of periodic one-dimensional wave propagation, modified only due to the presence of the additional term Q:

$$\dot{u} = f(u,v) + Q\frac{\partial u}{\partial \varphi} + k^2 \frac{\partial^2 u}{\partial \varphi^2}, \quad \varepsilon^{-1}\dot{v} = g(u,v). \tag{7}$$

Such propagation is described by the modified dispersion relation $\omega=\Omega(k,Q)$.

Comparing Eqs.(5) and Eqs.(7), we find

$$\dot{\chi} = \Omega(k,Q). \tag{8}$$

Note that since k and Q depend on χ_r and χ_{rr}, Eq.(8) is the nonlinear differential equation that determines temporal evolution of the SW shape.

If the spiral wave rotates steadily with some constant frequency ω_{sw} then $\dot{\chi}=-\omega_{sw}$=const since $\chi(r,t)=\chi(r)-t\omega_{sw}$, and Eq.(8) gives

$$\omega_{sw} + \Omega(k,Q) = 0. \tag{9}$$

Eq.(9) is the first-order differential equation for χ_r. A similar equation for reaction-diffusion system with two equally diffusive components was derived earlier by J.M.Greenberg[15].

1) First we consider a SW inside the ring $R_1 \leqslant r \leqslant R_2$. When the ring boundaries are impermeable, the wave front should be perpendicular to them and therefore

$$\chi_r(R_1) = \chi_r(R_2) = 0. \tag{10}$$

Note that we have two boundary conditions (10) for the equation (9), while it is the first order differential equation. Hence, conditions (10) define the nonlinear eigenvalue problem for (9) and determine the rotation frequency ω_{sw} of a spiral wave inside the ring as the solution.

2) If SW rotates around a hole of the radius R in an infinite medium, then $\chi_r(R)=0$ on the hole boundary. Besides, it can be shown [12] that in the limit $r\to\infty$ the step of a spiral becomes constant so that $k\to k_\infty$ and $Q\to 0$ as $r\to\infty$. These remarks lead to the boundary conditions

$$\chi_r(R)=0, \quad \lim_{r\to\infty} \chi_r = \text{const} < k_{max}. \tag{11}$$

Being applied to Eq.(9) they determine the rotation frequency ω_{sw}.

3) Numerical simulations of full time-dependent Eqs.(1) reveal [16] that, if one starts to diminish the size of a hole around which SW is rotating, the rotation frequency increases until a certain value $R=R_o$ is reached. Below this value, any significant dependence of ω_{sw} on R_o vanishes. The critical radius R_o is usually considered to define the intrinsic core size of SW in a given medium. Hence, SW in homogeneous media behave as if there were some holes of the radius R_o at their centers. Only the outer region $r>R_o$ is characterized by the self-sustained propagation, while inside the core the propagation is damped and externally driven.

4. The Problem of the Core

How can we determine R_o, i.e. the SW core radius? To answer this question it is necessary to study the stability of SW solutions.

When the terms J in Eqs.(5) are neglected and k and Q are kept fixed, at every value of r we have a one-dimensional propagation problem (7). This periodic propagation is characterized by the modified dispersion relation $\omega=\Omega(k,Q)$. The propagation should be stable in the outer region ($r>R_o$) but should lose stability on the core boundary ($r=R_o$), since the propagation is externally driven inside the core.

The dispersion relation $\omega=\Omega(k,Q)$ defines a surface in the three-dimensional space (ω,k,Q); the planar sections of such a surface at Q=const are shown in Fig.3. Only some part of the surface corresponds to solutions that are stable under small perturbations. We have marked the stable segments on the planar sections at a constant Q by solid curves in Fig.3. Projection of the stability regions onto the plane (ω,k) is shown in Fig.4 (the dashed area). It is bounded by the dispersion curve for Q=0 and by the line Γ given by the function $k=k^*(\omega)$.

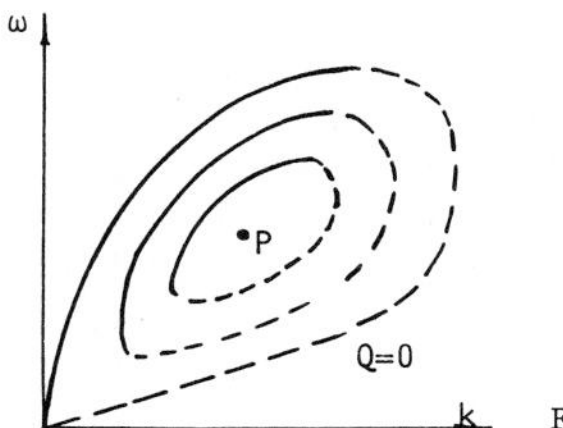

Fig. 3

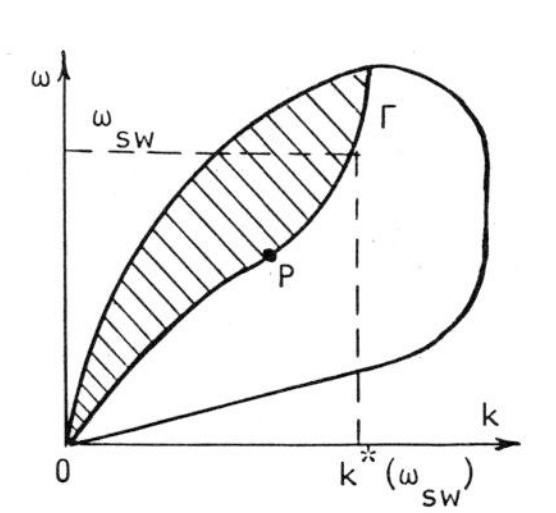

Fig. 4

We treat the core as an effective hole in the medium and, therefore $\chi_r(R_o)=0$. On the other hand, propagation should lose stability on the core boundary and, hence, $k(R_o)=k^*(\omega_{sw})$. By taking into account that $k^2=\chi_r^2 + r^{-2}$ we finally come to the conditions

$$\chi_r(R_o) = 0, \; R_o = \left[k^*(\omega_{sw})\right]^{-1}, \; \lim_{r\to\infty} \chi_r = \text{const} < k_{max}, \tag{12}$$

which set the nonlinear eigenvalue problem for the differential Eq.(9).

They determine the rotation frequency ω_{SW} and the core radius R_O of a spiral wave in the homogeneous infinitely extended active medium.

5. Discussion

This procedure was applied in [12] to the Rinzel-Keller model [17] where the dispersion relation for periodic nonlinear waves was found analytically. Fig.5 shows the dependence of the temporal period of one-armed spiral waves and the minimal period T_{min} of stable one-dimensional propagation on the relaxationability parameter ε in the Rinzel-Keller model. The dependence has an U-shaped form found in experiments with cardiac tissue as well as in numerical simulations. Near the critical value $\varepsilon \approx 1.4 \times 10^{-2}$, where one-dimensional pulse propagation becomes impossible, the SW period T is much greater than T_{min} and the rotation regime is controlled by the curvature limitations (see [14]). For the smaller values of ε, the SW period is close to the recovery time T_{min}, as it follows from the Wiener-Rosenblueth axiomatic model. Here both T and T_{min} increase as $\varepsilon^{-1/2}$.

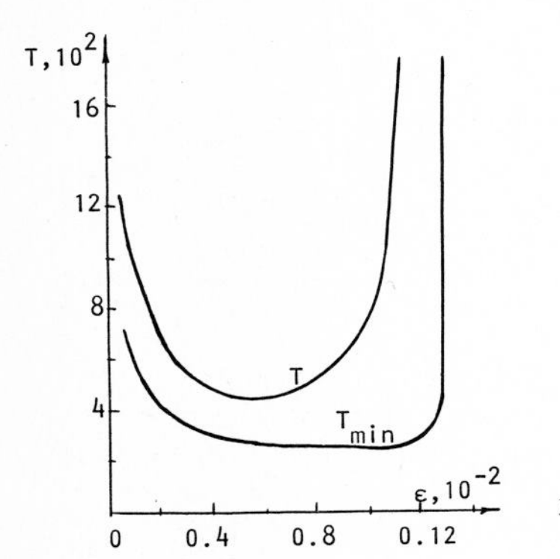

Fig. 5

The proposed procedure is approximate since we have neglected certain terms in Eqs.(5). The mathematical estimates given in [12] show that such terms can indeed be neglected if $k_{max} \ll 1$, i.e. if the minimal spatial period L_{min} of stable one-dimensional pulse propagation is great as compared with the width of the pulse front that was taken as a length scale unit. In this case the size R_o of the SW core is always much greater than the characteristic diffusion length ($R_o \gg 1$).

References

1. A.N.Zaikin and A.M.Zhabotinskii. Spatial effects in the self-oscillatory chemical system, in: Oscillatory Processes in Biological and Chemical Systems, vol.2, (Pushchino, 1971), p.279
2. A.T.Winfree. Science 175 (1972) 634
3. N.Wiener and A.Rosenblueth. Arch. Inst. Cardiol. Mex. 16 (1946) 205
4. I.S.Balakhovskii. Biofizika 10 (1965) 1063
5. G.R.Ivanitskii, V.I.Krinsky and E.E.Sel'kov. Mathematical Biophysics of the Cell (Nauka, Moscow, 1978)
6. T.Erneux and M.Herschkowitz-Kaufman. J. Chem. Phys. 66 (1977) 248

7. D.S.Cohen, J.C.Neu and R.R.Rosales. SIAM J. Appl. Math. 35 (1978) 536
8. A.S.Mikhailov and I.V.Uporov. Dokl. Akad. Nauk SSSR 249 (1979) 733
9. P.S.Hagan. SIAM J. Appl. Math. 42 (1982) 762
10. A.S.Mikhailov and V.I.Krinsky. Biofizika 27 (1982) 875
11. L.S.Polak and A.S.Mikhailov. Self-Organization in Non-Equilibrium Physical-Chemical Systems (Nauka, Moscow, 1983)
12. A.S.Mikhailov and V.I.Krinsky. Physica, 9D, (1983) 346
13. A.C.Scott. Active and Nonlinear Wave Propagation in Electronics (Wiley-Interscience, N.Y., 1970)
14. V.S.Zykov. Biofizika 25 (1980) 888
15. J.M.Greenberg. SIAM J. Appl. Math. 30 (1976) 199
16. A.M.Pertsov and A.V.Panfilov. Spiral waves in active media, in: Autowave Processes in Systems with Diffusion, M.T.Grekhova, ed. (Gorky, 1981) p.77
17. J.Rinzel and J.B.Keller. Biophys. J. 13 (1973) 1313.

One-Dimensional Autowaves, Methods of Qualitative Description

I.M. Starobinets and V.G. Yakhno
Institute of Applied Physics, Academy of Sciences of the USSR, SU-Gorky, USSR

The research of dynamics of nonequilibrium media involves the difficulty of obtaining typical spatial-temporal solutions for the known "basic" models. General methods to solve such equations in partial derivatives are still unavailable. However, some conclusions can be made using the considered examples describing typical nonlinear structures. In particular, the concepts of nonstationary processes in "point" systems can be applied to description of one-dimensional autowave processes. There are three main stages of investigation of possible solutions: obtaining of stationary solutions, study of their stability, determination of transitions from unstable stationary solutions to more stable ones. In this paper the examples of one-dimensional strongly relaxational autowave processes are presented. However, the proposed sequence of description seems to have a wider range of application. Recall that autowaves are a self-sustained wave process in a nonequilibrium medium which is independent of small but finite variations of both initial and boundary conditions [1].

What "images" are necessary to describe autowave processes? First, a great variety of experimental data on autowave processes calls for necessity to distinguish typical "basic" structures. Introduction of some autonomous space-time structures in physics of nonlinear processes is universally accepted today [2]. Solitons, regular and stochastic self-oscillations... This list can be naturally extended by a set of autowave processes (Fig.1a) [1,3-7]. Division of various autowaves into five types of nonlinear structures is to some extent conventional and cannot be regarded as strict. In particular, more complicated structures can be sometimes described by means of simpler ones. The profit of this division is obvious: the description of more complicated processes can be made using the peculiarities of the behavior of the introduced structures without going into details of their construction. Second, in order to describe these structures, "basic" models are built (some of them are shown in Fig.1b). Besides, there are model nonequilibrium media (for example, the Belousov-Zhabotinskii reaction). And, finally, a variety of solutions obtained for each "basic" model is available. The schemes of the solutions on the plane (x, t) are displayed in Fig.1c and enumerated in accordance with the order of nonlinear structures they describe. The positions of excitation fronts are shown in the schemes by a solid line, which bounds the hatched excited regions. Values $V=V_{cr}$, at which the front velocity is zero, are shown by a dotted line. The regions, where the slow variable exceeds V_{cr} are given by dots. A more detailed description of the schemes for typical solutions in one-dimensional systems is given, for example, in [5-7]. Comparison of the peculiarities of specific autowave structures with corresponding solutions definitely indicates these stages being necessary for a complete qualitative description. The first property - stability of spatial and temporal parameters of the structures and solutions - indicates the important role of the

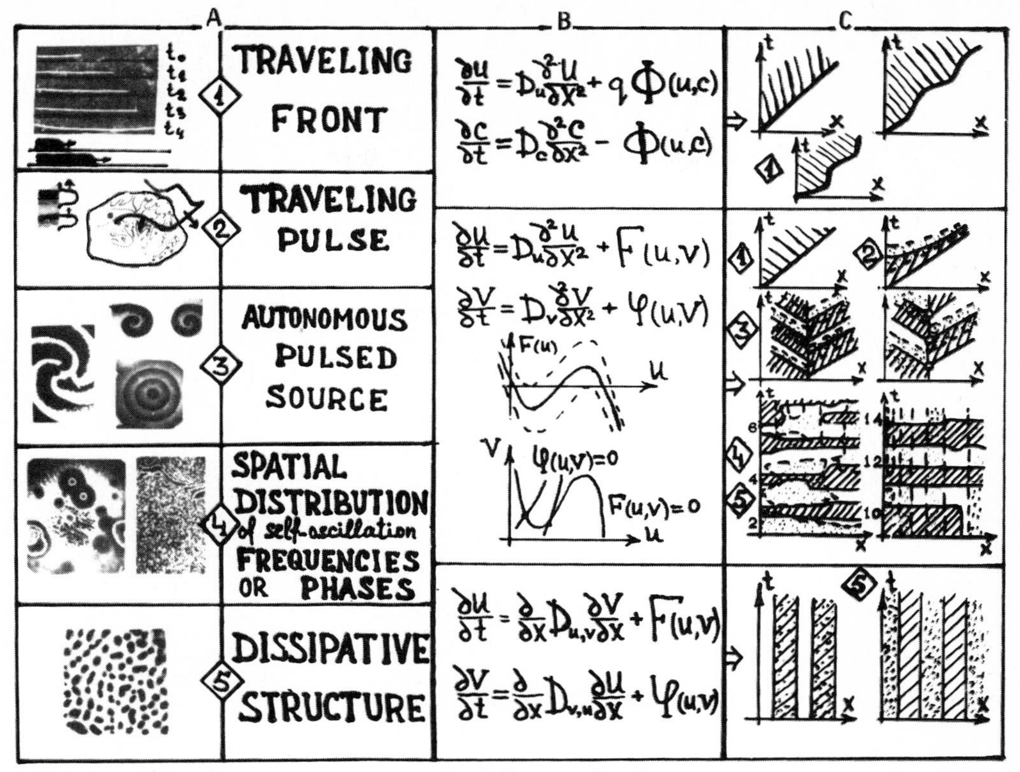

Fig. 1 Autowave structures, models, solutions.
a) Examples of experimental data illustrating five typical nonlinear structures constituting the "basis" of autowave processes;
b) Most popular types of "basic" two-component models;
c) Schemes of one-dimensional solutions obtained from these models

search for stationary solutions to the systems. The stability of the stationary state is investigated at the second stage. In distributed systems small perturbations are represented as a set of spatial modes. The stability type (focus, saddle, node, etc.) will depend on the spatial mode number. The third stage, which is evidently the most complicated one, is associated with determination of rules for transition from unstable stationary solutions to more stable ones. In real systems this stage reveals itself in the dependence of behavior of the transition mode on the form of initial perturbation. All these types are included directly or indirectly into the axiomatic description [3], into the method of consecutive approximations [9] and into methods of differentiation between slow and fast space-time processes [4,7,10] (see for details [5,7]). It is readily seen that all the three stages are present in considerations of practically all known autowave structures. The nonlinear structures displayed in Fig.2 and 3 are the simplest examples showing that all the three stages of the qualitative description are necessary. However, we omit this description here, since there is a more interesting question. Is it possible to apply this approach to search for unknown solutions? Let us deal with a simple "basic" model - a diffusion equation with a nonlinear N-shaped function in the form of a cubic parabola (Fig.4). Solutions are found in the finite interval $-L \leqslant x \leqslant L$. The boundary conditions are $\left.\frac{\partial U}{\partial x}\right|_{x=\pm L}=0$.

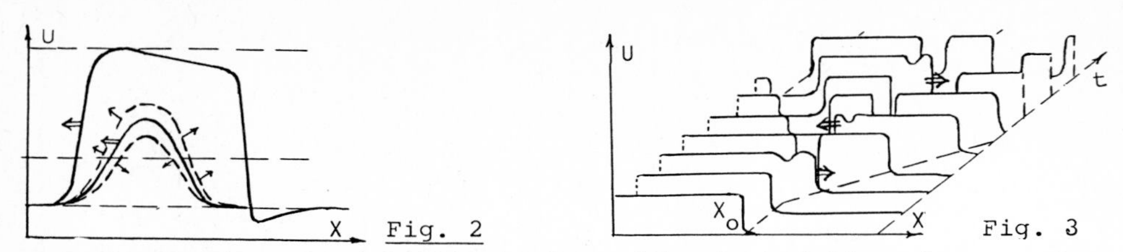

Fig. 2 Stationary travelling pulses (the dotted line shows perturbations near the unstable pulse)

Fig. 3 One-dimensional autonomous pulsed "echo"-type source

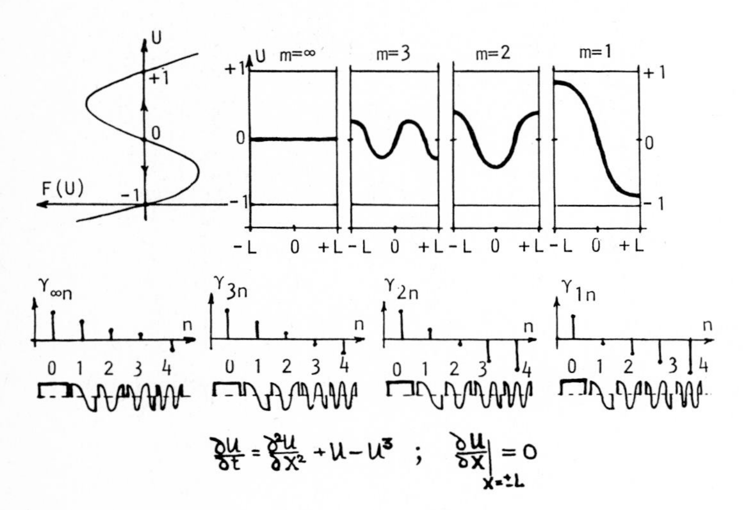

Fig. 4 Stationary solutions of equation $U_t = U_{xx} + F(U)$ and data on stability of three stationary waves and the central homogeneous state to perturbation modes with a different number of zero crosses

Now we demonstrate the possibility of a qualitative description of some transition processes in such an equation. Stationary solutions of this equation are obtained by conventional methods and regarded as analogs of equilibrium points in the "point" system. The upper part in Fig.4 displays the case of existence of three inhomogeneous stationary solutions and, as usual, three homogeneous solutions ($U_o=-1$, $U_o=0$, $U_o=1$). The small perturbations near the stationary solutions represent a set of spatial modes. The homogeneous solutions ($U_o=1$ and $U_o=-1$) are stable to all modes of small spatial perturbations. The central homogeneous solution ($U_o=0$) and three inhomogeneous stationary structures are unstable to some perturbation modes with a certain number of zeros [10]. In the lower part of Fig.4 the value and sign of the growth rate of instability γ_{mn} are plotted against the number of the spatial mode characterized by n, the number of zeros. The index refers to the stationary structure under study and defines the number of zeros in this solution (m=∞ corresponds to the homogeneous solution $U_o=0$, m=3 to the stationary structure with three zeros etc. (see Fig.4)). It is seen that inhomogeneous stationary solutions are stable only to perturbation modes having an equal or larger number of zeros compared with the stationary solution in question. These stationary solutions are unstable to perturbations with a smaller number of zeros. Comparison of the form of stationary solutions with the character of their stability to modes with a different number of zeros enables us to make the following assumptions: the unstable mode with K zeros can transform the unstable solution only to the stationary solution with the same number (K) of zeros. If this assumption is valid, there should exist a solution in the form of successive transitions from one stationary solution to another. Verification was performed by numerical calculations. Increasing perturbations really lead to the formation of a stationary solution with the same numbers of zeros. However, in a certain period of time, modes are found to appear and the stationary solutions are formed with a smaller number of zeros. In the initial distribution such perturbation modes are absent. We believe, that this behaviour is due to errors caused by discreteness of the numerical calculation scheme. Perturbations produced thereby grow up if they correspond to unstable modes. By now, we cannot unambiguously explain such behaviour of the solution. The existence of transitional process has been also found in the form of successive transformation of a less stable stationary solution to a more stable one. For example, Fig.5 shows the solution of the stationary equation with three zeros and then formations of profiles close to stationary solutions with two zeros, with one zero and the final transition of the system into the upper stable homogeneous state. Therefore, the character of the transition process agrees qualitatively with the earlier assumptions.

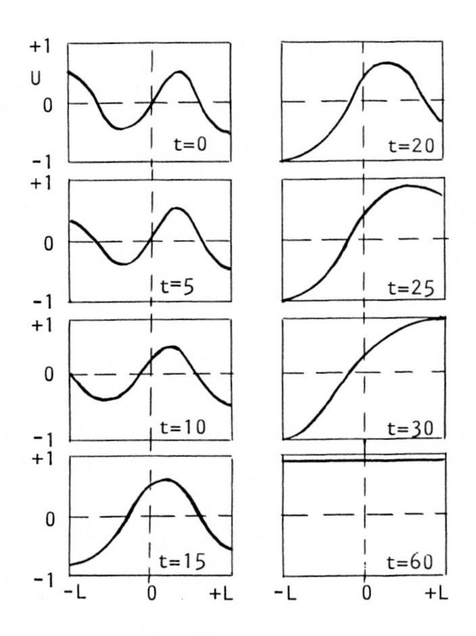

Fig. 5 The transition process in the system presented in Fig.4 under the initial condition $U_{in}=0.5\ \mathrm{Sin}(3\pi/2L)x + 0.05\ \mathrm{Cos}(\pi/L)x - 0.01017\ \mathrm{Sin}(\pi/2L)x$

The examples given here show that the distinction of staged in a qualitative description only not helps to fit the already known solutions into the general scheme of treatment but also allows finding the solutions previously unknown. Such a step-by-step approach is generally adopted in the oscillation theory. But as far as we know, no specific application of this general scheme to autowave processes has been reported as yet.

References

1. Autowave processes in diffusion systems (collected papers, in Russian), ed. M.T.Grekhova, Institute of Applied Physics, A. Sci.USSR, Gorky (1981)
2. A.V.Gaponov-Grekhov, M.I.Rabinovich. Nonlinear physics. Stochasticity and structures (to be published in "Physics of the twentieth century")
Nonstationary Structures - Chaos and Order, in "Synergetics" of Brain", ed. H.Haken, Springer (1983)

3. V.I.Krinsky. Fibrillation in excitable media.Problems of cybernetics (in Russian). M. Nauka (1968)
4. A.M.Zhabotinskii. Self-excited Concentration Oscillations (in Russian), Nauka, Moscow (1974)
5. V.A.Vasil'ev, Yu.M.Romanovskii, V.G.Yakhno. Autowave processes in distributed kinetic systems. Sov. Phys. Usp. 22(8), Aug. 1979 p.615-638
6. A.M.Zhabotinskii, V.I.Krinskii. Self-wave structures and perspectives of their investigation, in [1], p.6
7. V.G.Yakhno. Self-wave processes in one-dimensional relaxation systems, in [1], p.46
8. V.I.Talanov. Stimulated diffusion and cooperative effects in distributed kinetic systems. p.47, in "Nonlinear waves. Self-organization", (in Russian). M, Nauka (1983)
9. V.F.Pastushenko, V.S.Markin, Yu.Ya.Chizmadzhev. The foundations of the theory of excited media. Results of science and engineering, ser. Bionic, Biocybernetics, Bioengineering (in Russian) v.2, M. (1977)
10. B.S.Kerner, V.V.Osipov. Stationary and travelling dissipative structures in active kinetic systems. Microelectronics (in Russian) v.10, vyp.5, p.407 (1981)

Twisted Scroll Waves in Three-Dimensional Active Media

A.V. Panfilov, A.N. Rudenko, and A.M. Pertsov

Institute of Biological Physics of the USSR Academy of Sciences
SU-142292 Pushchino, USSR

The discovery of rotating waves (vortices) in the Belousov-Zhabotinsky (B-Z) reaction has initiated a great number of studies concerned with the reaction-diffusion equation which yields two-dimensional periodic solutions describing such waves. These studies have aroused interest among biologists because structures similar to those in B-Z reaction can also be observed in biological excitable media (cardiac and nerve tissues). The appearance of vortices in these tissues is associated with pathological changes such as paroxysmal tachycardia and fibrillation, waves of spreading depression in the brain cortex [1,2].

In chemical and biological excitable systems, vortices are, as a rule, three-dimensional. However, because of their complexity, only two-dimensional cases have been analysed in detail so far. Topological analysis suggests the existence of three distinct types of vortices in three dimensions - simple, twisted and knotted ones [3]. Only simple vortices were observed both experimentally and computationally [4,5].

In this work we obtained in numerical experiments the second topological type of vortices - a twisted scroll, which is a rotating wave with a screw-like surface. The conditions are specified under which such structures can be observed experimentally.

For the description of active medium, the system of a Fitz-Hugh-Nagumo type equations was used

$$\tau \frac{\partial E}{\partial t} = \Delta E + f(E) + I$$
$$\tau_1 \frac{\partial I}{\partial t} = (g_s E - I), \qquad (1)$$

where Δ is a three-dimensional Laplacian operator, f(E) is a non-linear N-shaped function and τ, τ_1 and g_s are parameters. This type of equations is much used to describe the distributed biological membranes, the meaning of the parameters being: f(E), the voltage-current characteristics of the fast inward current; g_s, the conductance of the slow outward current, τ and τ_1, the time constants of the fast and slow current, respectively ($\tau_1 = \tau \cdot \mu$, where $\mu >> 1$).

The calculations were performed for the piecewise linear approximation of f(E) of the type $f(E)=4E$ for $(E<E_1)$; $f(E)=-g_f(E-0.02)$ for $(E_1<E\leq E$ $f(E)=15E$ for $(E>E_2)$. The Neumann boundary conditions $(\partial E/\partial n)=0$ were used. System (1) was integrated in Cartesian coordinates by the simple Euler method (the mesh size being 30x30x30 elements).

Our numerical studies have shown that in homogeneous media twisted scrolls with open filaments are unstable. Whatever the initial conditions, they always degrade to simple (untwisted) scrolls during rota-

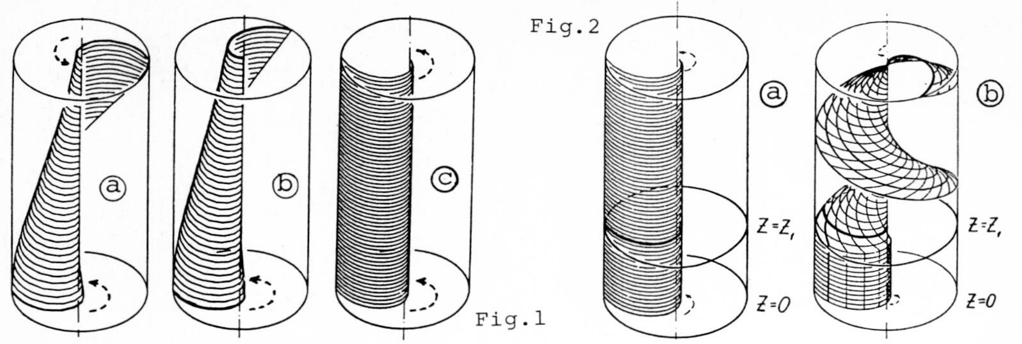

Fig. 1 Degeneration of a twisted scroll to a simple one in a homogeneous medium (g_f=0.9).
a - the initial shape of the scroll (t=0), the phase shift in the upper and lower cross-sections being equal to π. Dashed line is the wavefront, the arrow indicates the direction of rotation.
b - t=29.4, the phase shift reduced to $\pi/2$.
c - t=59.3, no phase shift is observable

Fig. 2 Generation of a twisted scroll.
a - initial conditions (simple scroll). At $z=z_1$, parameter τ is increased stepwise from $\tau = 1$ to $\tau = 2$.
b - the established shape of the twisted scroll

tion (Fig.1). This implies that such vortices are not physically realizable in homogeneous media.

However, active media are essentially inhomogeneous in parameters as a rule. The possibility of the occurrence of stable stationary twisted scrolls has been demonstrated for a broad class of inhomogeneous media. In particular, a twisted scroll can evolve from a simple one if there is a parameter gradient along the length of its filament. This phenomenon is illustrated by Fig.2*. Gradient in parameter τ in system (1) was set up stepwise. Fig.2a gives the initial conditions and Fig.2b the established form of a scroll. It is seen that the waves form a screw-like surface. In the cross-section z=const are shown spirals with the phase shift that increases with increasing z. Fig.3a (curve 1) plots the phase shift versus z for this case. It is seen that sufficiently far away from the boundary of inhomogeneity the phase shift gradient (the screw step) becomes constant (φ'=const). The same value of φ' = 0.36 is attained when the parameter τ is altered smoothly and not stepwise.

Twistedness of the scroll is due to the fact that the initial rotation velocities have different values because of parameter gradients along the length of the scroll filament. In this case the same rotation period (the minimal one) is established everywhere. In Fig.2, the initial gradient in the period along the z-axis was: T=30 at $z \leq z_1$, T=60 at $z > z_1$. During rotation, the shorter period (T=30) was established in the entire active medium. Twistedness of the scroll was observed not only with parameter τ gradient, but also with gradients in the other parameters of system (1), g_f and g_s. Also in that case the minimal period was established.

The step of the screw depended on the specific inhomogeneity produced. We studied the established shape of the twisted scroll with

*This figure was drawn by A.V.Samarin and V.N.Kochin using the graphic package "Atom" [6].

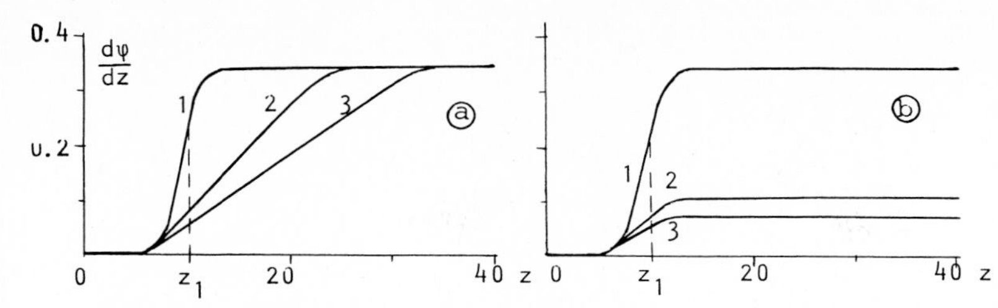

Fig. 3 The degree of the scroll twistedness as a function of z.
a - different gradients in parameter τ. 1, abrupt increase in τ, $\tau=1+\theta(z-z_1)$, where $\theta(z)$ is the Heaviside function. 2,3, smooth increase in τ, from τ=1 to τ=2 with different gradients: $\alpha=d\tau/dz=0.1$ (curve 2), α=0.05 (curve 3).
b - set up of different inhomogeneities (gradients in τ, g_f, g_s). Curve 1, $\tau=1+\theta(z-z_1)$, $g_f=0.9$, $g_s=1.0$. Curve 2, $g_f=0.9-0.12\theta(z-z_1)$, $g_s=1.0$, τ =1, Curve 3, $g_s=1.0+0.5\theta(z-z_1)$, $g_f=0.9$, τ=1

gradients in g_f, g_s and τ yielding the same period gradient $T=30+30\cdot\theta(z-z_1)$. In that case the established value of the phase shift with parameter τ was three times as great as with parameters g_f and g_s (Fig.3b).

The effects described were observed only with not too steep parameter gradients. In the case of strong inhomogeneity, the twisted scroll wave did not occur because the regions with long periods cannot be synchronized by shorter-period regions. In this case complicated periodic regimes may appear (in particular, when parameter is altered stepwise more than three-fold).

It can be concluded that the stable stationary twisted scrolls exist in a wide range of parameter values and do not depend on the kind of inhomogeneity prescribed. There is a good reason to believe that these structures will be observed experimentally in active media of chemical and biological nature. In particular, they are likely to occur in sufficiently thick layers of B-Z reagent as a result of the transverse inhomogeneity due to the inhibiting effect of oxygen on the reaction surface [7]. Significant transverse inhomogeneities exist also in myocardial tissue and therefore they may cause twistedness of the scroll waves occurring in the heart wall in the case of the appropriate filament orientation.

The authors are deeply indebted to Professor V.I.Krinsky for interest in the work and helpful discussions.

References

1. V.I.Krinsky, Pharmac. Ther., 3, No 4, 539 (1978)
2. N.A.Gorelova, J.Bures. Neurobiology 14 No 5, 353 (1983)
3. A.T.Winfree, S.H.Strogatz. Physica 9D, No 1, 65 (1983)
4. A.T.Winfree. Science 175, 634 (1972)
5. A.V.Panfilov, A.M.Pertsov. Doklady Akademii Nauk 274, No 6, 1500 (1984)
6. S.V.Klimenko, V.N.Kochin, A.V.Samarin. Upr.Syst. i Mash. No 3, 104 (1983)
7. K.I.Agladze. Characteristics of Two-dimensional Autowave Processes in Two-dimensional Medium. Thesis, Pushchino, Inst. of Biophysics, (1983).

On the Complex Stationary Nearly Solitary Waves

L.A. Beljakov and L.P. Šil'nikov

Scientific Research Institute for Applied Mathematics and Cybernetics
Gorky State University, SU-Gorky, USSR

When studying the mathematical models of various strongly nonlinear physical, chemical and other processes, the solutions of travelling stationary wave type are of special interest. The search for such solutions is specific and it is usually reduced to a study of peculiar trajectories, viz. heteroclinic and homoclinic orbits of finite-dimensional self-similar systems which may be investigated by the methods of qualitative theory of dynamical systems. Heteroclinic trajectories going from one equilibrium state to another correspond to travelling waves of the wave-front type which are typical for the Kolmogorov-Petrovsky-Piskunov equation (KPP-equation) as well as for equations of combustion, gas dynamics, and so on. Homoclinic trajectories, running from an equillibrium state and back to it, correspond to travelling waves called impulses or solitons (the Korteweg-de Vries equation, Fritz-Hugh-Nagumo equation, etc.).

The results which follow concern the existence problem for solitary travelling waves of complex form.

1. Complex Solitons. Suppose that a distributed system supports a travelling wave corresponding to the separatrix loop Γ_0 of a saddle-focus O in an n-dimensional self-similar system. As a saddle-focus we define the equilibrium state in which only one root λ_n of the characteristic equation lies in the right half-plane* and all others $\lambda_1, \lambda_2, \ldots \lambda_{n-1}$ have negative real parts, λ_1 and λ_2 are complex conjugated roots and $\mathrm{Re}\,\lambda_i < \mathrm{Re}\,\lambda_1$, where $i=3, n-1$. From earlier papers by Sil'nikov [1-4] (see also [5]), the following conclusions can be drawn.

1. If the saddle parameter $\sigma = \mathrm{Re}\,\lambda_1 + \lambda_n < 0$, then upon destruction of the loop Γ_0 no other loop remains within its neighbourhood.

This proposition is also valid if the negative root with the least absolute value is real.

Suppose now that 1) $\sigma > 0$, 2) Γ_0 enters O, touching the leading manifold, and 3) the separatrix parameter A differs from zero. These conditions distinguish the bifurcation submanifold with unity codimension in the space of dynamical systems.

2. When the conditions 1)-3) are fulfilled in the neighbourhood of the described submanifold, there exists a countable set of bifurcation submanifolds of the same codimension corresponding to homoclinic

---

* From the stability viewpoint, this case, according to [6], is of special interest.

curves of increasing number of rounds*, i.e. to multi-humped solitons of a distributed system.

Recently, this approch has been extended by American scientists [7-10] in connection with investigations of multipulse waves of Fitz-Hugh-Nagumo equation. Note that a similar situation occurs also for the equation

$$U_t + 4U_x U + (1-n)U_{xx} + U_{xxxx} = 0$$

obtained for the problem of a thin layer of liquid flowing over an inclined wall (here n is a parameter describing the wall slope) [11-14].

A question arises how the bifurcation submanifolds associated with complex solitons appear and vanish. This occurs due to violation of conditions 1)-3). Since only three-dimensional systems are considered below, there are only three basic cases [15], a two-parametric family of systems being involved in each case. The parameters μ_1, μ_2 are introduced in the following way: μ_1 controls the system behaviour in the neighbourhood of the equilibrium state O, whereas μ_2 controls splitting of the separatrix $\Gamma(\mu_1,\mu_2)$ so that $\Gamma(\mu_1,0)$ corresponds to a homoclinic curve at every $|\mu_1|\leqslant\bar{\mu}_1$. Now let us formulate some propositions about the structure of bifurcation subsets Π^2 and Π^3 corresponding to two- and three-rounding loops of separatrices.

A) Let $\lambda_1(0) = \lambda_2(0) = \gamma<0$ and $\sigma>0$ if $\mu_1=\mu_2=0$. In this case the parameter μ_1 is taken such that $\lambda_{1,2}=\gamma\pm\sqrt{-\mu_1}$ (the case of Jordan cell). Then, in the parameter plane μ_1, μ_2 the bifurcation subset Π^2 becomes a bundle of curves $\{\Pi^2_n\}_{n=1}^{\infty}$ converging at the point $\mu_1=\mu_2=0$ and the main asymptotics of these curves looks similar to that presented in [16] (see Fig.1). The same behaviour in the Fitz-Hugh-Nagumo model was found by Kuznetsov and Panfilov [17,18].

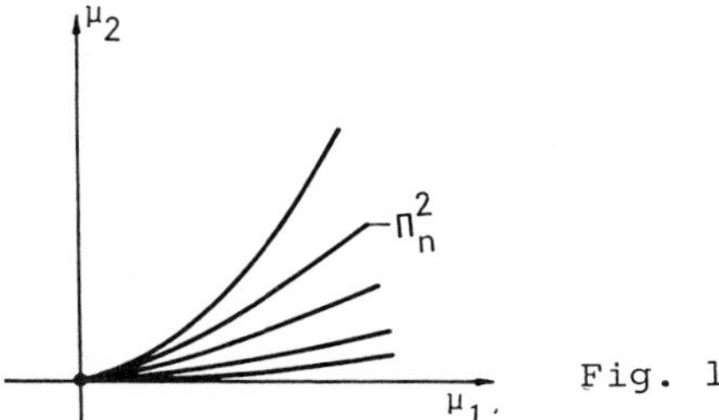

Fig. 1

B) Let $\sigma=0$, $\mathrm{Im}\lambda_i\neq0$, i=1,2 when $\mu_1=\mu_2=0$. This yields $\mu_1=\sigma$ and in the parameter plane μ_1,μ_2 there exists a countable set of curves Π^2_n of a parabolic type, their vertices tending to the origin as $n\to\infty$. Between every two curves Π^2_{n+1},Π^2_n there appears a set of curves Π^3_{nk} (k depends upon n), each curve consisting of two branches and the number of curves in this set increasing with the growth of n. The branches may be either oppositely directed or nested. In Fig.2 we show only a single pair of curves Π^2_n, Π^2_{n+1} and a single set between them.

---

* This result has been used in the Lorentz system [5] for constructing similar bifurcation curves.

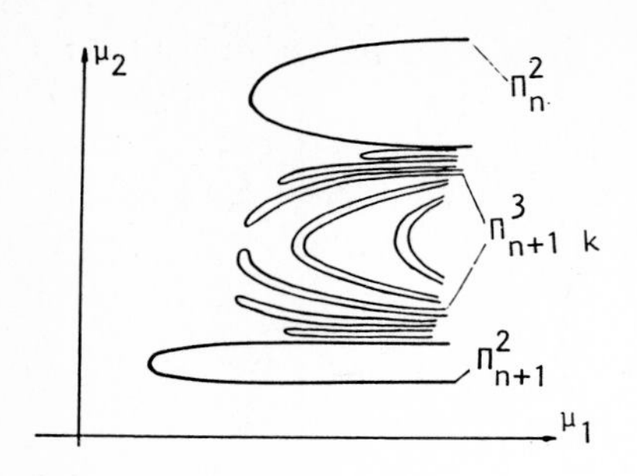

Fig. 2

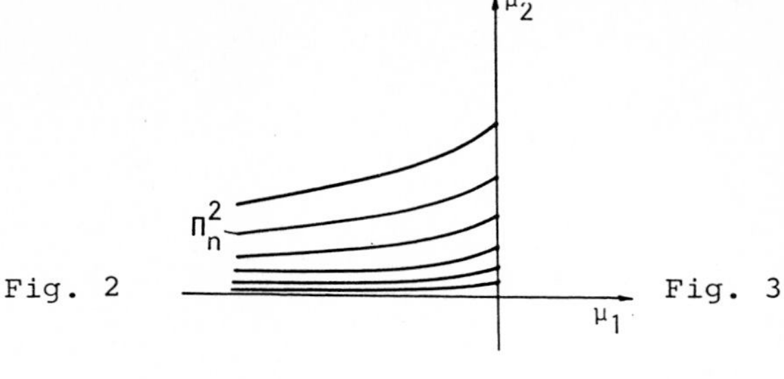

Fig. 3

C) Let $Re\lambda_i=0$, $Im\lambda_i\neq 0$, $i=1,2$, when $\mu_1=\mu_2=0$, and let the first Lyapunov value be negative, i.e. O is a complex saddle-focus. Suppose for this case that $\mu_1=Re\lambda_1$. Then on the plane μ_1,μ_2 we obtain a countable set of curves Π^2_n and Π^3_{nk} defined within $-\bar{\mu}_1\leqslant\mu_1\leqslant 0$ converging to $\mu_2=0$, $-\bar{\mu}_1\leqslant\mu_1\leqslant 0$ when $n\to\infty$ (see Fig.3).

Here, in contrast to dissipative systems, the existence of a homoclinic curve is generic since the manifolds W^s and W^v of a saddle point lie at the same energy level, and therefore they can intersect transversally. We remind that the equilibrium state of a Hamiltonian system with n degrees of freedom is said to be a saddle point if among the roots $\lambda_1,\dots\lambda_n$, $\lambda_{n+1},\dots\lambda_{2n}$ there is no root with a zero real part; then we assume O to be a saddle, if λ_1 and λ_2 are real, $\lambda_1=-\lambda_2$ and $|\lambda_1|<|Re\lambda_i|$ $(i\geqslant 3)$, and a saddle-focus, if $\lambda_{1,2,3,4}=\pm\rho\pm i\omega$ $(\omega\neq 0)$ and $|\rho|<|Re\lambda_i|$ $(i\geqslant 5)$.

3. Suppose now that a Hamiltonian system with two degrees of freedom (n=2) contains a nontransversal homoclinic curve Γ_o of a saddle-focus. The set of homoclinic curves of the saddle-focus, of which all lie within a small neighbourhood of Γ_o, is then countable and is in one-to-one correspondence (except Γ_o) with a set of segments $[j_1,\dots j_k]$ $(k\geqslant 1)$ composed of symbols 1,2,... n... .

This statement remains also valid for the multi-dimensional case under some conditions listed in [3,19]. Note that from the point of view of symbolic dynamics, a complete description of hyperbolic set of trajectories, located entirely within a sufficiently small neighbourhood of Γ_o on the energy level of the saddle-focus, is fully similar to the description of all trajectories in a small neighbourhood of the Poincaré homoclinic curve; the latter description was given in [20]. Also note that the existence of a countable set of homoclinic curves can be established (under definite conditions) in the case of a saddle, provided it has two homoclinic curves.

2. Complex fronts. For the equation $U_t=DU_{xx}+f(u)$ the problem concerning the front structure is known to be simple, since there are no limit cycles in a self-similar system. However, periodical motions are typical for multi-dimensional systems, therefore, in R^n (at $n\geqslant 3$) the problem is much more complex and informative. In this case, we restrict ourselves to the class of self-similar systems with a finite number of periodic motions. Only two statements are to be made.

1. Let O_1 be a saddle of the type (n-1,1), i.e. W_1^v is one-dimensional, O_2 a saddle of the type (1,n-1), i.e. W_2^s is (n-1)-dimensional, and L a periodic motion of the saddle-node type, i.e. all the multiplicators, except for that equal to 1, lie inside a unity circle and

the first Lyapunov value differs from zero. Suppose also that one of the trajectories Γ_1, outgoing from O_1, at $t\to\infty$ tends to L, and W_L^v transversally intersects with W_2^s. Then a countable set of bifurcations takes place, if the periodic motion vanishes and $\Gamma_1 \subset W_2^s$. Such heteroclinic curves result in the appearance of fronts with an arbitrary large number of motions.

It is interesting to note that such a case can also occur in R^2 [21] when, together with Γ_1, L is rounded by a stable separatrix of the saddle O_2 (see Fig.4). We want to stress that the bifurcation submanifolds are not achievable.

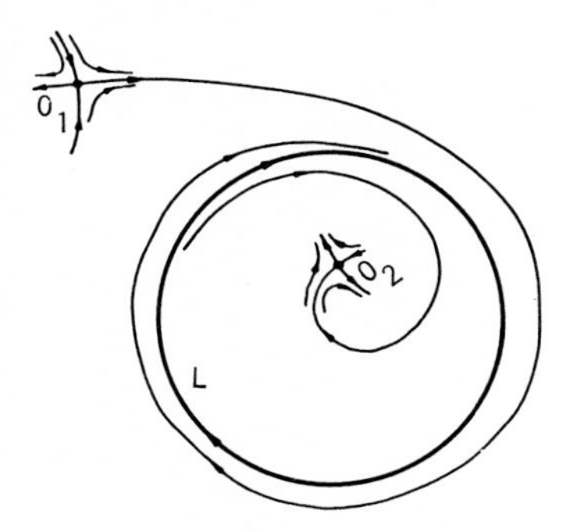

Fig.4

2. Suppose that W_1^v of the saddle O_1 of the type (p,q) transversally intersects with W_2^s of the saddle O_2 of the type (p-1; q+1). Their intersection may contain both limited and countable set of heteroclinic trajectories. In the latter case, in the closure of W_1^v a finite number (at least one) of saddle periodic motions of the type (p,q+1) is contained*.

According to the number and locations of these cycles in the closure, the trajectories connecting O_1 and O_2 can have various structure and their associated wave profiles can have any large number of oscillation.

Finally we would like to note that solitons and fronts are usually unstable motions of the saddle type. Their stable and unstable manifolds and also the manifolds of periodic waves and of the waves, corresponding to Poisson stable trajectories of self-similar systems, may intersect generating homoclinic structures of the initially distributed system. This behaviour is responsible for various chaotic wave motions and also for strange attractors or, at least, for "metastable chaos".

1. L.P.Šil'nikov. Mat. Sb., (1963), v.61(104), No 4, p.433-466
2. L.P.Silnikov. 3 Dokl. Akad. Nauk USSR, (1965) v.160, No 3, p.558-561
3. L.P.Šil'nikov. Mat. Sb. (1970) v.81 (123) No 1, p.92-103
4. L.P.Šil'nikov. Mat. Sb. (1968) v.77 (119) No 3, p.461-472

---

*To exactly describe the special set, the application of topological Markov chains with a countable set of trajectories is needed [22].

5. L.P.Šilnikov. The appendix II to: J.E.Marsden, M.McCracken "Bifurcation of cycle generation and its applications", Moscow (1980)
6. J.W.Evane. Indianne Univ. Math. J. (1972) v.22, No 6, p.577-593
7. J.A.Feroe. Siam. J. Appl. Math. (1982) v.42, No 2, p.235-246
8. J.A.Feroe. J. Math. Biosc. (1981) v.55, No 3/4, p. 189-204
9. J.W.Evans, N.Fenichel ., J.A.Feroe. Siam. J. Appl. Math. (1982) v.42, No 2, p.219
10. S.P.Hastings. Siam. J. Appl. Math. (1982) v.42, No 2, p.247-260
11. P.Lin. J. Fluid Mech. (1974) v.63, p.3
12. A.A.Nepomnjashi. Izv. Akad. Nauk SSSR, Ser. MJG, (1974) No 3 p.28-34
13. B.Ja.Shkadov. Izv. Akad. Nauk SSSR, Ser. MJG (1977) No 1, p.63-67
14. O.Ju.Cvelodub. Izv. Akad. Nauk SSSR, Ser. MJG (1980) No 4, p.142-146
15. L.A.Beljakov. Proc. IX IGNO, Kiev (1981)
16. L.A.Beljakov. Math. Zametki (1980) v.28, No 6, p.911-922
17. Ju.A.Kuznetsov, A.V.Panfilov. Preprint, Pushchino (1982)
18. Ju.A.Kuznetsov. Preprint, Pushchino (1982)
19. R.Devaney. J.Differential Equations, (1976) v.21, p.431-438
20. L.P.Šilnikov. Mat. Sb. (1967) v.74 (116) p.378-397
21. A.A.Andronov, E.A.Leontovič. Dokl. Akad. Nauk USSR (1938) v.21, No 9, p.427-430
22. V.S.Afraimovič, L.P.Šilnikov. Proc.MMO (1973) v.28, p.181-214

Elements of the "Optics" of Autowaves

O.A. Mornev

Institute of Biological Physics of the USSR Academy of Sciences
SU-142292 Pushchino, USSR

1. Introduction. Active media (AM) with chemical reactions and diffusion and their analogs play an important role in various branches of biology, chemistry and physics [1-7]. A simple example of such media is furnished by a bistable AM [1,3,5,7] described by the equations

$$\partial_t u + \operatorname{div} \vec{j} = f(u), \qquad \vec{j} = -D \operatorname{grad} u \tag{1a}$$

or, similarly, by

$$\partial_t u = D\Delta u + \operatorname{grad} D \cdot \operatorname{grad} u + f(u). \tag{1b}$$

Here u is a concentration variable, $\vec{j}$ is the diffusion flux, $D=D(\vec{r})$ is diffusivity (in the general case, it is a function of the space points $\vec{r} \equiv \{x,y,...\}$), Δ is the Laplacian which acts on the space variables, f(u) is an N-shaped source function (Fig.1a). Two stable states of the medium are given by the left and right zeros of f(u). Spatial dimensionality of the medium will be denoted as dim AM.

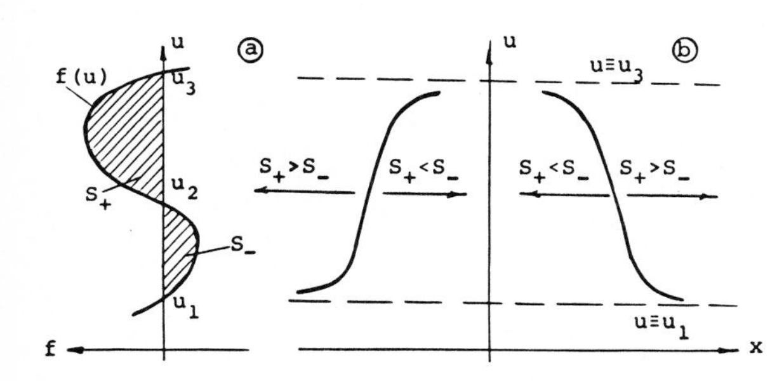

Fig. 1 a - the plot of the function f(u); b - space profiles of the travelling wavefront solutions to Eq.(2) (the arrows indicate the direction of propagation of the wavefronts)

The properties of Eq.(1) are studied in some detail for the case where the parameter D is independent of points of the medium. If D = const and dim AM = 1, Eq. (1b) after substituting $x \to x \cdot D^{\frac{1}{2}}$ becomes

$$\partial_t u = \partial_x^2 u + f(u). \tag{2}$$

This equation is known to have autowave solutions

$$u = u_s(\zeta), \qquad \zeta = t - x/c_s$$

of the type of travelling wavefronts which switch the medium from one stable state to another *) (see Fig.1b). The case D = const, dim AM>1 presents no difficulties either: Eq.(1b) is rewritten in the form

$$\partial_t u = D\Delta u + f(u) \tag{4}$$

and the travelling wavefront solutions to Eq.(4) are obtained from (3) by the formulae

$$u = u_s(\zeta), \qquad \zeta = t - \vec{\nu}\cdot\vec{r}/c, \qquad c = c_s\cdot D^{\frac{1}{2}}, \qquad |\vec{\nu}| = 1 \tag{5}$$

(it is evident that vector $\vec{\nu}$ in Eq.(5) may have any direction).

The aim of this communication is to present some new autowave solutions to Eqs.(1) for the case of spatially inhomogeneous AM with the piecewise constant profile of D.

We consider an infinite medium (dim AM = 2) which consists of two semi-planes $\Pi_1 = \{y|y>0\}$ and $\Pi_2 = \{y|y<0\}$ stuck together along the Ox-axis in Oxy space coordinates. It is assumed that y=0 is the line of an abrupt change of parameter D:

$$D = D_i \quad \text{if} \quad r \equiv \{x,y\} \in \Pi_i \qquad (i=1,2) \tag{6}$$

$$D_1 = \text{const}, \qquad D_2 = \text{const}, \qquad D_1 \neq D_2.$$

Eqs.(1) for such a medium reduce to the relationships**)

$$\partial_t u = D_i \Delta u + f(u) \qquad (\vec{r} = \{x,y\} \in \Pi_i, \qquad i=1,2) \tag{7a}$$

$$u\big|_{y=+o} = u\big|_{y=-o} \tag{7b}$$

$$D_1 \partial_y u\big|_{y=+o} = D_2 \partial_y u\big|_{y=-o} \tag{7c}$$

and admit two interesting types of solutions. The first type of solutions describes the steady state regime of the autowave refraction on the line of D jump. In some respects this regime differs radically from the refraction of linear electromagnetic waves. However, also in this case the well-known optical "sine-condition " applies. This allows the concept of the autowave refractive index to be introduced. The second type of solutions corresponds to a stand--still of the wavefront running transversally to the line of D jump. The stand-still condition can be described in terms of the autowave refractive index. Let us describe the solutions.

---

*) The wavefronts have a nonzero velocity only when $S_+ \neq S_-$, S_+ and S_- being moduli of the dashed areas on the plot f(u), see Fig.1a. When $S_+ > S_-$ ($S_+ < S_-$), the travelling wavefront switches the medium from the state of $u \equiv u_1$ to the state of $u \equiv u_3$ (from $u \equiv u_3$ to $u \equiv u_1$) (see, for example, [7]).

**) Eq.(7) results from Eq.(1b). Eq.(7b) stems from the natural requirement of continuity of the concentrational variable u. Eq.(7c) can be derived from Eqs. (1a) applying condition (7b).

2. Autowaves propagating along the line of D jump. The steady state regime of refraction. The solutions to problem (7) which describe this regime are found as follows. Let $u = u_s(\zeta)$, $\zeta = t - x/c_s$ is an autowave solution of "one-dimensional" equation (2) (the solution is assumed to be known). Consider the functions

$$u = u_s(\zeta_i), \quad \zeta_i = t - \vec{\nu}_i \cdot \vec{r}/c_i, \quad c_i = c_s \cdot D^{\frac{1}{2}}, \quad |\vec{\nu}_i| = 1$$
$$(\vec{r} \equiv \{x,y\} \in \Pi_i, \quad i=1,2) \tag{8}$$

where $\vec{\nu}_i = \vec{x}^o \cos\varphi_i + \vec{y}^o \sin\varphi_i$, $\vec{x}^o$ and $\vec{y}^o$ are orts of the axes Ox and Oy in Oxy coordinates, φ_i is the angle between $\vec{\nu}_i$ and $\vec{x}^o$.

Functions (8) automatically obey Eqs.(7a) (just as function (5) obeys Eqs.(4))and therefore, the solution is obtained by concatenation of the functions on the line y=o taking into account conditions (7b,c). The concatenation can be shown possible only for certain (eigen) vectors $\vec{\nu}_i$ in (8), which are determined by the formulae

$$\cos\varphi_1 = \pm [D_1/(D_1+D_2)]^{\frac{1}{2}}, \tag{9a}$$
$$\sin\varphi_1 = \pm [D_2/(D_1+D_2)]^{\frac{1}{2}}, \tag{9b}$$
$$\cos\varphi_2 = \pm [D_2/(D_1+D_2)]^{\frac{1}{2}}, \tag{9c}$$
$$\sin\varphi_2 = \pm [D_1/(D_1+D_2)]^{\frac{1}{2}}. \tag{9d}$$

Table

	I	II	III	IV
$\cos\varphi_1$	+	−	−	+
$\sin\varphi_1$	+	+	−	−
$\cos\varphi_2$	+	−	−	+
$\sin\varphi_2$	+	+	−	−

Placing correctly the signs in (9) is done by one of the four ways shown in the vertical columns of the Table. As a result, we obtain four autowave solutions to Eq.(7) corresponding to four pairs of eigenvectors $\{\vec{\nu}_1^N, \vec{\nu}_2^N\}$ (N = I,II,III,IV) (Fig.2a). Let us discuss the properties of these solutions.

Fig.2b shows the lines ζ_i=const of the solution relating to the pair $\{\vec{\nu}_1^{IV}, \vec{\nu}_2^{IV}\}$. It is seen that the travelling wave consists of the incident and the refracted components moving along the eigenvectors $\vec{\nu}_1^{IV}$ and $\vec{\nu}_2^{IV}$ with rates $c_1 = c_s \cdot D_1^{\frac{1}{2}}$ and $c_2 = c_s \cdot D_2^{\frac{1}{2}}$. These vectors can traditionally be characterized by the angles of incidence

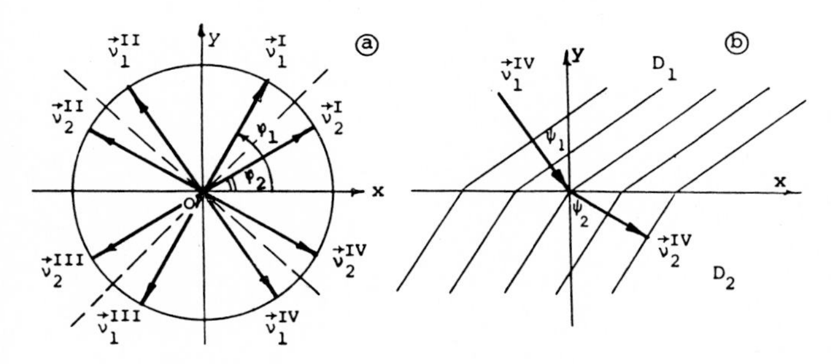

Fig. 2 a - eigenvectors of the wave normals of the piecewise-plane autowave solutions of Eq. (8). The pair $\{\vec{\nu}_1^N, \vec{\nu}_2^N\}$ is in the N-th quadrant and is symmetrical to the bisector of the appropriate coordinate angle; all the pairs can be obtained from each other by reflection onto the axes Ox and Oy; b - steady state refraction of an autowave on the line of D jump

and refraction (ψ_1 and ψ_2 in Fig.2b). It is easy to see (Fig.2b, Eqs.(9a,c) and column IV in the Table) that in this case

$$\sin \psi_i = \cos \varphi_i = \left[D_i/(D_1 + D_2)\right]^{\frac{1}{2}} \quad (i = 1,2)$$

and therefore

$$\sin \psi_1/\sin \psi_2 = a_{12}, \qquad a_{12} = c_1/c_2 = (D_1/D_2)^{\frac{1}{2}}. \tag{10}$$

(Similar relationships can be obtained also for other pairs $\{\vec{v}_1^{\,N}, \vec{v}_2^{\,N}\}$).Eq.(10) shows that the steady state refraction of autowaves meets the optical sine-condition. By analogy, it is natural to consider the parameter a_{12} as the autowave refractive index of the homogeneous regions Π_1 and Π_2. The principal difference between the refraction of nonlinear concentrational autowaves and the refraction of linear electromagnetic waves is as follows.

i. In optics, the incident and the refracted waves always co exist with the reflected wave. In active media with diffusion, the reflected waves are not observable.

ii. In optics, the angle of incidence of the light wave can have any value in the range from 0 to $\pi/2$, the angle of refraction being determined by the formula similar to Eq.(10). In the case of the *steady state* refraction of autowaves, these angles are uniquely determined by the diffusivities of the homogeneous regions from Eq.(9)*).

iii. In optics, the total internal reflection takes place when the angles of incidence exceed Bruster's angle. This phenomenon is not observed in active media; the refracted component of the autowave propagating in *steady state* conditions exists with any value of ψ_1 within the interval $0<\psi_1<\pi/2$.

3. "Opacity" of Autowaves and the Conditions for the Formation of Statinary Dissipative Structures. In this section we consider the interface of homogeneous regions be exactly zero**). In this case the wavefront running to the interface from the region with a smaller value of D (the semi-plane $\Pi_1 = \{y|y>0\}$) will be affected by the alternate region with a greater value of D (the semi-plane $\Pi_2 = \{y|y<0\}$) with the resulting decrease in its velocity. If the refractive index $a_{12} = (D_1/D_2)^{\frac{1}{2}}$ is less than a certain critical value a_{12}^{cr}, the wave cannot pass across the interface and comes to a stand-still - the line of D jump becomes "opaque" for on-running plane autowaves. The critical value a_{12}^{cr} can be found when it is considered that the standstill is equivalent to the formation of dissipative structures described by the steady state ($\partial_t \equiv 0$) solutions of Eq.(7a). In addition to conditions (7b,c), these solutions must satisfy one of the following two conditions

$$u = u_1 \text{ ahead of the wavefront } (y \to -\infty),$$
$$u = u_3 \text{ behind the wavefront } (y \to +\infty), \quad \partial_y u = 0 \text{ at } y \to \pm\infty \tag{11a}$$

---

*) This statement means that the initial position of the wavefront on both sides of the line of D jump can be preset arbitrarily; then,as time elapses,the position of the propagating wavefront will be determined by Eqs.(9).

**) Again,we consider a two-dimensional inhomogeneous AM described by the relationships (6)and(7). An additional assumption is that $D_1<D_2$.

or

$u = u_3$ ahead of the front $(y \to -\infty)$,

$u = u_1$ behind the wavefront $(y \to +\infty)$, $\partial_y u = 0$ at $y \to \pm\infty$ (11b)

Condition (11a) ((11b)) is used at $S_+ > S_-$ $(S_+ < S_-)$, see footnote *) p.114. The value of a_{12}^{cr} can now be found considering that the standing wavefronts are nonexistent at $a_{12} > a_{12}^{cr}$.

Let us determine a_{12}^{cr} for the case of $S_+ > S_-$ (the steady state problem (7), (11a)). By introducing the flux variables $j_i = -D_i \partial_y u$ $(\vec{r} \equiv \{x,y\} \in \Pi_i,\ i=1,2)$ and neglecting the term $\partial_t u$ in (7a), we reduce (7a) to $j_i \partial_u j_i + D_i f(u) = 0$ $(\vec{r} \equiv \{x,y\} \in \Pi_i,\ i=1,2)$. Integration of these equations with due regard to (11a) yields

$$j_1(u) = -D_1 \partial_y u = (2D_1)^{\frac{1}{2}} \cdot [S_+ - \sigma(u)]^{\frac{1}{2}} \qquad (y>o,\ u>u_o) \qquad (12a)$$

$$j_2(u) = -D_2 \partial_y u = (2D_2)^{\frac{1}{2}} \cdot [S_- - \sigma(u)]^{\frac{1}{2}} \qquad (y<0,\ u<u_o) \qquad (12b)$$

where

$$S_- = -\int_{u_1}^{u_2} f(u)\,du, \qquad S_+ = \int_{u_2}^{u_3} f(u)\,du, \qquad \sigma(u) = \int_{u_2}^{u} f(u)\,du$$

and

$u_o = u|_{y=+o} = u|_{y=-o}$ is the value of u on the line y=o to be determined. u_o can be found graphically from the relationship

$$K(u_o) = a_{12} \qquad \left\{K(u_o) \equiv [S_- - \sigma(u_o)]^{\frac{1}{2}} / [S_+ - \sigma(u_o)]^{\frac{1}{2}},\ a_{12} \equiv (D_1/D_2)^{\frac{1}{2}}\right\} \qquad (13)$$

obtained by substituting (12) into (7c) (the steady state solutions to the problem (7), (11a) are now obtained by integrating the equations $D_i \partial_y u = -j_i(u)$ (i=1 if y>o, i=2 if y<o) with the boundary condition $u|_{y=o} = u_o$). Examination of the function $K(u_o)$ shows that $K(u_1) = K(u_e) = 0$ (u_e being found from the equation $\sigma(u_o) = S_-$), $K(u_o) > 0$ at $u_1 < u_o < u_2$ and $\max_{u_1 < u_o \leq u_e} K(u_o) = K(u_2) = (S_-/S_+)^{\frac{1}{2}}$, see Fig.3a,b. Therefore, both Eq.(13) and the steady state problem (7)(11a) have two solutions at $a_{12} < (S_-/S_+)^{\frac{1}{2}}$. The solutions are represented in Fig.3a,c as points u_{o1} and u_{o2} and profiles 1 and 2; only profile 1 is stable to small perturbations (analysis for stability is omitted). If $a_{12} = (S_-/S_+)^{\frac{1}{2}}$, then $u_{o1} = u_{o2} = u_2$ and profiles 1 and 2 merge into one semi-stable profile (dashed line in Fig.3c). At $a_{12} > (S_-/S_+)^{\frac{1}{2}}$ both Eq. (13) and steady state problem (7),(11a) have no solutions and, therefore,

$$a_{12}^{cr} = (S_-/S_+)^{\frac{1}{2}}$$

when $S_+ > S_-$. The case of $S_+ < S_-$ (the steady state problem (7),(11b)) is studied similarly. Here

$$a_{12}^{cr} = (S_+/S_-)^{-\frac{1}{2}}$$

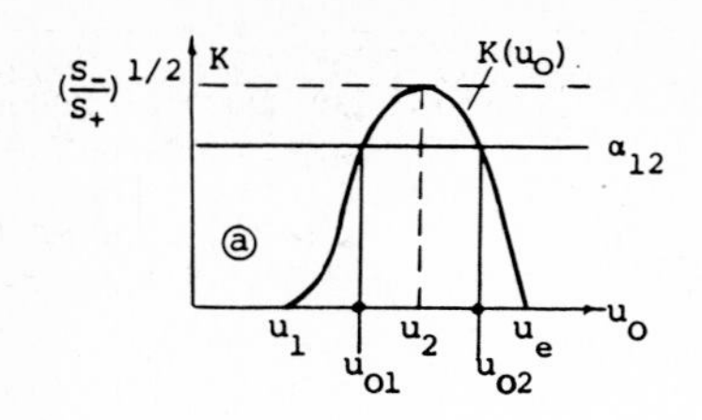

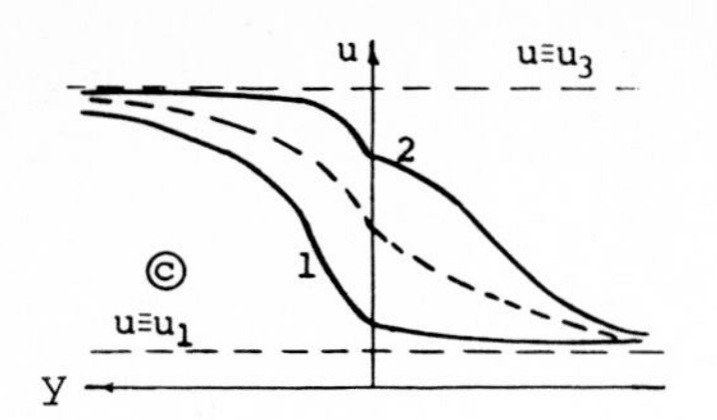

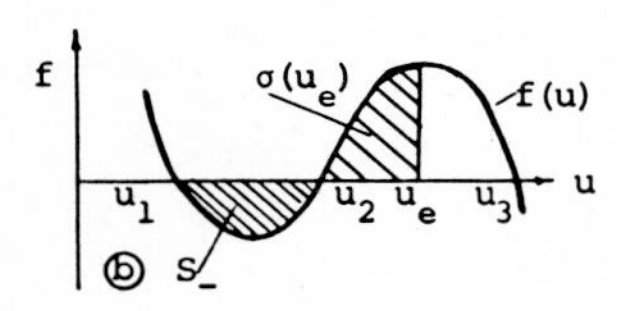

Fig.3.
(a) The plot of the function $K(u_o)$;
(b) the graphical solution to the equation $\sigma(u_o) = S_-$;
(c) inhomogeneous steady state solutions to the problem (7), (11a)

It is of particular interest to consider the case of $S_+ = S_-$. It turns out that in this case the steady state problem (7) has non-trivial steady state solutions with either of the conditions (11) if and only if $a_{12}=1$, i.e. $D_1=D_2$ (homogeneous medium). In inhomogeneous media, the standing wavefronts do not exist at $S_+=S_-$.

To conclude this section, note that the effects of autowave opacity are not observed if the autowave runs to the interphase of homogeneous regions from the region of greater diffusivity. It can be shown that the occurrence of standing wavefronts is impossible in this case.

4. One-dimensional Branching Active Media. Conditions for Opacity of the node of Branching. The propagation of autowaves in branching active media*) is characterized by the effects similar to those considered above. Indeed, the velocity of an autowave approaching a node of branching along one of the medium appendages decreases due to the shunting effects of other appendages [8-10]. If the effect is rather strong, the node becomes opaque to the incoming autowaves [7]. In the case of bistable media, such a node becomes a centre of the formation of dissipative structures. Below, the condition is determined for the existence of the dissipative structure that occurs when n wavefronts, entering simultaneously a node of branching along appendages numbered I,..., m, come to a stand-still. (It is assumed that the node consists of n one-dimensional homogeneous appendages of infinite length (Fig.4); the appendages are characterized by the diffusivities D_i=const (i=1,...,n) and are described by Eqs.(1) with the same function f(u). This condition is given by the inequality

$$(a_1 + \dots + a_m)/(a_{m+1} + \dots + a_n) \leq Q$$

($a_i = D_i^{1/2}$ (i=1,...,n), $Q = (S_-/S_+)^{1/2}$ at $S_+>S_-$ and $Q = (S_+/S_-)^{1/2}$ at $S_+<S_-$) which is obtained by the methods described in the previous section. When all D_i (i = 1,...,n) are identical and the wavefront enters the node along only one of the appendages, the condition for opacity of the node of branching is $n>n^{cr}$. The critical number of appendages n^{cr} is determined by the relationship $n^{cr}=1+Q^{-1}$,

*) The biological examples of branching AM are provided by neuron syncytium, dendrites, Pourkinje fibre of heart muscle, etc.

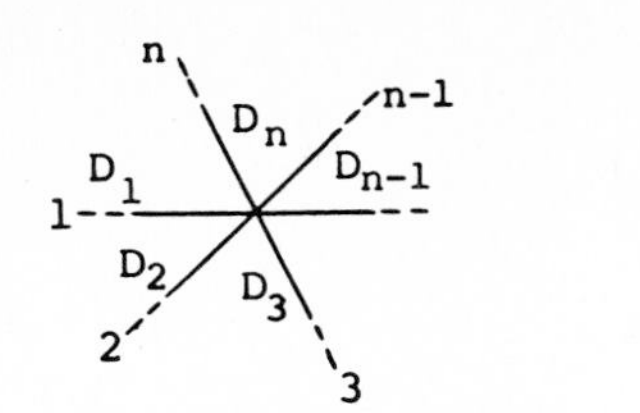

Fig. 4 Geometry of the node of branching

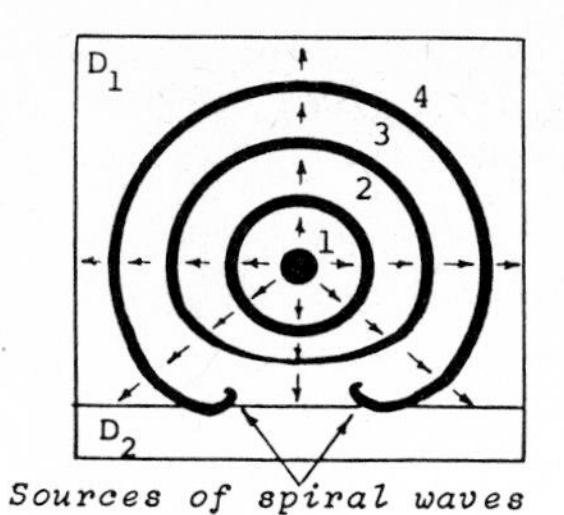

Fig. 5 Initiation of reverberators in the case of partial extinction of a concentric autowave of an impulse type on the line of D jump ($D_1 < D_2$; numbers indicate the consecutive position of the autowave)

if Q^{-1} is an integer, and $n^{cr} = [2+Q^{-1}]$, if Q^{-1} is a composed fraction (here $[x]$ stands for the integer part of x).

5. Generalizations. Consider a multicomponent AM with one diffusion component described by the system

$$\partial_t u + \operatorname{div} \vec{j} = F(u,\underline{v}), \qquad \vec{j} = -D \operatorname{grad} u, \tag{14}$$

$$\partial_t \underline{v} = \underline{\Phi}(u,\underline{v})$$

(Here $\underline{v} \equiv \{v_1,\ldots,v_k\}$, $\underline{\Phi} = \{\Phi_1,\ldots,\Phi_k\}$). We assume also that Eq.(14) has autowave solutions in the form of travelling or periodic impulses. Let us describe the alterations in the results of sections 2-4 induced by passing from (1) to (14).

Refraction. The results of section 2 remain unaltered by introducing Eq. (14).

Opacity Effects. In AM described by the systems of type (14), these effects are manifested by extinction of impulses on the line of D jump and at the nodes of branching [10]. The conditions for extinction are easy to obtain if the variable u in (14), responsible for the evolution of the impulse wavefront, is a fast one and the variables v determining the dynamics of the plateaux are slow ones. The extinction in this case proceeds as follows. The fore front of an impulse (it is described by the equations

$$\partial_t u + \operatorname{div} \vec{j} = f(u), \qquad \vec{j} = -D \operatorname{grad} u \tag{15}$$

where $f(u) = F(u,\underline{v}^*)$, $\underline{v}^*$ being a fixed value of the slow variables of the fore front [11]) slows down sharply when approaching a node of branching or the line of D jump from the region of a smaller value of this parameter. The moving rear front overtakes the fore front of the impulse and the impulse vanishes. The conditions for extinction in this case are approximately determined by the stand-still conditions for the fore front of the impulse. They may be easily found by the methods described in section 3 if we take into account that Eq.(15) coincides formally with Eq.(1a) and the function $f(u)=F(u,v^*)$ in Eq.(15) is usually an N-shaped one (the statement holds at least for the equations cited in footnote *).

*) The important particular cases of (14) are the well-known Hodgkin-Huxley and Fitz-Hugh -- Nagumo equations and their analogs used to describe the waves of excitation in nerve tissue [7] and cardiac muscle [6].

Hypothesis. Partial extinction of concentric autowaves on the lines of D jump must result in the occurrence of sources of spiral waves, the so-called reverberators (see Fig.5); the inequality $a_{12}<a_{12}^{cr} = (S_+/S_-)^{\frac{1}{2}}$ $(a_{12}=(D_1/D_2)^{\frac{1}{2}}$, S_+ and S_- being determined by the function $f(u)=F(u,v^*)$) is the condition for the occurrence of reverberators in the case where the region of initiation of concentric autowaves is distant from the line of D jump by more than a wavelength.

References

1. A.Scott. Active and Nonlinear Wave Propagation in Electronics (Wiley, New York, 1970)
2. A.M.Zhabotinsky. Concentrational Self-Oscillations ("Nayka", Moscow 1974) (in Russian)
3. W.Ebeling. Strukturbildung bei irreversiblen Prozessen (BSB B.G.Teubner Verlagsgesllschaft, 1976)
4. G.Nicolis and I.Prigogine. Self-Organization in Nonequilibrium Systems (Wiley, New York, 1977)
5. H.Haken. Synergetics (Springer-Verlag, Berlin, 1978)
6. G.R.Ivanitsky, V.I.Krinsky, E.E.Sel'kov. Mathematical Biophysics of the Cell("Nauka", Moscow, 1978) (In Russian)
7. A.C.Scott. The Electrophysics of a Nerve Fiber. Rev. of Modern Phys., v.47, No 2 (1975)
8. Yu.I.Arshavsky, M.B.Berkenblit, S.A.Koval'ov, V.V.Smolyaninov, L.M.Chaylakhyan. The Analysis of Functional Properties of Dendrites depending on their Structure. In: The Models of Structural and Functional Organization of some Biological Systems ("Nauka", Moscow, 1966) (In Russian)
9. M.B.Berkenblit, S.A.Koval'ov, V.V.Smolyaninov, L.M.Chaylakhyan. Electrical Behaviour of Myocardium as a System and Characteristics of Heart Cell Membranes. In: The Models of Structural and Functional Organization of some Biological Systems ("Nauka", Moscow, 1966) (In Russian)
10. V.S.Markin, V.F.Pastushenko, Yu.A.Chismadzhev. Theory of Excitable Media ("Nauka", Moscow, 1981) (In Russian)
11. L.A.Ostrovsky, V.G.Yakhno. Formation of Impulses in the Excitable Medium. Biofizika, v.20, No 3, pp. 489-493 (1975) (In Russian)

Numerical Simulation and Nonlinear Processes in Dissipative Media

A.A. Samarskii

Keldysh Institute of Applied Mathematics USSR Academy of Sciences
SU-Moscow, USSR

1. Introduction

A series of works on studying self-maintained dissipative structures in plasma (the T-layer) have been performed in our Institute and registered as Scientific discovery No 55. Later on these works have been experimentally confirmed. References for the "T-layer" problem can be found in [34,14]. Numerical techniques in combination with analytical research are an efficient tool for investigating many problems concerned with nonlinear media. These investigations are of fundamental importance since they give way to new conceptions of the phenomena in nonlinear media and contribute much to the solution of modern global problems of energetics (plasma physics problems), ecology, active biological media, cosmogony, universal field theory, and so on. They are also essential for the development of various technological problems.

In this review the investigation results are discussed describing the localization phenomena in diffusion processes and the appearance of nonstationary thermal structure in a simple, but very comprehensive model of nonlinear dissipative medium. The simplicity of the model allowed to develop some new mathematical means to study the problems of nonstationary nonequilibrium thermodynamics and of the theory of localized structures.

The above investigations are being carried out by the group of scientists working at the Keldysh Institute of Applied Mathematics of the USSR Academy of Sciences under guidance of the author and Prof. S.P.Kurdyumov. Among these scientists are V.A.Galaktionov, V.A.Dorodnitsyn, N.V.Zmitrenko, G.G.Malinesku, A.P.Mikhailov as well as G.E.Elenin and E.S.Kurkina from Moscow State University. Besides, a number of graduate students from IAM and Moscow State University participate in this work.

Unlike the traditional synergetic problems, the phenomena under study are strongly nonstationary (unlimited growth of basic parameters on a finite time interval, the so-called aggravation regimes). The aggravation regimes may be initiated, e.g., by cumulative effects (collapse of bubbles in a fluid), nonhomogeneity of a medium in the initial state (emergence of a shock at the surface of a star) or by anomalous properties of a medium (negative viscosity in the atmosphere physics problems), nonlinearity of a process (strong temperature dependence of the thermonuclear reaction rate, nonlinear dependence of the ohmic heating on gradients of magnetic field and temperature of plasma, etc.).

A number of latent features which were not observed under usual conditions become apparent in the diffusion processes, in aggravation regimes. One of them is the metastable localization of the medium mass in certain regions (diffusion inertia). In application to heat conduction and burning processes, the inertia means that any amount of energy may be concentrated and confined in limited zones of the medium for a finite time without dissipation outside the localization zones. Due to the localization effect, the temperature field is not smoothed out during heating or burning of a heat-conducting medium; on the contrary, it becomes more inhomogeneous. The medium breaks into localized (heat insulated), independently burning regions within which the temperature grows in the aggravation regime (strongly nonstationary dissipative structure). The scale of spatial structures (the fundamental heat length) is determined by the diffusion coefficient, energy release rate and character of the initial heat.

If the fundamental lengths of two or more structures are overlapped, the structures merge into one growing according to the fastest law. With a proper choice of fundamental lengths and initial temperatures, the process of "coherent" burning of a medium in the form of a complex structure (a localized formation with several peaks of temperature and complex architectonics) is maintained for a long time. The number of possible complex structures, their properties and spatial organization depend only on nonlinearity parameters of the medium. They are known a priori if the medium is specified. One would say that here we have the incarnation of the ideas of ancient philosophers that the architecture of the world is predestinated by the primordial medium.

The difficulties in studying nonlinear processes are overcome by using the numerical simulation in combination with the techniques especially developed for qualitative analysis of nonlinear parabolic equations. Specifically, the capabilities of the method of self-similarity have been considerably extended. As a result, a fairly general theory of the heat inertia effect and of nonstationary heat structures has been formulated for example, in the case of media with arbitrary nonlinearity, the conclusions being valid for any diffusion processes - the diffusion of matter, magnetic field,etc.

The above phenomena have been observed also in gas dynamics and in the motion of a high temperature gas with a large number of various dissipative processes. It testifies to the generality of the effects of localization in transport phenomena with the development of strongly nonstationary structures in objects of different physical nature.

2. The heat inertia effect

In order to gain more insight into the results given below we first analyze the localization effect for a simple case of the influence of the thermal boundary aggravation regime on a heat-conducting medium:

$$u_t = (k(u)u_x)_x, \qquad 0<t<T, \qquad x\in R^1_+ = (0,\infty), \tag{1}$$

$$u(t,0) = u_1(t)>0, \qquad 0<t<T, \qquad u_1(t)\to+\infty, \quad t\to T^-, \tag{2}$$

$$u(0,x) = u_0(x)\geq 0, \qquad x\in R^1_+, \qquad \sup u_0<+\infty. \tag{3}$$

Here $k\in C(R^{-1}_+)\cap C^2(R^1_+)$ is such that $k(u)>0$, $u>0$, $k(u)=0$ for $u=0$, $k(u)$ is the thermal diffusivity. The solution of problem (1)-(3) will be called localized if it goes to infinity when $t\to T^-<+\infty$ in a limited region of localization; otherwise it will be nonlocalized. The quantity

$$l_* = \text{mes}\{x \in R^1_+\} \mid u(T^-, x) = +\infty\}$$

will characterize the degree of localization. In relation to the introduced definition the problem arises to find out the classes of boundary aggravation regimes resulting in localization (or nonlocalization) in problem (1)-(3) for a rather arbitrary value of diffusivity k(u).

In the case of equation (1) where the diffusivity is a power function

$$u_t = (u^\sigma u_x)_x, \qquad 0<t<T, \qquad x \in R^1_+; \qquad \sigma = \text{const}>0 \tag{4}$$

and the aggravation regimes have the form

$$u_1(t) = (T-t)^n, \qquad 0<t<T; \qquad n=\text{const}<0, \tag{5}$$

the problem has been studied in detail by constructing and analyzing respective self-similar solutions [1-3]. The asymptotic stability of the solutions and the theorems of their comparison in boundary values were also studied [1-2]. It was shown that under the "slow" regimes, when $n \in [-1/\sigma, 0)$ the solutions were localized, and for $n<-1/\sigma$ ("fast" regimes) there was no localization with $u(t,x) \to +\infty$ when $t \to T^-$ everywhere in R^1_+. The analytical solution (the S-regime) of problem (4), (5) at $n=-1/\sigma$ first constructed in [5]

$$u(t,x) = (T-t)^{-1/\sigma}[(1-x/x_0)^+]^{2/\sigma}, \qquad x_0 = [2(\sigma+2)/\sigma]^{1/2},$$

demonstrates the localization property of nonlinear diffusion processes.

Generalization of the formulated results to the case of an arbitrary form of equation (1) was made in two ways. The first one was to construct a theory of operator (function) comparison between solutions of various nonlinear parabolic equations [6-8]. In the framework of this technique based on special point-to-point estimates for the higher derivative of the majorizing solution from the lower ones, the comparison can be done between the solutions of equations (1) and (4) with different diffusivities. Hence, some results obtained for the case of equations (4) may be "extended" to the case of (1) with an arbitrary k(u). Specifically, the following result has been obtained as to the heat localization effect in any media [9].

Theorem 1. Let there exist a constant $\alpha>0$ such that $[k^\alpha]'(0)<+\infty$. Then the class of boundary aggravation regimes resulting in localization in problem (1)-(3) may be determined.

Nearly the same general result indicating the absence of localization was obtained in [9].

Another approach is to construct the so-called approximate self-similar solutions (a.s.s.) which do not satisfy equation (1), but to which the solution of the problem (1)-(3) asymptotically converges. It was found in [10-12] that such solutions might be constructed in the case of rather arbitrary values of k(u) and u(t), when equation (1) had no appropriate exact self-similar or other invariant solutions. (Note that the latter exist very seldom [13,14]). Complete systems of a.s.s. were constructed in [10]. Below we give only one example of their application to the localization problem [11,12].

Let equation (1) be linear, i.e.

$$u_t = u_{xx}, \qquad 0<t<T, \qquad x \in R^1_+ \tag{6}$$

and the boundary regime has the form

$$u(t,0) = \exp\left[(T-t)^n\right]-1, \quad 0<t<T; \quad n<0. \tag{7}$$

The problem (6), (7) has no self-similar solutions [13,14]. It is proved in [11,12] that the properties of solutions to this problem are described by the a.s.s. u_a which satisfies the first order Hamilton-Jacobi equation

$$(u_a)_t = \frac{(u_a)_x^2}{1+u_a}, \quad 0<t<T, \quad x\in R_+^1$$

and which is given by

$$u_a=\exp\left[(T-t)^n\theta_a(\xi)\right]-1, \quad \xi= \frac{x}{(T-t)^{(1+n)/2}}, \tag{8}$$

where $\theta_a(\xi)\geqslant 0$, $\theta_a(0)=1$, $\theta_a(+\infty)=0$ is solution of the equation

$$(\theta_a')^2 - \frac{1+n}{2}\theta_a'\xi + n\theta_a=0, \quad \xi\in R_+^1, \quad ('= \frac{d}{d\xi}).$$

An estimate for the convergence rate was obtained in a special norm: $u\to u_a$ when $t\to T^-$ [10-12]. Then from the form of a.s.s. (8) it immediately follows that if $n\in |-1,0)$ the solution is localized, and if $n<-1$, it is not. At $n=-1$ (the S-regime) one can easily obtain the localization depth $l_*=2$ [11,12].

3. Thermal Structures and Localization Effect in Nonlinear Media with Volume Energy Release

Now let us turn to the study of essentially nonstationary structures whose evolution is described by the quasilinear parabolic equation

$$u_t = \nabla(k(u)\nabla u) + Q(u), \tag{9}$$

where $Q\geqslant 0$ is the power of the volume heat sources being a function of the medium temperature, $u=u(t,x)\geqslant 0$, $\nabla(\cdot)=\text{grad}_x(\cdot)$. When analyzing equation (9) three main problems arise. The first one is to determine the conditions for development of a strongly nonstationary process (the aggravation regime existing in a finite interval of time, $0<t<T<+\infty$). The second problem is to determine the conditions for k,Q under which the localized dissipative structures, infinitely growing when $t\to T^-$, develop in the limited localization region

$$\Omega_L = \{x|u(T^-,x) = +\infty\}.$$

Finally, the third problem (perhaps, the most important one) is to find specific "laws" of development of arising structures, in particular, "the laws" of resonance mergence of simple (elementary) structures into more complex ones with several extrema on the spatial profiles. The methods for analyzing these problems and the results obtained are briefly considered below.

3.1. Equations with Power Coefficients. The Cauchy Problem

The nonstationary dissipative structures whose development in time is described by the equation with the power coefficients k and Q of the form

$$u_t = \nabla(u^\sigma\nabla u) + u^\beta; \quad \sigma>0, \quad \beta>1 \tag{10}$$

$$u(0,x) = u_0(x)\geqslant 0, \quad x\in R^N; \quad u_0\in C(R^N). \tag{10'}$$

have been studied thoroughly. Equation (10) has unlimited self-similar solutions (see [15-19]) in the form

$$u_A(t,x) = (T-t)^{-1/(\beta-1)}\,\theta(\xi), \qquad \xi = \frac{x}{(T-t)^{[\beta-(\sigma+1)]/2(\beta-1)}}, \tag{11}$$

where the function $\theta(\xi)\geqslant 0$ satisfies the elliptic equation obtained as a result of substitution of (11) into (10) as

$$\nabla_\xi(\theta^\sigma \nabla_\xi \theta) - \frac{\beta-(\sigma+1)}{2(\beta-1)}\vec{\nabla}_\xi \theta \vec{\xi} - \frac{1}{\beta-1}\theta + \theta^\beta = 0, \quad \xi \in R^N. \tag{12}$$

The solution (11) develops in the aggravation regime, for example, if $\theta(0)\neq 0$ then $u_A(t,0)\to+\infty$ when $t\to T^-$. The existence of a self-similar solution (11) allows a constructive description of all the most typical features of arising structures. Let us emphasize that the solutions (11) of a particular form are asymptotically stable [18,19]. A number of numerical calculations were made [15-20] to show that the structure of these solutions describes the behavior of all classes of unlimited solutions of equation (10). Theoretically, the question on the variety of forms of self-similar structures of the solutions of equation (12) limited in R^N is of fundamental importance. It was studied by various methods and from various viewpoints [16-19,21,22]. In the LS-regime ($\beta>\sigma+1$) the number M of different structures is finite and may be obtained from the approximate relation $M\simeq(\beta-1)/[\beta-(\sigma+1)]+1$ (see [16-19,21]). They differ from each other by the number of extrema. A wide class of complex, radially non-symmetric self-similar structures was studied in multidimensional geometry in [22]. Note that in this study a remarkable relationship was obtained between heat structures and a class of solutions of the Schrödinger equation describing the electron structure of atoms [17].

Let us refer back to formula (11). The first thing to catch one's eye is the possibility of determining the condition for localization of an unlimited solution in the sense of the above definition. It is immediately seen that for $1<\beta<\sigma+1$ the solution (12) is not localized: $u_A(t,x)\to\infty$, $t\to T^-$ everywhere in R^N. The burning wave front envelopes the whole space, when $t\to T^-$.

On the contrary, when $\beta\geqslant\sigma+1$ the self-similar solution describes the process with localization. As seen from (12), at $\beta=\sigma+1$ the burning wave front is fixed in space during the entire time of aggravation $0<t<T$ (the S-regime). The localization property is extremely representative for the analytical solution of equation (10) in the one-dimensional case [15]. Here

$$u_A(t,x) = (T-t)^{-1/\sigma}\left[\frac{2(\sigma+1)}{\sigma(\sigma+2)}\cos^2\left(\frac{\pi x}{L}\right)\right]^{1/\sigma}, \qquad |x|<\frac{L}{2};$$

$$u_A = 0, \quad |x|\geqslant\frac{L}{2}, \qquad A \tag{13}$$

where $L=2\pi(\sigma+1)^{1/2}/\sigma$ is the so-called fundamental length of the S-regime. As a result of detailed analysis [15-19] it was convincingly shown that for any initial values $u_0(x)$ the unlimited solution at $\beta=\sigma+1$ goes to infinity when t T^- on a set with a measure divisible by L, i.e. L is really a fundamental (independent of external conditions) characteristic of a nonlinear medium.

In the case $\beta>\sigma+1$ unlimited solutions go to infinity when $t\to T^-$ on a set of measure zero (for example, at a point). As can be easily seen, for $\beta>\sigma+1$ the self-similar solution $u_A(t,x)$ leaves in space "a trace" (the limit distribution) of the following form:

$$u_A(T^-,x) = C_{\sigma,\beta} \,||x||^{-2/(\beta-\sigma-1)}, \qquad x\in R^N/\{0\}, \tag{14}$$

where $C_{\sigma,\beta}>0$ is a constant in an asymptotic distribution of the solution $\Theta(\xi)$ for large $||\xi||$

$$\Theta(\xi) \simeq C_{\sigma,\beta}\, ||\xi||^{-2/(\beta-\sigma-1)}, \qquad ||\xi|| \to \infty \tag{15}$$

The condition (14) means that $u_A(t,x)$ tends to $+\infty$, provided $t\to T^-$ only at the point $x=0$, everywhere in $R^N/\{0\}$ the solution is bounded above regularly in t by the limit distribution (14).

Now we formulate some statements about the conditions for appearance of unlimited solutions (see [18,23]). The sufficient conditions for appearance of aggravation regimes under arbitrary initial functions are obtained from the following theorem.

Theorem 2. Let $1<\beta<\sigma+1+2/N$ and $u_0(x)\not\equiv 0$. Then the solution to the Cauchy problem (10),(10') is unlimited.

Theorem 3. Let $\beta>\sigma+1+2/N$. Then for a sufficiently "large" u_0 the solution is unlimited, and for u_0 sufficiently "small" the problem has a global solution. In the latter case

$$\sup_{x\in R^N} u(t,x) = O(t^{-1/(\beta-1)}), \qquad t\to+\infty .$$

Theorems 2 and 3 are proved by constructing and analyzing some special kind of higher and lower solutions of equations (10) [23].

3.2. Equation with Power Nonlinearities. The Boundary Value Problem in a Limited Region.

A close relationship proved to exist between the localization property in the Cauchy problem of (10) and the solvability conditions for the boundary value problem in a limited region. We shall consider it briefly below. Let Ω be a limited region in R^N with a smooth boundary $\partial\Omega$. We shall consider for (10) the boundary value problem with the conditions

$$u(0,x) = u_0(x)\geqslant 0, \qquad x\in\Omega; \qquad u_0^{\sigma+1}\in H_0^1(\Omega), \tag{16}$$
$$u(t,x) = 0 \qquad t>0, \qquad x\in\partial\Omega .$$

The following theorems are valid (see [24]).

Theorem 4. Let $\beta<\sigma+1$. Then for any initial functions $u_0(x)$ there exists a global solution, the following inclusions being valid

$$u^{1+\sigma/2}\in L^\infty(0,T;\ L^2(\Omega)), \qquad \partial u^{1+\sigma/2}/\partial t\in L^2(0,T;\ L^2(\Omega)),$$
$$u^{1+\sigma}\in L^\infty(0,T;\ H_0^1(\Omega)).$$

Theorem 5. Let $\beta=\sigma+1$. If $\lambda_1>1$, where $\lambda_1>0$ is the first eigenvalue of the problem,

$$\Delta w + \lambda w = 0; \qquad w|_{\partial\Omega} = 0 \qquad (w\in H_0^1(\Omega)),$$

then (10), (16) have a limited solution for any $u_0(x)$.
But if $\lambda_1<1$, then all the solutions with $u_0\not\equiv 0$ are unlimited.

Theorem 6. Let $\beta>\sigma+1$. Then for sufficiently "small" functions $u_o(x)$ there exists a global solution, and for "large" u_o the problem is unsolvable as a whole.

Note that the above theorems may be proved also for eq.(9) with arbitrary nonlinearities [25].

3.3. Localization in a Medium with Uniform Heat Conduction. Construction of Approximate Self-Similar Solutions.

It is a fortunate feature of equation (10) that it allows the exact self-similar solutions to be obtained. Thereby, the condition is determined for localization of unlimited solutions of the Cauchy problem. In a general case, however, such invariant solutions do not exist, as shown by the group classification of equation (9) [26]. The latter takes place for the equation

$$u_t = \Delta u + (1+u)\ln^{\beta}(1+u), \tag{17}$$

where $\beta>1$ is constant. Equation (17) describes the heat diffusion and burning in a medium with constant diffusivity. It appears that an unusual form of the source in (17) allows to describe the most interesting features of burning in a medium with constant diffusivity, in particular, to observe the exchange (with variation of parameter β) of the HS-, S- and LS-burning regimes with aggravation. In this brief presentation of the results we mostly adhere to those outlined in [11,27].

Thus, equation (17) has no invariant (self-similar) solutions. Below we apply the approach of constructing approximate self-similar solutions (a.s.s.) which do not satisfy equations (17) but, nevertheless, they correctly describe an asymptotic behavior of nonlimited solutions. The a.s.s. approach [10] infers that from the operator on the right-hand side of (17), the part is extracted that basically determines the spatial-temporal structure of unlimited solution at the developed stage of aggravation regime. For this, according to [10,11], we introduce, instead of $u(t,x)$ a new function $U(t,x)\geq 0$ given by:

$$U(t,x) = \ln|1+u(t,x)|, \quad t>0, \quad x\in R^N.$$

Then for U we obtain the parabolic equation

$$U_t = \Delta U + (\nabla U)^2 + U^{\beta}, \quad t>0, \quad x\in R^N. \tag{18}$$

Like (17), equation (18) has no appropriate invariant solutions. Let us truncate it, i.e. omit the first term in the right-hand side, and consider the resulting equation

$$V_t = (\nabla V)^2 + V^{\beta}, \quad t>0, \quad x\in R^N \quad (\beta>1). \tag{19}$$

This equation allows one to obtain the self-similar solution

$$V_a(t,x) = (T-t)^{-1/(\beta-1)}\Theta_a(\xi), \quad \xi = \frac{x}{(T-t)^{(\beta-2)/2(\beta-1)}}, \tag{20}$$

where $\Theta_a(\xi)\geq 0$ satisfies the equation

$$(\nabla_{\xi}\Theta_a)^2 - \frac{\beta-2}{2(\beta-1)}\vec{\nabla}_{\xi}\Theta_a\vec{\xi} - \frac{1}{\beta-1}\Theta_a + \Theta_a^{\beta} = 0 \tag{21}$$

(the solvability of (21) in R^N was proved in [27]). Now one should see that the transition from equation (18) to (19) is correct, i.e.

the relation (20) is the a.s.s. of equation (17). It is easy to show that in the functions (20) the omitted term is "much less" than those remained, $(\nabla U)^2$ or U^β [11,27]. Evidence for this comes from estimates obtained when constructing the unlimited lower solutions of equation (18) [11,27] and also from some general results of the a.s.s. theory [10] and of numerous computations which prove the "structural" stability in the sense of a.s.s. (10) [18,19,27]. Thus, as shown in [11,27], the a.s.s. (20) correctly describe the behavior of the unlimited solution at times close to the beginning of aggravation. Then the structure of a.s.s. (20) allows one to determine the localization condition for unlimited solutions: if $1<\beta<2$ (the HS-regime) the localization is absent and $U(t,x)\to+\infty$ (and, hence, $u(t,x)\to+\infty$) when $t\to T^-$ everywhere in R^N. On the contrary, if $\beta\geqslant 2$ the solutions are localized; in the case $\beta>2$ (the LS-regime) the solution U (and also u) goes to infinity on a set of measure zero (at a point), which directly follows from (20). In the "critical" case $\beta=2$ the S-regime takes place and equation (21) has in R^N the generalized solution

$$\Theta_a(x) = \begin{cases} \cos^2\left(\frac{||x||}{2}\right), & ||x|| < \pi, \\ 0 & ||x|| \geqslant \pi \end{cases}$$

(here $\xi=x$). Thus, in the S-regime the localization region is a sphere $\{||x||<\pi\}$ of diameter $D=2\pi$ (the validity of this conclusion was verified in a series of computations |27|).

It should be noted that the a.s.s. $u_a(t,x)$ of the initial equation (17) obtained from (19) through the transformation

$V_a = \ln\,(1+u_a)$ has the form

$$u_a(t,x) = \exp\,\{(T-t)^{-1/(\beta-1)}\,\Theta_a(\xi)\} - 1, \quad \xi = \frac{x}{(T-t)^{\frac{\beta-2}{2(\beta-1)}}}$$

and satisfies the first-order Hamilton-Jacobi type equation

$$\frac{\partial u_a}{\partial t} = (\nabla u_a)^2/(1+u_a) + (1+u_a)\ln^\beta(1+u_a), \quad t>0, \quad x\in R^N.$$

It differs essentially from the initial parabolic equation (17), but, nevertheless, it correctly describes the properties of unlimited solutions of (17). Such an unusual feature of "degeneration" (reduction in the order of equation) is typical only of aggravation regimes.

The conditions for development of aggravation regimes in the Cauchy problem for (17) were studied in detail in [11,27] by constructing the lower solutions. General classes of "degenerate" a.s.s. satisfying the first-order equation were constructed in [28] for equations (9) with arbitrary nonlinearities by using a general technique.

3.4. On a General Approach to the Problem of Localization

Earlier we presented two methods for investigation of the localization effect by constructing exact and approximate self-similar solutions (for equations with power nonlinearities and more general equations). Below we shall make some sufficiently general and mathematically strict statements concerning the localization problem. Here we shall consider equations (9) with arbitrary coefficients k,Q.

The proof of the theorem is based on the comparison between the unlimited solution of equation (9) and the family (U) of stationary solutions of this equation

$$\nabla(k(U)\nabla U) + Q(U) = 0, \qquad x \in R^N. \tag{22}$$

(U is assumed to be limited in R^N).

The proposed method (see [39,30]) is based on the fact that the behavioral features of solutions of equation (9) are described by the family of stationary solutions {U}, but they are "parameterized" in a special way (the role of a parameter may be played by $U_0 = \sup U(x)$). In other words, the construction of the family of solutions for equation (22) allows one to describe all the most important features (for example, the spatial-temporal structure) of arising nonstationary dissipative structures. Note that equation (22) is of the elliptic type and the properties of its solutions have been studied in some detail.

By using the above approach, we can prove the following statement [29].

Theorem 7. Let the functions k and Q be such that

$$\int_0^u k(\eta)d\eta / Q(u) \to +\infty, \qquad u \to +\infty. \tag{23}$$

Then any solution of the Cauchy problem (9), (10') is not localized, and $u(t,x) \to +\infty$ everywhere in R^N.

Thus the condition (23) is sufficient for the absence of heat localization. In the case of equation (23) it yields $\beta < \sigma+1$, this being in agreement with the calculations made above. It turns out [29] that everywhere, except for a special class of "weakly nonlinear" functions (which shall be discussed later), the condition (23) is also necessary, i.e. when it is violated, the solution is localized. Moreover, if

$$\int_0^u k(\eta)d\eta / Q(u) \to 0, \qquad u \to +\infty, \tag{24}$$

then the LS-regime takes place ($\mathrm{mes}\,\Omega_L = 0$); if

$$\int_0^u k(\eta)d\eta / Q(u) \to \mu > 0, \quad u \to +\infty, \tag{25}$$

then the S-regime develops when the structure infinitely grows on the set Ω_L with a nonzero measure (but $\mathrm{mes}\,\Omega_L < \infty$). Particularly, in order to estimate the size of the localization region under the S-regime the following theorem is to be proved.

Theorem 8. Let the condition (25) be fulfilled. Then the solution cannot be localized in the region Ω_L of the diameter less than

$$D^* = 2\mu^{-1/2} z_N^{(1)},$$

where $z_N^{(1)} > 0$ is the first (least) root of the Bessel function $J_{\frac{1}{2}(N-2)}$.

The theory of stationary solutions allows the description of some features of the structures when the condition (24) [29] is fulfilled, i.e. under the LS-regime.

Now let us consider the class of "weakly nonlinear" functions k,Q where the condition (23) is not necessary (but it is always sufficient) for the absence of localization. Most typical representatives of this conditions are the family of the functions

$$k=k(u), \quad Q=Q_\alpha(u) = \frac{u+1}{k(u)} \left\{ \int_0^u \frac{k(\eta)}{\eta+1} d\eta \right\}^\alpha, \tag{26}$$

where $\alpha \geqslant 0$ is a parameter and $k(u) \geqslant 0$ is such that

$$\int_0^\infty \frac{k(\eta)}{\eta+1} d\eta = +\infty, \qquad \lim_{u\to+\infty} \left(\frac{k}{k'} \right)(u) = \infty.$$

For such k,Q, the problem proves to have a.s.s. satisfying the first-order equation

$$(u_a)_t = \frac{k(u_a)}{1+u_a}(\nabla u_a)^2 + Q_\alpha(u_a), \qquad t>0, \qquad x \in R^N.$$

Therefore, when $t \to T^-$ the solution u(t,x) does not satisfy the parabolic equation. Then the approach based on the analysis of the family of solutions of stationary parabolic problems cannot yield optimal results. By constructing a.s.s., it can be shown that with the coefficients (26) the localization occurs if $\alpha > 2$, but if $\alpha < 2$ the solutions are not localized [11,27,29].

4. On some Extensions of the Theory of Essentially Nonstationary Dissipative Structures

Now the aggravation regimes and the conditions for their localization have been studied in application to parabolic systems of quasilinear equations describing the burning processes in multicomponent reacting media [30,31]. This study was carried out by various methods including the construction of self-similar solutions, numerical calculation [31], and the analysis of the family of stationary solutions [30]. The existence of the localization effect and the appearance of structures in gasdynamic processes [34,35] have been proved, specifically, in terms of a complete nonlinear MHD-system [34]. Applications of the theory of aggravation regimes and localization effect in some problems are discussed in [17,32]. Most of the results presented in this paper were verified by series of computational runs. It is of interest to study the conditions for development of finite-difference aggravation regimes arising in discrete evolution problems. The investigation of unlimited solutions of implicit (nonlinear) difference schemes for evolution equation of type (9) was performed in [33].

References

1. A.A.Samarskii, N.V.Zmitrenko et al. Dokl. Akad. Nauk SSSR, (1975), v.223, No 6, p.1344-1347.
2. N.V.Zmitrenko, S.P.Kurdjumov et al. Preprint No 103, Keldysh Inst. Appl. Math. Acad. Sci. USSR, Moscow (1977).
3. V.A.Galaktionov, A.P.Mikhailov. Preprint No 53, Keldysh Inst.Appl. Math. Acad. Sci. USSR, Moscow (1977).
4. V.A.Galaktionov, S.P.Kurdjumov et al. Differencial'nye Uravnenija (1980) v.16, No 7, c.1196-1204.
5. A.A.Samarskii, I.M.Sobol'. Z.Vyčisl.Mat. i Mat.Fiz. (1963) v.3., No 4, p.703-719.

6. V.A.Galaktionov, S.P.Kurdjumov S.P. et al. Z.Vyčisl. Mat. i. Mat. Fiz. (1979) v.19, No 6, p.1451-1461.
7. V.A.Galaktionov, S.P.Kurdjumov et al. Dokl. Akad. Nauk SSSR (1979) v.248, No 3, p.586-589.
8. V.A.Galaktionov. Dokl.Akad.Nauk SSSR, (1980) v.251, No 4, p.832-835.
9. V.A.Galaktionov, S.P.Kurdjumov et al. Differencial'nye Uravnenija (1981) v.17, No 10, p.1826-1841.
10. V.A.Galaktionov, A.A.Samarskii. Matem.Sbornik (1982) v.118, c.292-322, c.435-455 (1983), v.120, c.3-21, v.121, c.131-155.
11. A.A.Samarskii. In: Trudy Mat.Inst.Steklov (1981) v.158, c.153-162.
12. A.A.Samarskii, V.A.Galaktionov et al. Dokl.Akad.Nauk SSSR, (1979) v.247, No 2, p.349-353.
13. L.V.Ovsjynnikov. Dokl.Akad.Nauk SSSR (1959) v.125, No 3, p.492-495.
14. N.X.Ibragimov. Groups Transformation in Mathematical Physics. Moscow: Nauka (1983).
15. A.A.Samarskii, N.V.Zmitrenko et al. Dokl.Akad.Nauk SSSR, (1976) v.227, No 2, p.321-324.
16. A.A.Samarskii, G.G.Elenin et al. Dokl. Akad.Nauk SSSR, (1977) v.237, No 6, p.1330-1333.
17. S.P.Kurdjumov. In: Modern Problems of Mathematical Physics and Computational Mathematics. Moscow: Nauka (1982) p.217-243.
18. G.G.Elenin, S.P.Kurdjumov. Preprint No 106, Keldysh Inst.Appl. Math.Acad.Sci.USSR, Moscow (1977).
19. G.G.Elenin, S.P.Kurdjumov, A.A.Samarskii. Z.Vyčisl.Mat. i Mat. Fiz. (1983) v.23, No 2, p.380-390.
20. S.P.Kurdjumov, G.G.Malinetskii et al. Dokl.Akad.Nauk SSSR (1980) v.251, No 4, p.836-839.
21. M.M.Adjutov, Y.A.Klokov, A.P.Mikhailov. Differencial'nye Uravnenija, (1983) v.19, No 7, p.1107-1114.
22. S.P.Kurdjumov, E.S.Kurkina, A.B.Potapov. Preprint No 75, Keldysh Inst.Appl.Math.Acad.Sci.USSR, Moscow (1982).
23. V.A.Galaktionov, S.P.Kurdjumov et al. Dokl.Akad.Nauk SSSR (1980) v.252, No 6, p.1362-1364.
24. V.A.Galaktionov. Differencial'nye Uravnenija (1981) v.17, No 5, p.836-842.
25. V.A.Galaktionov. Z.Vyčisl.Mat. i Mat.Fiz. (1982), v.22, No 6, p.1369-1385.
26. V.A.Dorodnicyn. Z.Vyčisl.Mat. i Mat.Fiz. (1982) v.22, No 6, p.1393-1400.
27. V.A.Galaktionov, S.P.Kurdjumov et al. Preprint No 161, Keldysh Inst. Appl.Math.Acad.Sci.USSR, Moscow (1979).
28. V.A.Galaktionov. Preprint No 16, Keldysh Inst.Appl.Math.Acad.Sci. USSR, Moscow (1981).
29. V.A.Galaktionov. Dokl.Acad.Nauk SSSR (1982) v.264, No 5, p.1305-1040.
30. V.A.Galaktionov, S.P.Kurdjumov, A.A.Samarskii. Differencial'nye Uravnenija (1983) v.19 No 12.
31. S.P.Kurdjumov, E.S.Kurkina et al. Dokl.Akad.Nauk SSSR, (1981), v.258, No 5, p.1084-1088.
32. N.V.Zmitrenko, S.P.Kurdjumov et al. Pisma v JETP (1977) v.26, p.620-624.
33. V.A.Galaktionov, A.A.Samarskii. Z.Vyčisl.Mat. i Mat. Fiz. (1983) v.23, No 3, p.646-659, No 4, p.831-838.
34. S.P.Kurdjumov, N.V.Zmitrenko. Z.Prikl.Mech. i Teor. Fiz. (1977) No 1, p.3-23.
35. M.A.Anufrieva, A.P.Mikhailov. Differencial'nye Uravnenija (1983) v.19, No 3, p.483-491.

The Onset and the Development of Chaotic Structures in Dissipative Media

A.V. Gaponov-Grekhov, M.I. Rabinovich, and I.M. Starobinets
Institute of Applied Physics, Academy of Sciences of the USSR, SU-Gorky, USSR

1. Introduction

The appearance of space structures in dissipative media results from the development of space-inhomogeneous instabilities and their subsequent stabilization due to the balance between the internal dissipative loss in the medium and the pumping from the source of nonequilibrium. This balance, in the average, may take place both the regular, completely ordered states and in nonstationary chaotic structures.

Regular and chaotic structures are generally opposed. However,both chaos and order in a nonlinear dissipative medium unaffected by external noises, originate from the nonlinear dynamics of the system. Only chaos (turbulence) occurs in systems with unstable dynamics (nearly all trajectories are unstable). Regular structures (order) develop in the systems with stable dynamics. Regular structures may transit to irregular ones even with a very small variation of the parameters in the evolving system. This is a transition across the critical point*. Therefore, it is only natural to consider chaotic and regular structures as a general problem, characterizing them, in particular, by the level of order or the degree of disorder [2],rather than to oppose them. The more so, as even far enough from the boundary of the "order-disorder" transition, the established chaos still has some features of the order that has already disappeared. This is demonstrated by coherent structures in turbulent flows, chemical turbulence in the form of an ensemble of cylindrical or spiral waves,etc.

In this report, we consider a possibility of describing nonstationary dissipative structures in space-confined systems by finite-dimensional models and discuss possible transitions of regular structures to chaotic ones. We also give an example to illustrate continuous transformation of a regular structure in the dissipative medium into a chaotic one with a high degree of disorder.

2. Structures and Finite-Dimensional Attractors

The nonlinear equations with diffusion are traditional models of non-equilibrium dissipative media. The point analogies of these equations either demonstrate a self-oscillating behavior or exhibit a stationary pulse response to an external impulse(excitable media). The processes in these media and the established structures are primarily defined by the characteristics of the media and weakly depend on the initial and boundary conditions. Nevertheless, in order to make the analysis more correct, we discuss the structures in limited nonequilibrium media with trivial or periodic boundary conditions.

*The evolution of a real system in its mathematical model may be often taken into account only under the assumption that the parameters of the model are time-dependent [1].

The attractor (an attracting set of trajectories that corresponds to the behaviour of a system at $t \to \infty$) is a mathematical image of the structures in an infinite-dimensional phase space of the dissipative media (systems) under study [2]. Note, that we do not speak about transient structures, whose time of development is of the order of the characteristic time of the evolution of the system.

The attractors in space-confined dissipative systems of the general type have a wonderful property: they are finite-dimensional, i.e. the motion on this attractor and, hence, the corresponding structure, are described by a finite-dimensional dynamic system. This statement has been proved mathematically. Rigorous results have been obtained, in particular, for two-dimensional dissipative media described by the Navier-Stokes equations [3-5] and for various model media described by the Burgers equations. For example, Iliashenko has shown that the attractor in a ring one-dimensional medium described by the equation

$$\frac{\partial U}{\partial t} = - \sigma \frac{\partial^2 U}{\partial x^2} + \left(\frac{\partial U}{\partial x}\right)^2 - \frac{\partial^4 U}{\partial x^4}, \tag{1}$$

that is met in models of chemical turbulence [6], is a finite-dimensional object.

Thus, for a rather wide class of limited dissipative media, the mathematical study of the appearance and transformation of the structures in the evolutionary process consists in the analysis of the finite-dimensional attractors, their transformation and bifurcations.

3. Scenarios of the Onset of Chaotic Structures

It has been recently found out (which is a remarkable fact!), that the number of typical scenarios in which regular structures transit to chaotic ones - - turbulence - - is small. This is explained by the fact that the number of the bifurcations, as a result of which regular attractors are replaced by stochastic ones, is also limited. The typical bifurcations observed in the evolution are given in Table 1. The onset of chaos is most often concerned with the following bifurcations: a) the disappearance of the regular attractor when it converges with its unstable trajectory (usually, a cycle), i.e. intermittency; b) the destruction of the torus (a quasiperiodic motion); c) the appearance of a strange attractor as a result of an infinite sequence of period doublings that bunch at a critical point. The spectral transformation of the nonstationary structures (Taylor vortices for a Couette flow between the cylinders and the rollers for the Rayleigh-Benard convection) is given elsewhere (see, for example, [14]). The space structures in the close vicinity of the critical point have very specific features for every transition. In particular, if chaos appears via intermittency, successive variation of regular and disordered structures may be observed before the transition point. This transition ends at $t \to \infty$, where complete order is established. A similar picture was observed by P.Berge and M.Dubois [7] in a convective cell (see Fig.1).

This alternation of chaos and order at $t \to \infty$ does not disappear immediately after the transition point. Such an intermittency that does not vary with $t \to \infty$ is called an established regime. It corresponds to the appearance of a strange attractor in the phase space.

What happens to the chaotic structures that replace complete order in the process of further evolution of the system? How do they develop, interact with each other and the external fields and, perhaps,

Table 1. Scenarios of Chaotization of the Structures

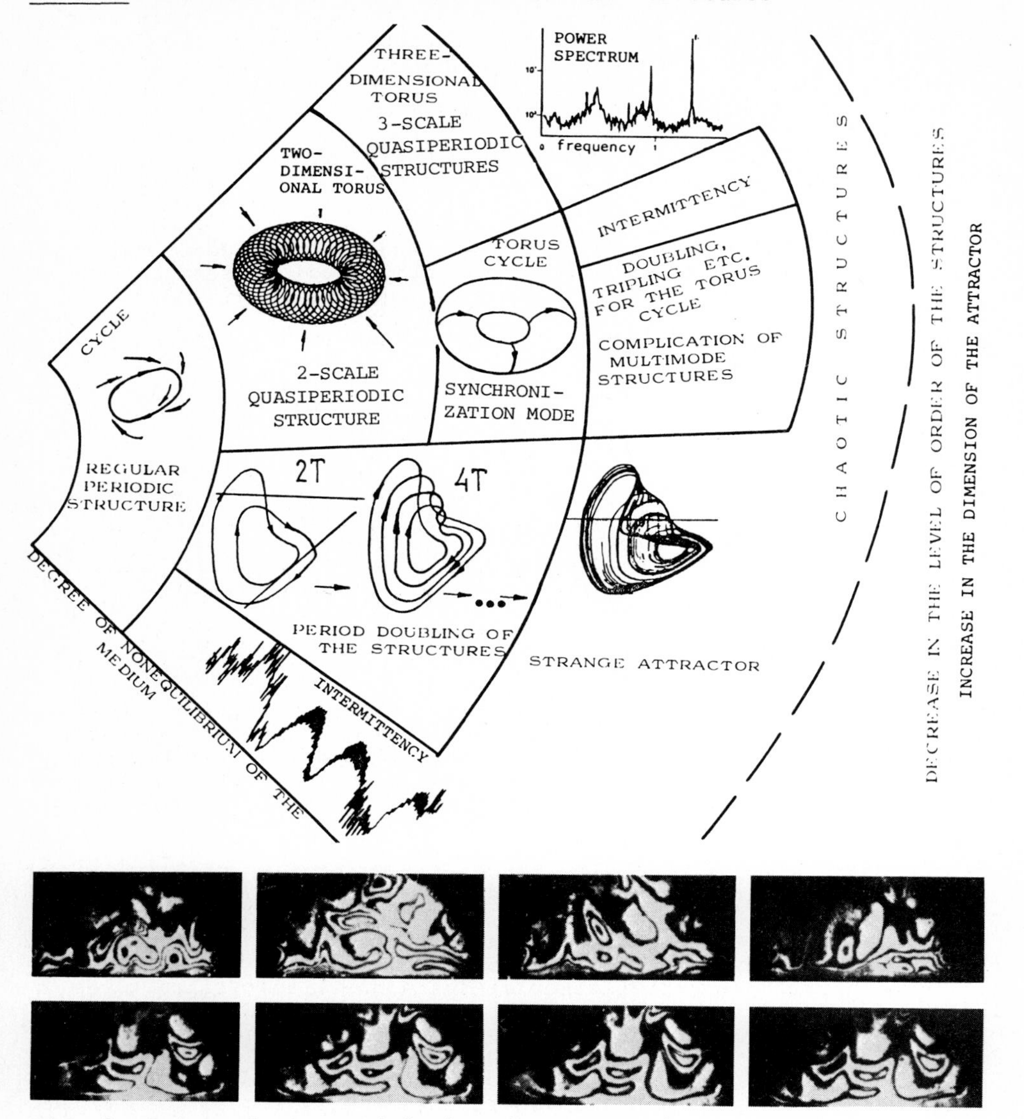

Fig. 1 A transient spatial chaotic state: a) strong chaotic state; b) transient lockings on different well-defined spatial arrangements. Pictures, showing temperature isogradient lines $\nabla_z T$ in the vertical plane and along the greatest extension of the cell ($R_a/R_{a_c} \approx 540$)(from[7]).

disappear? These questions seem to be most important in the field under study. We believe that the main ways of the development of chaotic structures in the process of evolution (i.e. when the system changes its parameters), as well as the scenarios in which chaos appears, are not infinitely diverse. We shall discuss them using the example of the structures described by a sequence of self-oscillators coupled dissipatively, a simple model in the theory of self-organization.

4. The Characteristics of the Model

Let us consider a sequence of coupled self-oscillators of the Van der Pol type. Each oscillator has a simple periodic behavior. The equation for such an oscillator has the following form

$$\frac{dw}{dt} = w-(1+i\beta)|w|^2 w. \tag{2}$$

This equation has a single stable periodic solution - $-w(t) = I_0 \exp|i\omega(I_0)t|$. Figure 2 gives a schematic of the sequence of oscillators (2) coupled dissipatively. The sequence is described by a set of differential-difference equations

$$\frac{dw_j}{dt} = w_j-(1+i\beta)|w_j|^2 w_j + e(1-ic)(w_{j+1}-2w_j+w_{j-1}), \quad j=1,2,\dots . \tag{3}$$

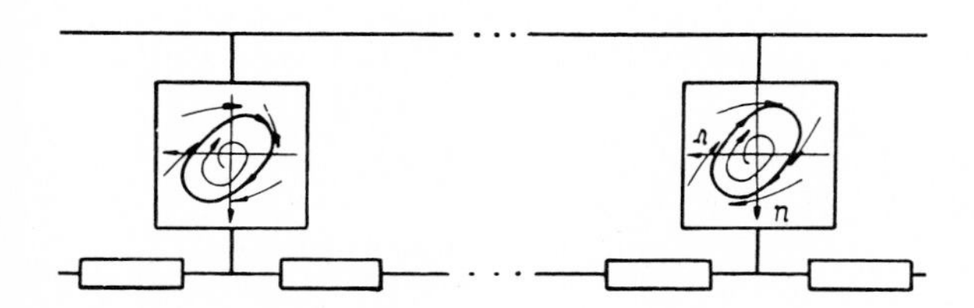

Fig. 2 Scheme of the sequence of coupled self-oscillators. Each oscillator has an independent periodic behavior

Here all parameters are assumed to be positive. The parameter e defines the coupling of oscillators, as well as the degree of the dissipation in the sequence; β characterizes the dependence of the oscillation frequency of a single oscillator on the oscillation amplitude; c "controls" the space dispersion of the medium that is also defined by the discreteness of the sequence. The sequence (3) has stationary periodic solutions, similar to a single oscillator. In this case, they are waves

$$w_j(t) = I(k)\exp\left[i(\omega(I^2,k)-kj)\right], \tag{4}$$

where $k=2\pi/\Lambda$ (Λ is the length of the stationary wave expressed in terms of the number of the oscillators). The dispersion law for such waves and the dependence of their amplitude on the propagation constant are shown in Fig.3.

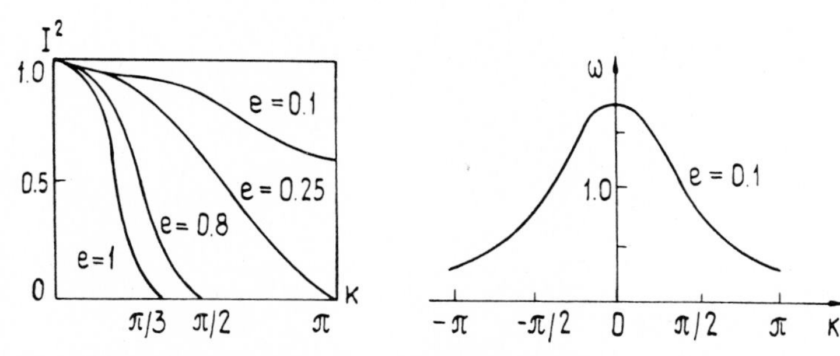

Fig. 3 The characteristics of stationary waves of system (3): a) the amplitude versus the wave number; b) the frequency versus the wave number

One can readily see, that system (2) coincides, in the longwave limit ($\Lambda >> 1$), with the one-dimensional Ginzburg-Landau equation that can be met in various fields of nonlinear physics:

$$\frac{dw}{dt} = w - (1+i\beta)|w|^2 w + D\nabla^2 w \tag{5}$$

(here the discrete coordinate j is replaced by the space coordinate x). System (3) in the real form is, at the same time, a discrete analog of the generalized $\lambda-\omega$ system ($W=U+iv$, $R=U^2+v^2$)

$$\begin{aligned} \frac{\partial u}{\partial t} &= \lambda(R)U-\omega(R)v+D_1\nabla^2 U+D_2\nabla^2 v, \\ \frac{\partial v}{\partial t} &= \lambda(R)v+\omega(R)U+D_1\nabla^2 v-D_2\nabla^2 U, \end{aligned} \tag{6}$$

which is rather popular in the theory of self-organization.

In order to give an approximate finite-dimensional description of a moving continuous medium, we usually either present the field in the form of a finite set of modes (some elementary collective excitations of the system), or use differential-difference equations, such as (3), instead of partial differential ones. We believe that for the study of the transition of regular structures to irregular ones and of the transformation of chaotic structures, the second approach is more favorable. On the one hand, we can readily conceive the spatial pattern of the structures connected with the natural choice of variables. On the other hand, when studying a discrete analog, it is easier to vary the characteristics of the model, in particular, to take the inhomogeneity of the medium, the change of conditions at the boundaries and other features into account. It should be also noted, that the analysis of the discrete system (3) is of independent physical interest. That is associated, for example, with the dynamics of the Taylor vortices in the Couette flow between the cylinders, when the inner cylinder rotates [9].

We consider here one-dimensional structures that are described by (3) under the following periodic boundary conditions, $w_j(t)=w_{j+N}$. In such a "medium" closed in a ring, there are solutions in the form of stationary travelling harmonic waves, such as (4), with the wave number $k=2\pi n/N$ (n is an integer, $n \leqslant N/2$). The amplitude and the frequency of these waves depend on their wavelength. The long waves (small k) have the highest amplitude and frequency (see Fig.3). The elementary structure-modes, such as (4), are stable, i.e. they are established at $t\to\infty$ either at a rather strong or a very weak coupling of the oscillators. When the coupling is strong (e is large), the most intense longwave, that is weakly inhomogeneous, structures ($n\sim 1$) survive as a result of mode competition. When the coupling is weak (e is small), the interaction of oscillators affects only strongly inhomogeneous structures that are characterized by a maximum shift of the oscillation phases of the neighbouring oscillators. As a result, in spite of a, in comparison with the longwave structures, smaller intensity, shortwave π-oscillations or close to them ($n \leqslant N/2$)-oscillations survive.

As shown by the numerical experiments with a homogeneous sequence in a wide range of e and β parameters, regular structures are transformed to chaotic ones. Note, that simple structures, such as (4), prevail among the observed regular structures. As to chaotic structures, they are rather diverse and differ in their space pattern, intensity, peculiarities of the time spectrum, level of order [2], and in other characteristics.

5. The Appearance of Chaotic Structures

Our experiments were carried out with the sequences of different lengths, N. The main results were obtained at N=50 and N=9. The amplitude and phase distributions of the field along the sequence, the space and time spectra, the characteristic Lyapunov exponents [10] averaged along the trajectory of the stochastic attractor, the Kolmogorov-Sinai entropy [11], the fractal dimension of the strange attractor that corresponds to the chaotic structure, and the level of order in the structures were calculated at N=9.

It was found, that the character of transition at small e does not depend on the length of the sequence. The order-chaos transition was qualitatively the same at N=50 and N=9. The critical values of the parameters at which the transition took place, were also close. The rearrangement (at N=50) in the space pattern of the nonstationary structures, the evolution of their oscillograms in some point of the sequence, and the variation of the time spectra with an increase in diffusion (parameter e) are shown in Fig.4. One can see, that a periodic structure close to π-oscillations (e=0.10800) appears immediately before the critical point. A simple limit cycle corresponds to these structures in the phase space of system (3). With a slight increase in diffusion (e=0.10850), the pattern of the regular structures gets more complicated, they become space and time modulated (see Fig.4b). The attracting winding on the two-dimensional torus* is the image of these structures in the phase space. Finally, in transition across

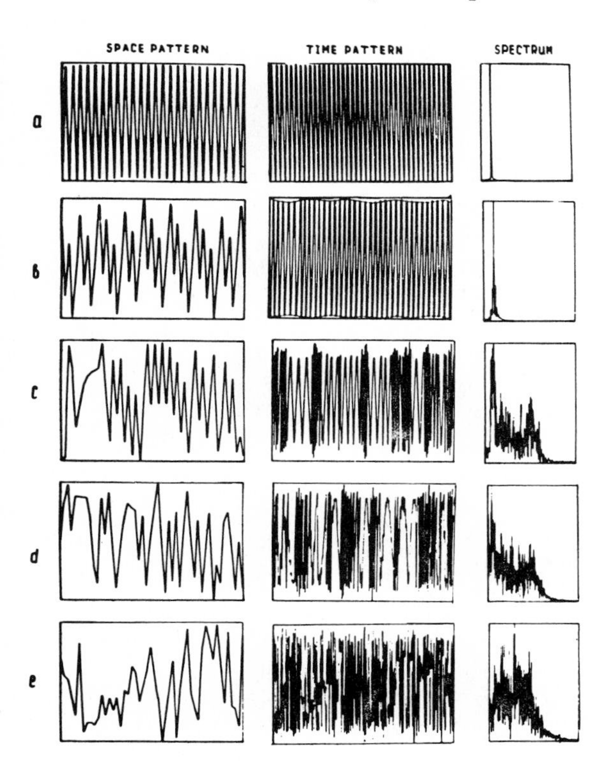

Fig. 4 Pattern evolution in a sequence (N=50) with an increase in diffusion: a) a homogeneous periodic structure (e=0.10800); b) a modulated structure (e=0.10850); c,d) chaotic structures with intermittency (e=0.1086, e=0.1088); e) developed chaotic structures (e=0.1330)

*This winding, in the general case, may be closed, then this is a cycle on a two-dimensional torus.

the critical point ($e^*\approx 0.10855$), chaos abruptly emerges in the medium. The time spectrum widens essentially, the dependences of the field (amplitude, I, and phase, θ) on the coordinate and the time become irregular. One can easily note, however, that in the immediate vicinity of the transition point ($e=0.1086$ and $e=0.1088$), both the time and the space patterns of the chaotic structures still have a trace of the order, for example, turbulent regions (bursts) intermit with almost periodic (laminar) regions.

Note, that the mathematical nature of the discovered "order-chaos" transition via intermittency differs from the known mechanism of intermittency observed in the systems that are described by a one-dimensional map [12]. In our case, the transition via intermittency is connected with the disappearance of the two-dimensional torus. At $e\geqslant e^*$, the system stays for a rather long time in the former vicinity of the torus (laminar regions in the form of modulated wave trains correspond to this situation). Then, the system gets into the developed stochastic set that appeared at $e<e^*$ and was not attracting before the torus disappeared. With a further increase in e, the memory of the regular structures escapes (the laminar regions decrease) and it vanishes at $e=0.1330$ (see Fig.4e). Note, that the dimension of the stochastic set, in contrast to the known cases, very rapidly grows at $e\geqslant e^*$ with an increase in e, and is rather large at $e=0.1330$ ($D>>1$, see Section 6).

The "order-chaos" transition via intermittency for N=9 was also observed at larger values of diffusion, in particular, at $e\approx 0.94$. The bifurcations associated with the disappearance of the tori seem to be quite common for multidimensional systems.

6. The Development of Chaos

How are the chaotic structures transformed and how do they develop with a further change of parameters, in particular, as the degree of their nonequilibrium grows? The most generally accepted concept is as follows. New modes, that were not excited before, get involved in the process of formation of chaotic structures, as the degree of nonequilibrium in the system grows. Since the transition of such modes across the neutral curve (i.e. their transformation from damped to unstable modes) takes place at discrete values of parameters of the system, the structures should get complicated in a discrete manner. The quantitative characteristics (the fractal dimension of the stochastic attractor, the level of order, the topological entropy, etc.) that correspond to the structures, should change abruptly together with the number of supercritical modes.

The detailed study of system (3) at N=9 showed that this almost obvious concept is wrong.

In order to elucidate the peculiarities of the developing chaotic structures on the corresponding attractors, we calculated the spectrum of the average characteristic Lyapunov exponents, $\lambda_1\geqslant\lambda_2\geqslant..>\lambda_{2N}$ ($\sum^{2N}\lambda_j<0$)*, and the parameters of chaos that are determined by them. We found, in particular, the dimension of the stochastic attractor, $D=M+d$, where M is the number of the exponents whose sum is positive

*The algorithm of our calculations coincided, principally, with that described in [10].

($\sum^{M}\lambda_j \geq 0$), but the sum of M+1 exponents is negative ($\sum^{M+1}\lambda_j < 0$) and $d=\sum^{M}\lambda_j/|\lambda_{M+1}|$ [13]. We also determined the level of order in irregular structures |2|, defined by $P=(2N-D)/(2N-1)$ and the Kolmogorov-Sinai entropy, $H=\sum^{m}\lambda_j$ (where m is the number of positive exponents). Note that as the parameters e and β change, all these values change continuously; only their derivatives are discontinuous (in the critical points). Figure 5 plots D,P and H versus r=1/e that characterizes the degree of non-equilibrium in the system. The region of small r was not studied in detail.

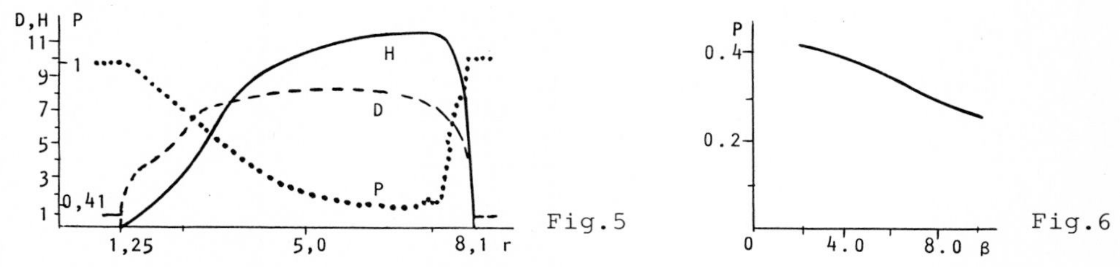

Fig. 5 The dependence of the entropy, H, the dimension of the stochastic set, D, and the level of order, P, on the back diffusion, r=1/e, for the sequence with N=9 (β=c = 1.71)

Fig. 6 The decrease in the level of order in the structures with the increase in nonisochronism of single oscillators ($\sim\beta$)

The longwave chaotic structures are replaced by the regular ones at $r>r_1$. Then, at $r=r_2=1.25$, partial stochastization of the structures takes place. This means that the level of order, P, monotonically decreases from 1; the dimension, D, of the corresponding attractor and the entropy, H, first monotonically grow with the increase in r up to $r_3=6.4$, after this D and H slightly decrease and at $r_4=8.1...$ abruptly drop down to 1 and 0, respectively. At $r>r_4$, the strongly disordered structures ($P\approx 0.3$) are replaced by the strictly regular structures in the form of ¶-oscillations (see Section 5). All these results were obtained for the sequence with N=9. It is important, that D and H changed substantially, when the number of oscillators was changed. At a larger N, the entropy, and the dimension increased, while the level of order, P, in the chaotic structures remained practically constant.

The spectrum of the Lyapunov exponents also changed very little with the increase in N. When N was increased by a unity, one positive and one negative exponent were added, with the largest and the smallest exponents changing very slightly. This is due to the fact, that the exponents characterize the unstable manifold of the periodic solution that corresponds to ¶-oscillations, and the stable manifold that corresponds to the homogeneous oscillations of the medium as a point system, respectively. The existence of both types of oscillations does not depend on the number of oscillators in the ring system (at $N\geq 2$).

We would like to add, that the increase in the degree of nonisochronism of single oscillators (parameter β) resulted in the decrease in the level of order in the structures (see Fig.6).

All the observed effects: the continuous variation of the Lyapunov exponents (transition across the zero value included), the monotonic increase in the dimension and the entropy, and the continous decrease in the level of order, as the system evolved, proved to be rather stable (rough) with respect to various perturbations in the system, in particular, to the disturbance of its homogeneity, i.e. the identity of the oscillators in the sequence.

7. Conclusion

One cannot but wonder at the harmony of Nature that has taken care of the fact that there should not be a sharp transition from complete order to complete disorder. Both complete order and complete disorder are limiting cases that are not often met in nonequilibrium media. Order supplemented by chaos, or chaos with a substantial contribution of order are more typical. The fact that complete chaos may monotonically and continuously transit to developed chaos (see Fig.5) is one of our most important results.

We believe, that the results obtained for the λ-ω system can be applied in a rather wide range of nonequilibrium media. However, it seems important to undertake analogous study of the cases that can be directly compared with the experiment. For example, the analysis of the structures that appear as a result of the Rayleigh-Benard convection in the fluid layer heated at the bottom, or of the structures in the form of the ensemble of spiral or cylindrical waves in a two-dimensional Belousov-Zaikin reactor.

We have all grounds to hope that a qualitative theory of structures similar to the classical theory of bifurcations in dynamic systems, will be developed in the nearest future. Such a theory will make it possible to predict admissible reconstruction of the structures, as the system evolves, and to foresee their features based on the analysis of the evolutionary history of the system. Indeed, no structures appear from nothing and disappear to nowhere. The memory of their properties is "recorded" in the topology of the corresponding phase space and in the relationship of its basic elements (cycles, separatrices, etc.).

1. M.Eigen and P.Schuster. The Hypercycle. A Principle of Natural Self-Organization, Springer (1979)
2. A.V.Gaponov-Grekhov and M.I.Rabinovich. Nonstationary Structures - Chaos and Order. In: Synergetics of Brain, ed. H.Haken, Springer (1983)
3. Yu.S.Iliashenko. - Advances in Mechanics (1982) v.5, No 1-2, p.31-64
4. C.Foias and R.Temam. - J. Math. Pures et Appl. (1979), 58;3
5. A.V.Babin and M.I.Vishik. - Uspekhi Mat. Nauk, (1983), v.38, No 4 (232), p.133-187
6. Y.Kuramoto. Induced Chemical Turbulence. In: Dynamics of Synergetic Systems, ed. H.Haken, Springer (1980) p.134-146
7. P.Berge and M.Dubois. - Phys. Lett. (1983) v.93A, No 8, p.365-368
8. N.Kopell and L.N.Howard. - Studies in Appl. Math. (1973), v.52. p.291-328
9. V.S.L'vov and A.A.Predtechensky. Step by Step Transition to Turbulence in the Couette Flow. In: Nonlinear Waves. Stochasticity and Turbulence, Gorky, Inst. of Appl. Phys. Acad. Sci. USSR (1980) p.57
10. I.Shimada and T.Nagashima. - Progr. Theor. Phys. (1979) v.61, No 6, p.1605-1616
11. V.I.Arnold and A.Avez. Ergodic Problems of Classical Mechanics, Benjamin, New York (1968)
12. Y.Pomeau and P.Manneville. - Commun. Math. Phys. (1980) v.77, p.189
13. F.Lidrappier. - Commun. Math. Phys. (1981) v.81, p.229
14. J.P.Eckmann. - Rev. Modern. Phys. (1981) v.53, p.643.

Part IV

Autowaves and Auto-Oscillations in Chemical Active Media

Mathematical Models of Chemical Active Media

A.M. Zhabotinsky and A.B. Rovinsky

Institute of Biological Physics, Acad. Sci. USSR, Moscow Region
SU-142292 Pushchino, USSR

Autowaves and dissipative structures arise in different physical chemical and biological systems with dynamics governed by nonlinear parabolic equations

$$U_t = f(U, \nabla U) + D \Delta U. \quad (1)$$

The reaction-diffusion systems can be separated as a special class:

$$X_t = f(X) + D \Delta X. \quad (2)$$

The homogeneous isothermic liquid phase chemical systems are the simplest ones in some sense. In this case the nonlinear functions have the following form:

$$f_i(X_j) = \alpha_i^j X_j + \beta_i^{kl} X_k X_l, \quad (3)$$

where X_j are concentrations of species.

Two types of autowave structures have received much attention:

1) travelling waves (plane, circular, spiral, etc.):

$$X = X(\omega t + k\rho + N\varphi); \quad (4)$$

2) stationary space periodic structures (Turing structures):

$$X(\alpha\xi + \beta\eta + L) = X(\alpha\xi + \beta\eta). \quad (5)$$

The so-called basic two-component model (N-model) plays an important role in analysis of autowave processes:

$$\begin{aligned} X_t &= f(X,Y) + D_X \Delta X \\ Y_t &= g(X,Y) + D_Y \Delta Y \end{aligned} \quad (6)$$

with an N-shaped nullcline of Y-variable for the local system

$$\begin{aligned} X_t &= f(X,Y) \\ Y_t &= g(X,Y), \end{aligned} \quad (6a)$$

Fig. 1 shows the nullclines of system (6a) for the main cases: monostability, oscillation and bistability. In most theoretical studies particular versions of model (6) are used |3,4|.

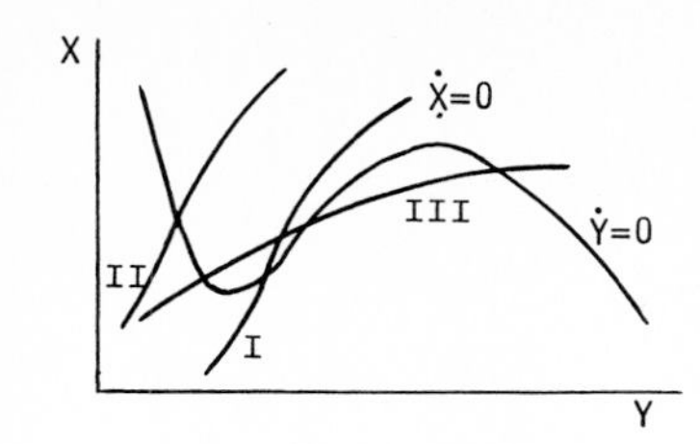

Fig. 1 Nullclines of model (6): I - oscillations, II - monostability, III - bistability

The spatial chemical system based on the modified Belousov reaction (BZ-medium) is very convenient for experimental work [5,6].

A variety of structures of type (4) has been found in different systems including homogeneous chemical systems [5-10]. Among the recent works most interesting results have been obtained with BZ-medium by Agladze and Krinsky:
1) the multi-armed spiral waves (N=2,3 and 4) [11];

2) a peculiar non-stationary behaviour in the core of spiral waves rather than a stationary rotation of the wave front [11];

3) the drift of wave breaks caused by external high frequency waves (when external influence is ceased the breaks develop into spiral waves) [12].

On the contrary, no unequivocal evidence of the Turing patterns (5) has been obtained in reaction-diffusion systems while their hydrodynamic analog - the Benard cells have been known for a long time [1,2]. The data by Showalter [13] and Agladze [14] indicate that the quasistationary periodic structures found in the BZ-medium [15] may be the Benard cells with convection forced by the surface tension gradient (Marangoni effect). If the BZ-system can be simulated by an N-model, the Turing structures are most unlikely to exist in this medium.

Thus, a reliable model of the BZ-system would be very useful for a comparison of some theoretical and experimental data on autowave structures.

Mechanism and Models of the Belousov Reactions

The generally accepted scheme of autocatalytic oxidation of metal ions by bromate inhibited by bromide ion (Fig.2) was suggested by Vavilin [16]. Then, Field, Koros and Noyes [17] gave the now-familiar mechanism for the cerium system which included Vavilin's scheme, Scrabal's scheme of the bromate-bromide reaction [18] and the assumption that the kinetics of bromide production can be described by the simple equation:

$$V_{Br^-} = qk[Red][Ce^{4+}], \quad (7)$$

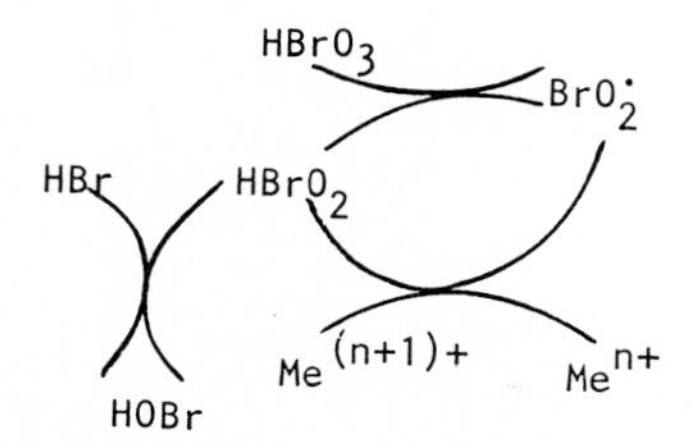

Fig. 2 Autocatalysis and inhibition of metal ion oxidation by bromate

where k is the rate constant of Ce^{4+} reduction and q is a stoichiometric factor. Fig. 3 shows the simplest scheme of the FKN mechanism. Apparently, in this system either HOBr or Br_2 is rapidly removed because of bromination of the reductant. Thus the following system of equations corresponds to the scheme:

$$
\begin{aligned}
\dot{X} &= k_2HU(C-X)-k_{-2}YX-k_6BX \\
\dot{U} &= 2k_1HAY-2k_{-1}U^2-k_2HU(C-X)+k_{-2}YX \\
\dot{Y} &= -k_1HAY+k_{-1}U^2+k_2HU(C-X)-k_{-2}YX+k_3H^2AZ-k_4HYZ-2k_5Y^2 \\
\dot{Z} &= qk_6BX-k_3H^2AZ-k_4HYZ,
\end{aligned}
\tag{8}
$$

where $X=[Ce^{4+}]$, $U=[BrO_2^{\bullet}]$, $Y=[HBrO_2]$, $Z=[Br^-]$, $C=[Ce^{3+}]+[Ce^{4+}]$, $H=[H^+]$.

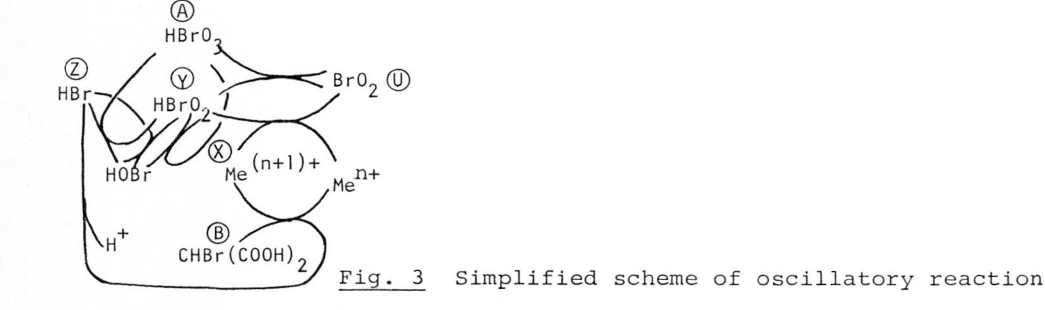

Fig. 3 Simplified scheme of oscillatory reaction

Field and Noyes have put forward a simplified phenomenological model known as Oregonator [19]

$$
\begin{aligned}
\dot{X} &= k_1AY-k_6X \\
\dot{Y} &= k_1AY+k_3AZ-k_4YZ-2k_5Y^2 \\
\dot{Z} &= qk_6X-k_3AZ-k_4YZ.
\end{aligned}
\tag{9}
$$

The Oregonator can be deduced as an asymptotics of system (8) under Vavilin's assumption that $U \ll Y$ [16], which is, however, not the case in the cerium-catalyzed reaction [23].

Tyson presented an elegant and simple N-model reducing the Oregonator by the methods of singular perturbation theory:

$$
\begin{aligned}
\frac{dx}{d\tau} &= y-x \\
\varepsilon\frac{dy}{d\tau} &= y(1-y) - 2qx\,\frac{y-\mu}{y+\mu},
\end{aligned}
\tag{10}
$$

where $x = \frac{k_6k_5B}{(k_1HA)^2}X$, $\quad y = \frac{2k_5}{k_1HA}Y$,

$\tau = k_6Bt$, $\quad \mu = \frac{2k_3k_5}{k_1k_4}$, $\quad \varepsilon = \frac{k_6B}{k_1HA}$.

Model (10) provides a correct qualitative description of the BZ reaction dynamics. However, the oscillatory behaviour simulated by system (10) is different from what was experimentally observed. The amplitude of X does not depend on C and may even exceed the C value. The parameters of the right hand side of the equations do not depend on A and B. This makes it impossible to transform monostability into oscillation by variation of A or B while it can be easily done in the real system.

Tyson [21] also suggested another N-model closely related to model (10) but free of its drawbacks. Unfortunately, this model cannot be obtained as an asymptotic approximation of model (8).

On the Mechanism of Cerium Catalyzed BZ Reaction

It has been shown in [22,23] that for a correct description of the kinetics of the reaction

$$HBrO_3 + 4Ce^{3+} \longrightarrow 4Ce^{4+} + HOBr + 2H_2O \qquad (11)$$

no term can be neglected in model (8, B=O). Besides, this model cannot, seemingly, be reduced to the second-order one. If it could, the right-hand sides of the reduced system would be complicated enough. On the other hand it was demonstrated earlier [24,25] that relationship (7) is not correct for the bromate-cerium-bromomalonic acid system. It means that for description of the bromide ion production at least the term qk_6BX in model (8) should be replaced by a more complex one. Moreover, probably some additional variables should be introduced into the model.

Thus, no simple and reliable model of the cerium-catalyzed BZ-reactions has, as yet, been proposed.

Model for the Ferroin-Catalyzed Oscillatory Reaction

Fortunately some peculiarities of the reactions involving ferroin ions allow formulation of a reasonably simple model for the bromate-ferroin-bromomalonic acid system, which is very convenient for observation of autowave phenomena.

There are two main factors accounting for the difference in the kinetics of the reactions with $Fe(phen)_3$ ions and of those with Ce. The first is the difference in the redox potentials of the $Fe(phen)_3^{2+}/Fe(phen)_3^{3+}$ (1.14 V) and of Ce^{3+}/Ce^{4+} (1.61 V) couples. The second is the presence of ligands in ferroin which themselves can take part in the oxidation-reduction process. As shown in [26,27], for the separate reactions of metal ion oxidation by bromate and of metal ion reduction by bromomalonic acid only the first factor is significant under conditions of oscillatory reaction. As a consequence of the low redox potential value of the ferroin/ferriin couple the equilibrium of elementary steps should be shifted to the oxidized state $Fe(phen)_3^{3+}$. Therefore in this case in (8) $k_2 >> k_{-2}$, while the rate of the $Fe(phen)_3^{3+}$ reduction by bromomalonic acid is described by the following equation [27]:

$$v_X = \dot{X} = - \frac{k_{15}BX}{1+\frac{k_{-15}}{k_{16}} h_o(C-X)}, \qquad (14)$$

where $X = [Fe(phen)_3^{3+}]$, $C = X + [Fe(phen)_3^{2+}]$; h_o - acidity function,

k_{15}, k_{-15} and k_{16} are the rate constants of the steps:

$$Fe(phen)_3^{3+} + CHBr(COOH)_2 \rightleftharpoons {}^{\bullet}CBr(COOH)_2 + Fe(phen)_3^{2+} + H^+ \quad (15)$$

$$CBr(COOH)_2 + H_2O \longrightarrow {}^{\bullet}COH(COOH)_2 + H^+ + Br^-. \quad (16)$$

In paper [27] the estimates are given: $k_{15}k_{16}/k_{-15} \approx 2\cdot10^{-5} M\cdot s^{-1}$, $k_{-15}/k_{16} > 5\cdot10^4 \cdot M^{-2}$. Since in chemical wave experiments usually $C \sim 10^{-3} M$, $h_o \sim 1M$, then

$$v_X = -\frac{k_7 BX}{C-X}, \text{ provided that } 1 - \frac{X}{C} >> 10^{-2} \quad (k_7 = \frac{k_{15}k_{16}}{k_{-15}h_o}). \quad (17)$$

The assumption (7) about the rate of bromide production for the ferroin case yields:

$$v_Z = /\dot{B}r^-/ = q\frac{k_7 BX}{C-X}. \quad (18)$$

Adding to the model of the ferroin oxidation by bromate [26] the terms (17) and (18), describing ferriin reduction and bromide production, brings to the following model for the ferroin catalyzed oscillatory system:

$$\begin{aligned}
\dot{X} &= k_2U(C-X) - k_{-2}YX - \frac{k_7BX}{C-X}, \\
\dot{U} &= 2k_1h_oAY - 2k_{-1}U^2 - k_2U(C-X) + k_{-2}YX, \\
\dot{Y} &= -k_1h_oAY + k_{-1}U^2 + k_2U(C-X) - k_{-2}YX + k_3h_oAZ - k_4h_oYZ - 2k_5h_oY^2, \\
\dot{Z} &= q\frac{k_7BX}{C-X} - k_3h_oAZ - k_4h_oYZ.
\end{aligned} \quad (19)$$

The U variable remains small and consequently rapid unless $1-(X/C) << 1$ [26]. Therefore the methods of singular perturbation theory can be used to reduce system (19). Besides, the term $k_{-1}U^2$ can be neglected. Thus system (19) becomes:

$$\begin{aligned}
\dot{X} &= 2k_1h_oAY - \frac{k_7BX}{C-X}, \\
\dot{Y} &= k_1h_oAY + k_3h_oAZ - k_4h_oYZ - 2k_5h_oY^2, \\
\dot{Z} &= q\frac{k_7BX}{C-X} - k_3h_oAZ - k_4h_oYZ.
\end{aligned} \quad (20)$$

We have made the following estimates from the experimental data of [22,25,27]: $k_1 \approx 10^2 M^{-1}s^{-1}$, $k_3 \approx 10 M^{-1}s^{-1}$, $k_4 \approx 10^7 M^{-1}s^{-1}$, $k_5 \approx 2\cdot10^4 M^{-1}s^{-1}$, $k_7h_o \approx 2\cdot10^{-5} M^{-1}s^{-1}$.

They are close to those evaluated by Tyson [28]. Further analysis reveals that at the values of the parameters $C \sim 10^{-3} M$, $A \sim 10^{-2}$, $B \sim 10^{-1}$, $h_o \sim 1M$, the variables Y and Z are fast compared with X and their time scales are approximately the same. Reduction to the

X,Y - variables yields a simpler N-model:

$$\frac{dx}{d\tau} = y - \alpha\frac{x}{1-x}$$
$$\varepsilon\frac{dy}{d\tau} = y(1-y) - 2q\alpha\,\frac{x}{1-x}\cdot\frac{y-\mu}{\mu+y}, \qquad (21)$$

where

$$X = Cx,\quad Y = \frac{k_1A}{2k_5}\,y,\qquad \varepsilon = \frac{k_1A}{k_5C},\qquad \alpha = \frac{k_5k_7B}{k_1^2A^2h_o},$$
$$\mu = \frac{2k_3k_5}{k_1k_4},\qquad t = \frac{k_5C}{k_1^2A^2h_o}\cdot\tau.$$

The nullclines of system (21) are shown in Fig.4.

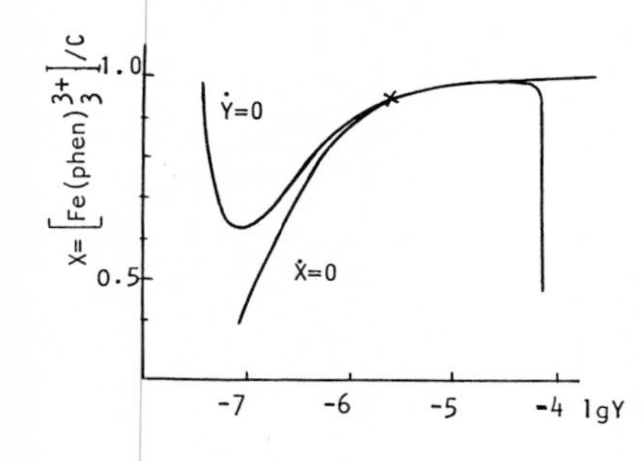

Fig. 4 Nullclines of model (21) for the oscillatory bromate-ferroin-bromomalonic acid system

Model (21) produces the oscillations of $Fe(phen)_3^{3+}$ with extremes proportional to the total ferroin concentration C, which is in good agreement with the experiment. The dependence of the extremes and period of oscillations on the other parameters A,B,h_o also agrees satisfactorily with the experimental data. Figures 5 and 6 show the oscillations in the real bromate-ferroin-bromomalonic acid system along with those simulated by model (21) at the corresponding values of the parameters.

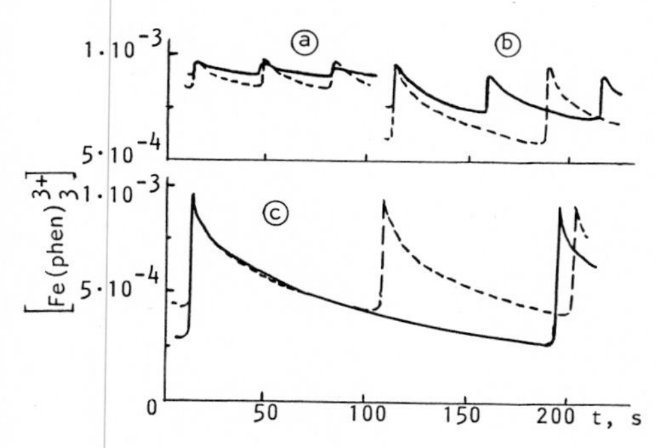

Fig. 5 Oscillations in the bromate-ferroin-bromomalonic acid system: experiment-solid line, model (21) - broken line A=0.025M, C=0.001M, q=0.5, $[H_2SO_4]$=1.25M (h_o=2.5M), t=40°C; a) B=0.05M, b) B=0.2 M, c) B=0.4M

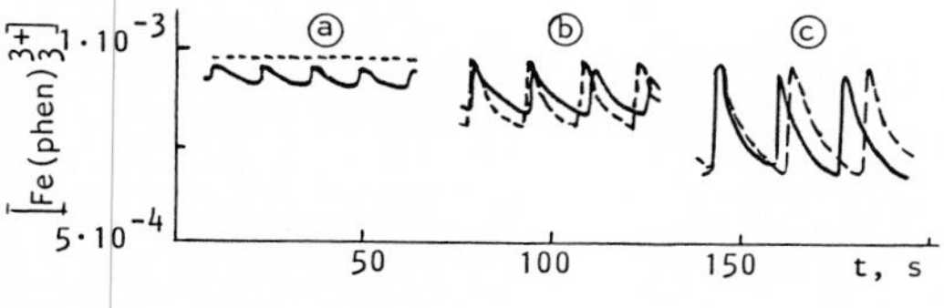

Fig. 6 The same as in Fig.5 A=0.05M, $[H_2SO_4]$= =1.1M (h_o=2.1M); a) B=0.1M, b) B=0.2M, c) B=0.37M

As already mentioned analysis of the phenomenological N-model indicates that the appearance of the Turing patterns seems to be unlikely in BZ media. Indeed, Turing's bifurcation may arise in system (6) with the local kinetics described by model (21) only if

$$\gamma D_Y < D_X \qquad (22)$$

with $\gamma > 1$.

Since $Fe(phen)_3^{3+}$ is a big multi-charged complex ion then, apparently, $D_Y > D_X$. The validity of condition (22) is still less probable.

References

1. G.Nicolis and I.Prigogine. "Self-Organization in Nonequilibrium Systems", Wiley, New York (1977)
2. H.Haken. "Synergetics", Springer, Berlin (1978)
3. V.Vasiliev, Y.M.Romanovskii, V.G.Yahno. Usp. Phys. Nauk, 128, 625 (1979)
4. Autowave processes in diffusion systems. Ed. Grekhova, Gorky (1981)
5. A.N.Zaikin and A.M.Zhabotinskii. Nature, 225, 535 (1970)
6. A.T.Winfree. Science, 175, 634 (1972)
7. A.T.Winfree. Science, 181, 937 (1973)
8. P.De Kepper. In: "Nonlinear Phenomena in Chemical Dynamics", C.Vidal and A.Pacault Eds., Springer, Berlin (1981) p.192
9. K.W.Pehl, L.Kuhnert and H.Linde. Nature, 282, 198 (1979)
10. M.Orban. J. Am. Chem. Soc., 4311 (1980)
11. K.I.Agladze and V.I.Krinsky. Nature, 269, 424 (1982)
12. V.I.Krinsky, K.I.Agladze. Physica, 8D, 50 (1983)
13. K.Showalter. J. Chem. Phys., 73, 3735 (1980)
14. K.I.Agladze and V.I.Krinsky. In: Synerg. and Coop. Phen., Tallin (1982)
15. A.M.Zhabotinsky and A.N.Zaikin. J. Theor. Biol., 40, 45 (1973)
16. V.A.Vavilin and A.N.Zaikin. Kinet. catal., 12, 309 (1971)
17. R.J.Field, E.Kőrös and R.M.Noyes. J. Am. Chem. Soc., 94, 8649 (1972)
18. A.Scrabal, H.Schreiner. Monatsch. Ch., 65, 213 (1935)
19. R.J.Field and R.M.Noyes. J. Chem. Phys., 60, 1877 (1974)
20. J.J.Tyson. An. N.Y. Acad. Sci., 316, 279 (1979)
21. J.J.Tyson. J. Math. Biol., 5, 351 (1978)
22. V.A.Vavilin and A.M.Zhabotinskii. Kinet. catal., 10, 83 (1969)
23. A.B.Rovinskii and A.M.Zhabotinskii. Theor. exper. khim., 14, 183 (1978)
24. A.M.Zhabotinskii and A.B.Rovinskii. Theor. exper. khim., 16, 386 (1980)
25. A.M.Zhabotinskii, A.N.Zaikin, A.B.Rovinskii. React. Kinet. Catal. Lett., 20, 29 (1982)
26. A.B.Rovinskii and A.M.Zhabotinskii. Theor. exper. khim., 15, 25 (1979)
27. A.B.Rovinskii. J. Phys. Chem. (to appear)
28. J.J.Tyson. Oscillations and Traveling Waves in Chemical Systems, R.J.Field and M.Burger, Eds., Wiley, New York (1983)

On the Mechanism of Target Pattern Formation in the Distributed Belousov-Zhabotinsky System

K.I. Agladze and V.I. Krinsky

Institute of Biological Physics of the USSR Academy of Sciences, Moscow Region
SU-142292 Pushchino, USSR

The wave patterns in the distributed Belousov-Zhabotinsky reaction are generated by two different types of sources. Those are spiral waves [1,2] and leading centers [1,3] producing the so called "target patterns". The nature of leading centers remains unclarified up to now. Target patterns may be generated either by some small auto-oscillatory regions of the medium or by a special type of sources (echo-sources) arising in a medium with non-oscillatory (excitable) kinetics [4].

Winfree [5] has found that leading centers can be drastically reduced in number or even completely eliminated by careful filtration of the reaction solution. This led him to conclude that the only cause of target pattern formation in the B-Z reagent are local inhomogeneities triggering the medium into an auto-oscillatory state. Tyson has added to the view by demonstrating the impossibility of echo-sources in a two-component mathematical model of the distributed B-Z system [6]. However, Zaikin and Kawczynski, using a three-component model, have concluded that echo-sources are, in principle, realizable [7].

We have developed a method which can be used to discriminate experimentally between the two alternatives. The method is based on the study of interaction between the autowave source under examination and waves from an external source.

1. Synchronization of a Leading Center by an External Source

The idea of the experiment was to synchronize the leading center by a higher frequency source and then to remove the synchronizing influence. The wave pattern produced thereby can characterize the underlying mechanism. The following three cases are possible. 1) The leading center will disappear due to the fact that the topological charge of the target pattern is zero. 2) The source will reappear exactly at the same place. 3) It will appear again but in some other place (the induced drift of the source [8]). If the first case is realized, it means that the wave source is a priori not due to the soft auto-oscillatory kinetics. On the contrary, the second case is an indication of the auto-oscillatory kinetics being involved. The existence of the drift in the third case means that the source is not linked with the local geometry of the medium (e.g., holes, inhomogeneities). Such a source arises from singularities in phase distribution. This may be either a source of the echo type or a source of the phase waves in an auto-oscillatory medium.

The experiments were run in a medium with the composition approaching the standard one: $NaBrO_3$, 0.3 M; $CHBr(COOH)_2$, 0.1 M; $CH_2(COOH)_2$, 0.1 M; H_2SO_4, 0.45 M; ferroin, 0.005 M.
4 ml of the solution were poured into a 9 cm Petri dish, t^o=20°C. This medium displays excitable kinetics [9].

The leading centers in the B-Z reaction usually appear when the reaction approaches an auto-oscillatory state (with either H_2SO_4 concentration exceeding 0.45 M or with the solution layer thicker than 1 mm at about 0.3 M H_2SO_4). We used as a synchronizing source a pair of spiral waves which were induced by local stirring of the wavefront.

The spiral waves with frequences higher than those of leading centers always suppressed these latter [1] (Fig.1a,b). After the leading centers disappeared, the spiral waves were eliminated (Fig.1c) by a drop of 4M KBr solution which produced an inexcitable bridge between the wave cores. As a result, the leading centers emerged again at exactly the same place (Fig.1d). This evidences that the second case is realized. Neither the first nor the third case was ever observed in these experiments. Thus, we are led to conclude that the leading centers observed in the B-Z chemical reagent are small loci of the medium with auto-oscillatory kinetics, while the rest of the medium displays non-oscillatory (excitable) kinetics. Indeed, when external synchronization is switched off, such a locus continues generating oscillations with the period determined by its intrinsic characteristics.

2. Densitometric Estimation of the Parameters of the B-Z Chemical Active Medium

Axiomatic model analysis gives the following existence condition for an echo-source [4]:

$$\tau/R > 1/2 \qquad (1)$$

where τ is the duration of the excited state, and R is the medium refractory period.

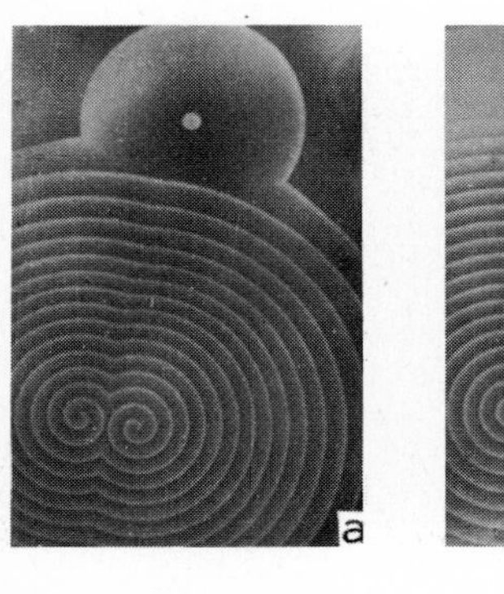

Fig. 1 Suppression of a leading centre by reverberators and its re-establishment

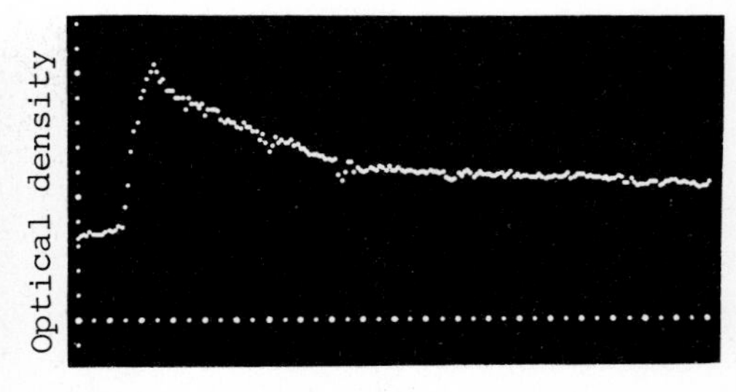

Fig. 2 A wave profile obtained by densitometrical analysis. The interval between points is 25 μm

We can evaluate these parameters from densitometric processing of the photos of a wave pattern which yields the optical density distribution. The optical density in a given region corresponds to $Fe(phen)_3^{3+}$ concentration. Its distribution across the wavefront is shown in Fig.2.

The wave profile obtained allows the duration of the excited state to be estimated. The excited state of the medium corresponds to a peak of increased $HBrO_2$ concentration |10|. The length of a pulse of $HBrO_2$ concentration can be estimated from the wave profile of $Fe(phen)_3^{3+}$.

According to [10],

$$[HBrO_2] \sim \frac{d[Fe(phen)_3^{3+}]}{dt} \qquad (2)$$

where $[HBrO_2]$ and $[Fe(phen)_3^{3+}]$ are concentrations.

As seen from Fig.2, the length of the forefront of $Fe(phen)_3^{3+}$ is about 0.12 mm. At the wave velocity of about $3 \cdot 10^{-2} mm \cdot s^{-1}$ this value corresponds to the duration of the excited state ~ 4s. The refractory period for this system, evaluated in [9] from the maximum frequency of wave propagation, is ~30s.

Thus, for the B-Z reaction the ratio $\tau/R \sim 0.13$, this being a priori insufficient for the generation of an echo-source. The results obtained by the method of re-establishment of an autowave source after the termination of external synchronization as well as the densitometric evaluation of the B-Z reaction parameters, indicate that the target patterns are generated by small loci of the medium with auto-oscillatory kinetics.

References

1. A.M.Zhabotinsky. Concentrational Oscillations. Moscow, "Nauka", Publ. (1973) (in Russian)
2. A.T. Winfree.Science 175, 634,(1972)
3. A.N.Zaikin, A.M.Zhabotinsky, Nature 225, 535,(1970)
4. V.I.Krinsky, Biofizika 11, 676,(1966)
5. A.T.Winfree, Theor. Chem. 4, 1,(1978)
6. J.J.Tyson, P.C.Fife, J.Chem. Phys. 73, 2224,(1980)
7. A.N.Zaikin, A.L.Kawczynski, J. Non-Equilibr. Thermodyn. 2,39,(1977)
8. V.I.Krinsky, K.I.Agladze, Physica 8D, 50,(1983)
9. K.I.Agladze, Thesis, Pushchino,(1983)
10. A.B.Rovinsky, A.M.Zhabotinsky A.M., Theor. Chem. 15, 25,(1979) (in Russian)

Iodide-Induced Oscillation in Uncatalyzed Bromate Oscillators

E. Kőrös, M. Varga, and T. Pauló

Institute of Inorganic and Analytical Chemistry, L. Eötvös University
H-Budapest, Hungary

1. Introduction

At least fifty of such organic compounds (mostly phenol- and analine-derivatives) are known at present which under certain concentration conditions in their reaction with acidic bromate show chemical oscillations. These systems are called uncatalyzed bromate oscillators [1,2] and have been investigated from various aspects in many laboratories. Although the skeleton mechanism suggested [3] seems to account for the behaviour of these systems properly [4], we are rather ignorant of the chemical prerequisites of the oscillation.

In one of our earlier publications we have demonstrated that the course in time of chemical oscillation - its length in time, period time, amplitude and damping factor - is different in character for every single organic substrate [5]. We also pointed out that the reaction between bromate and the aromatics is a rather complex one: large number of intermediates and products are formed [6]. Our knowledge on the chemistry of bromate oscillators is rather vague and attempts to find some relation between the chemical composition of the aromatics and the occurrence of oscillations failed so far. Chopin-Dumas [7] has suggested a selection criteria for the aromatics based on their electrochemical behaviour and tendency to enter into coupling reaction. Although some of her statements seem to point at correct direction, her rules have shortcomings, because these are irrespective of the fact that at the onset of chemical oscillation the initial compound has already been converted (partly or completely) into oxidation and bromination products [6].

2. Experimental Results and Discussion

2.1. Uncatalyzed Systems

With these facts in mind we have decided to start a systematic experimental study trying to reveal the conditions of chemical oscillation. As a first step we selected three p-substituted phenols with closely similar side chains in the para position. These are shown in Fig.1.

The chemical oscillation was followed by immersing a smooth Pt electrode and a $Hg/Hg_2SO_4/K_2SO_4$ reference electrode into the stirred reacting system, and the potential trace was recorded. Temperature: $25.0 \pm 0.2^{\circ}C$.

With HMP and HME the domains of existence are very wide, e.g. at 1 M H_2SO_4 it is between 0.04 and 0.15 M $KBrO_3$, and between 0.015 and 0.06 M aromatics. With Tyr the domain is narrower as shown in Fig.2.

The highest number of oscillations with Tyr is only 5; with HMP 8 and with HME about 30. Typical curves are shown for HME in Fig.3,

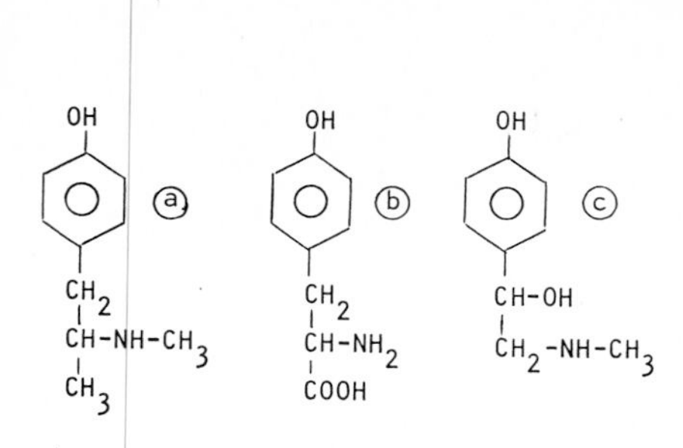

Fig. 1 a) 1-(p-hydroxyphenyl)-2--methylamine propane (HMP); b) 1-(p-hydroxyphenyl)-2-aminopropionic acid (Tyr); c) 1-(p-hydroxyphenyl)-1-hydroxy-2-methylamine ethane (HME)

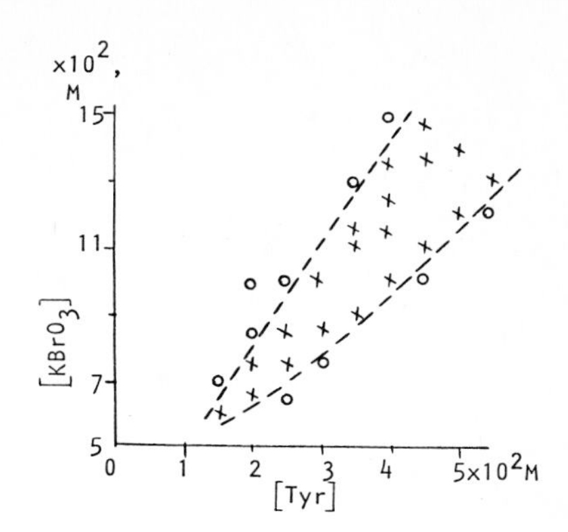

Fig. 2 The domain of existence of chemical oscillation at 1.5 M H_2SO_4 x - the system oscillates; o - the system does not oscillate

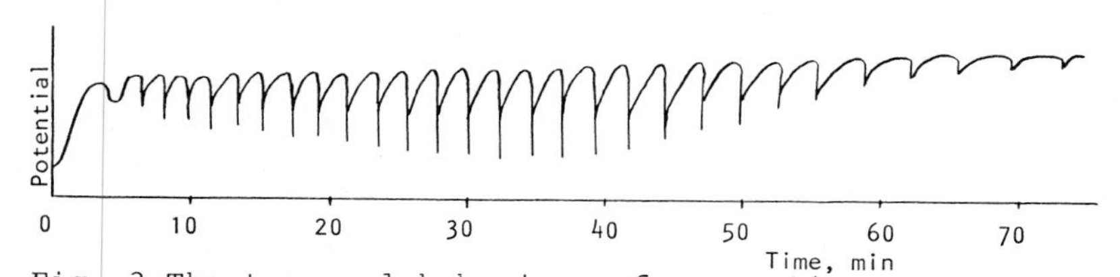

Fig. 3 The temporal behaviour of a reacting system with the following composition: 0.09 M $KBrO_3$, 0.03 M HME and 0.5 M H_2SO_4

for Tyr in Fig. 6.a and for HMP in Fig. 4.a. With Tyr the preoscillatory period is rather long.

One of the main products of the reactions is 1,2-quinone, its rate of formation during the oscillatory phase of the reaction is periodic.

It is salient that the chemical oscillation is the longest in the HME-containing reacting system. In our opinion in the oscillatory reaction an important part is played by the group in p-position, especially that part which is closest to the aromatic ring. This is $-CH_2-$with HMP and Tyr and the more reactive (towards bromate)-CHOH-with HME.

2.2. Effect of Iodide

It had been found earlier that iodide affects the Belousov-Zhabotinsky oscillatory system [8], and the iodide-induced high-frequency oscillation has been studied thoroughly in our laboratory [9].

The rather dramatic impact of iodide on the catalyzed bromate oscillators prompted us to investigate its effect also on uncatalyzed bromate oscillators. Already the preliminary experiments have shown that there are some uncatalyzed bromate oscillators which are not affected by iodide (only inhibition occurs at high iodide concentration), and there are such systems which are greatly affected by iodide. To the first group belong e.g. phenol, 1,2,3-trihydroxybenzene, and 3,4,5-trihydroxy benzoic acid; to the second one all the three aromatics (HMP, Tyr and HME) under study. With the latter group of aromatics iodide - in a certain concentration - induces the reacting system to a more pronounced oscillatory activity: the number of oscillations increases considerably and the preoscillatory period - if any - disappears. It is, however, no matter at what phase of the reaction iodide is added to the system. Fig. 4 shows the differences.

In Fig.5 the concentration decrease of $KBrO_3$ in time in the same reacting system is shown.

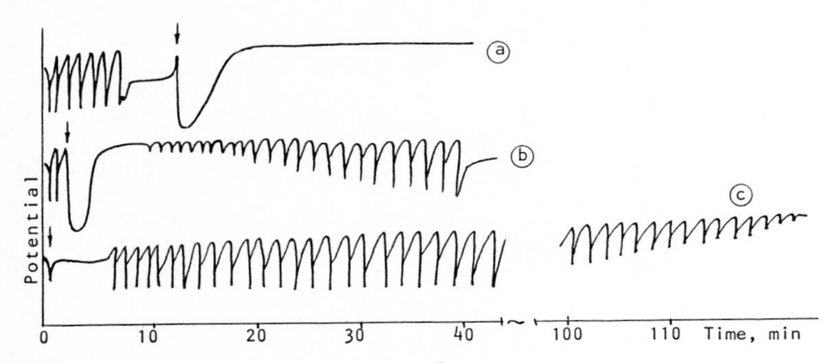

Fig. 4 The effect of $4 \cdot 10^{-3}$M iodide on a reacting system of the following composition: 0.05 M $KBrO_3$, 0.04 HMP, 1.0 M H_2SO_4. Iodide is added a) after the termination of oscillation, b) after the second oscillation; c) at the start of the reaction

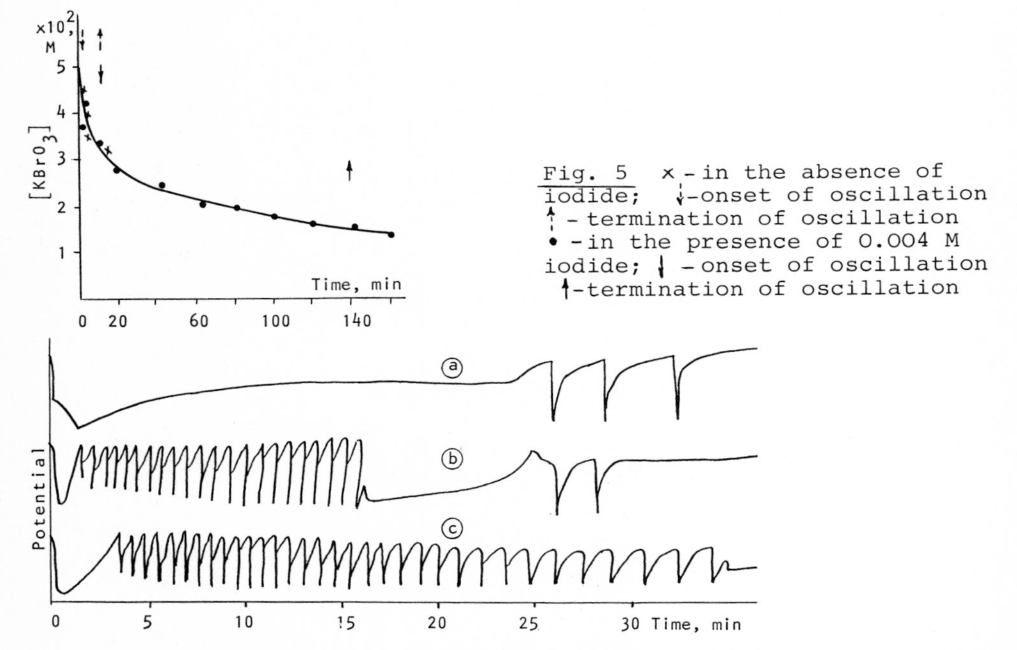

Fig. 5 × - in the absence of iodide; ↓ - onset of oscillation ↑ - termination of oscillation • - in the presence of 0.004 M iodide; ↓ - onset of oscillation ↑ - termination of oscillation

Fig. 6 The composition of the reacting system: 0.09 M $KBrO_3$, 0.02 M Tyr, 1.0 M H_2SO_4 a) without iodide; b) 0.010 M iodide; c) 0.011 M iodide. In both cases iodide was added to the system at the start of the reaction

From other aspect, Fig.6 shows how a slight change in iodide concentration influences the temporal behaviour of a reacting system.

The mechanism for the iodide-induced oscillatory reactions remains to be worked out. One probable scheme, however, can be envisioned that would lead to oscillation. The reaction of iodide with bromate in the presence of an aromatics produced bromide, bromaromatics, and iodide (+I). The latter appears as a -CHI- group by its reaction with the $-CH_2-$ group in HMP or Tyr. This hydrolyses readily disposing a -CHOH- group active towards bromate

$$-CHOH- + BrO_3^- + H^+ \longrightarrow -CO- + HBrO_2 + H_2O$$

The production of $HBrO_2$ drives the key process as suggested for the bromate oscillators [10]. (The reacting system containing HME, which has a -CHOH- group, exhibits a prolonged oscillation; see Fig.3).

It is our firm conviction that sustained oscillations in the reaction between acidic bromate and iodide in a stirred tank reactor observed by the Brandeis group [11] are closely connected with our recent observations on the iodide effect in closed systems, and the occurrence of chemical oscillations in both systems can be explained on the same basis.

The details will appear in a forthcoming publication.

References

1. E.Kőrös and M.Orbán. Nature, 273, 371 (1978)
2. M.Orbán and E.Kőrös. J. Phys. Chem., 82, 1672 (1978)
3. M.Orbán, E.Kőrös and R.M.Noyes. J. Phys. Chem., 83, 3056 (1979)
4. P.Herbine and R.J.Field. J. Phys. Chem. 84, 1330 (1980)
5. M.Orbán and E.Kőrös. Synergetics. Far from Equilibrium /Eds.: A.Pacault, C.Vidal/ Springer, Berlin, Vol. 3, 1979, p.43
6. M.Orbán and E.Kőrös. Kinetics of Physicochemical Oscillation, Vol. 1, Aachen, 1979, p.83
7. J.Chopin-Dumas. J. Chem. Phys., 78, 461 (1981)
8. R.J.Kaner and I.R.Epstein. J. Am. Chem. Soc., 100, 4073 (1978)
9. E.Kőrös and M.Varga. J.Phys. Chem., 86, 4839 (1982)
10. R.J.Field, E.Kőrös and R.M.Noyes. J. Am. Chem. Soc., 92, (1972)
11. M.Alamgir, P.De Kepper, M.Orbán and I.R.Epstein. J. Am. Chem. Soc., 105, 2641 (1983)

Topological Similarities in Dissipative Structures of Marangoni-Instability and Belousov-Zhabotinsky-Reaction

H. Linde

Akademie der Wissenschaften der DDR, Zentralinstitut für Physikalische Chemie
Rudower Chaussee 5, DDR-1199 Berlin, German Democratic Republic

In the early investigation of Marangoni-instability (MI), the view dominated about the irregular behaviour of the spontaneous interfacial convection (called interfacial turbulence), due to both strong driving forces and the irregularities of additional convections |1,2|. The transition from the foregoing regular dissipative structures to different kinds of chaotic behaviour with increasing driving forces and changing of other parameters is now again of interest from the point of generalization of the turbulence problem [3,4]. Even in the case of MI, there exist at least three different kinds of transitions to different kinds of macroscopic chaotic behaviour: from a sharp one-peak size distribution in systems of regularly shaped "stationary" roll cells to a wide band distribution of the size of irregularly shaped roll cells([5-8], Fig.1); from systems of rather regular "stationary" roll cells (three-dimensional convection cells) to unharmonic relaxation oscillations, which interact at the interface chaotically with each other; from the regular standing waves of harmonic longitudinal capillary waves to their chaotic interaction caused by unharmonic behaviour under very strong driving forces [9-11].

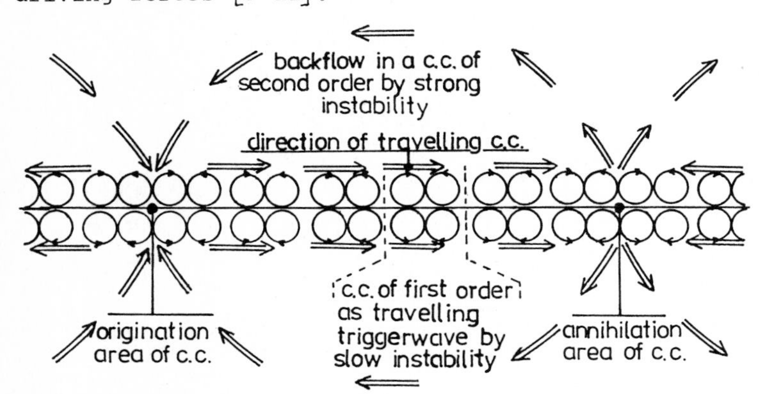

Fig. 1 Convection cells c.c. of first and second order in Marangoni-instability

Avoiding the sources of irregularities, i.e. using a quiet system with lower driving forces of heat or mass transfer through the interface, the so-called "stationary" regime (with the above mentioned rather regular and theoretically predictable roll cells of the first order [5-7,11-14] and with the less understandable roll cells of higher order (Fig.2) and their transition to relaxation oscillation) and the oscillatory regime (longitudinal capillary waves) [9-11] dominate.

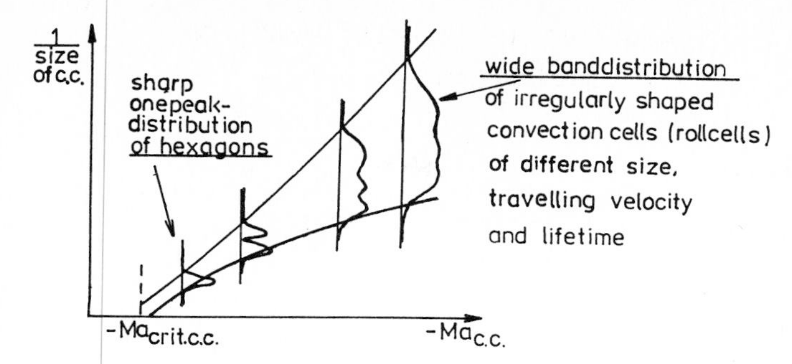

Fig. 2 Size distribution of c.c. in the "stationary" regime of MI and transition to spatial and temporal chaos

Comparing the patterns of the "stationary" regime of MI with those of chemical dissipative structures of reactions like the Belousov-Zhabotinsky-reaction (BZR) in quiet thin layers (see Fig.3) [15,16], we find topological similarities under conditions with very low (supercritical) driving forces immediately after the nucleation: in both systems there exist ripples, which travel as concentric circles from a pointlike leading centre or an elongated leading centre. In MI, the ripples are the image of four coupled traveling convection cylinders, in the cross-section with streamlines according to Fig.2; in BZR the "ripples" are traveling concentration waves. In both cases, the diffusion cooperates with nonlinear convective transport and nonlinear chemical reaction. Both kinds of waves behave

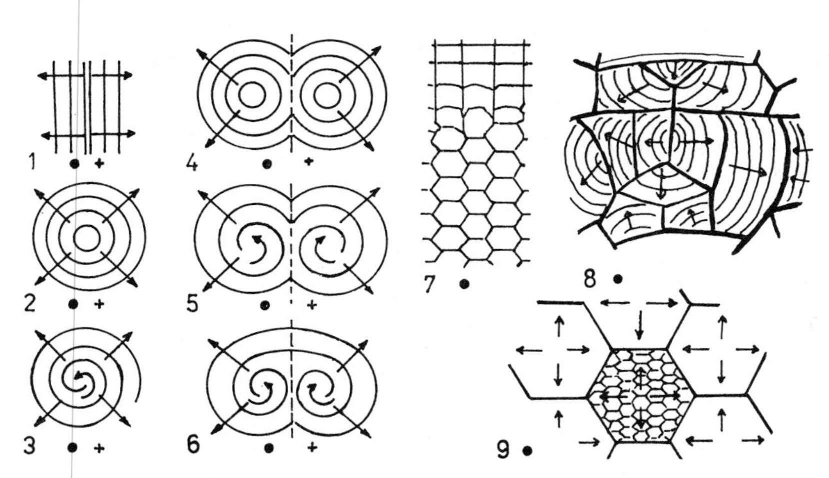

Fig. 3 Topological map of realized dissipative structures of the "stationary" regime of MI· and of BZR +
Traveling trigger waves from
1. elongated, 2. point-like, 3. spiral leading centre
Confrontation area between
4. circular, 5. independent spiral, 6. coherent spiral trigger waves
7. Transition from convection cylinders via rectangular roll cells to hexagonal roll cells in MI.
Formation of structures of second (or higher) order
8. with convection cylinders in MI
9. with polygonal roll cells in MI

as trigger waves and it is common for these, that there exist no interference and reflection (in strong contrast to the behaviour of the longitudinal capillary waves of MI): the trigger waves disappear at a wall or annihilate each other in the confrontation areas of waves of different leading centres.

The question is: what can we learn from the similarities in topology and kinetics of both instabilities for the further investigation in this field? (Fig.3). Firstly, the relaxation oscillation in MI (i.e. periodic new amplification of convection cells), the traveling of trigger waves in both instabilities and especially the spiral leading centres in the BZR are strictly concerned with the behaviour of refracterity. Consequently we looked for spiral leading centres in MI.

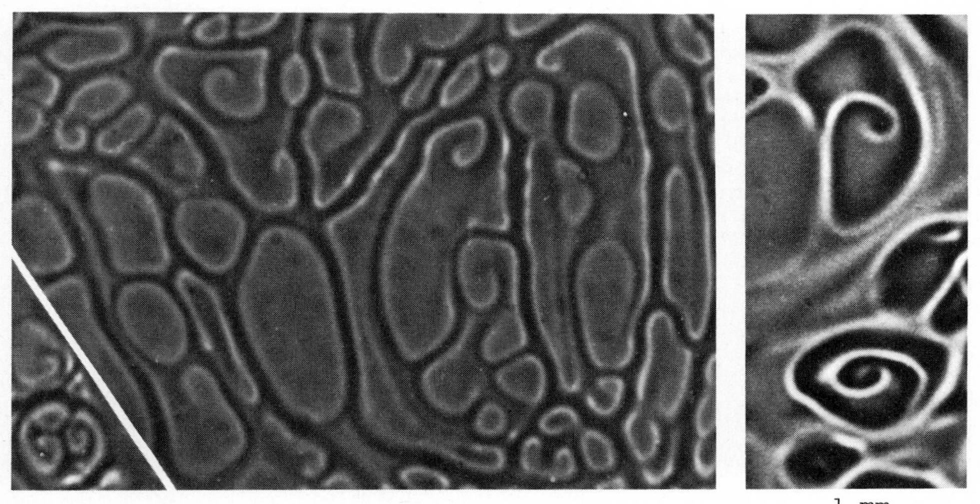

Fig. 4 Spiral trigger waves in Marangoni-instability 1 mm

Spiral waves can be observed in MI under conditions of small instability in the early steps of nucleation of convection cells. They seem to be very unstable: 180° to 360°-spirals and double-spirals are rare but the first steps of spiralization are more frequent (Fig.4). They are not of long lifetime and are annihilated by expanding of traveling roll cells resp. roll cell systems. We found them recently by a systematic proof of the complete series of pictures, which gave the basis for the publication of 1963 with E.Schwarz [7]. However after our cognizance of this pattern we are able to recognize them in Fig.5 and 6 of paper [7] too. In this time we were not encouraged to such prognostic deductions.

Secondly, there arises the question, whether the confrontation areas of a lot of concentrically circular or spiral trigger wave-systems can compose stable structures of higher order with characteristic shapes and wavelengths (average diameters) like in MI, resp. whether the structures of higher order in MI can be the result of a transition to another hydrodynamic regime which is not due to the trigger wave only. This and the related three-dimensional task too is concerned with the problem of the existence of real stationary structures in BZR.

Further it can be looked for chemical waves of higher modes and for a periodic breakdown and new amplification resp. for a periodicity in the activity of circular or spiral trigger wave systems.

For the kinetics of the structure elements of MI, i.e. the traveling kinetic of roll cylinders and roll cells, their refracterity and their trigger wave behaviour under conditions of low instability, we can expect some theoretical understanding in adapting the theory of chemical waves.

References

1. I.B.Levis. Chem.Engng.Sci. 3, 248-278 (1954)
2. K.Sigwart, H.Nassenstein. Naturwissenschaften 42, 458 (1955)
3. R.Krishnamurti. Fluid.Mech. 42, 295-307 (1978)
4. J.A.Yorke, E.D.Yorke. Chaotic behaviour and fluid dynamics in: Hydrodynamic instabilities and the transition to turbulence. Springer-Verlag (1981)
5. H.Linde, E.Schwarz. Mber.Dt.Akad.Wiss. 3, 554, 563, 569 (1961)
6. A.Orell, J.W.Westwater. AIChE J. 8, Nr.3, 350 (1962)
7. H.Linde, E.Schwarz. Z.Phys.Chem. 224, 331-352 (1963)
8. H.Linde. Marangoni-Instabilities in: "Convective transport and instability phenomena", ed. J.Zierep and H.Oertel jr. G.Braun-Verlag Karlsruhe 1982
9. H.Linde, K.Loeschcke. Chem.Ing.Techn. 39, 65-74 (1966)
10. H.Linde, M.Kunkel. Wärme-Stoffübertragung 2 60-64 (1969)
11. K.Loeschcke, E.Schwarz, H.Linde. Abhandlung Dt. Akad. der Wiss. Berlin, 6b, 720-733 (1966)
12. L.E.Scriven, C.V.Sternling. J.Fluid Mech. 19, 321-340 (1964), AIChE J. 5 514-523 (1959)
13. H.Linde, P.Schwartz. Chem.Tech. 26, 455-456 (1974)
14. H.Linde, J.Reichenbach. J.Coll.Interf.Sci. 84, 433 (1981)
15. A.M.Zhabotinsky, A.N.Zaikin. J.Theor.Biol. 40, 45-61 (1973)
16. V.I.Krinsky, K.I.Agladze. Doklady Akad. Nauk SSSR 263 (1982) 335

A Generalized Mechanism for Bromate-Driven Oscillators

L. Kuhnert

Zentralinstitut für Physikalische Chemie der Akademie der Wissenschaften der DDR,
Rudower Chaussee 5, DDR-1199 Berlin, German Democratic Republic

This class of chemical oscillators was first discovered by Belousov |1| in the reaction of citric acid with acidic bromate catalyzed by cerium salts. Zhabotinskii [2] extended this system to other organic substrates and catalysts (so-called classical BZ-reaction). In this system also other unusual nonlinear chemical phenomena like bistability and propagating waves were discovered stimulating a great deal of theoretical work. Therefore, the understanding of the general nonlinear mechanisms of bromate-driven oscillators is of general interest for recent progress in nonlinear kinetics and chemical dissipative structures. Besides the classical BZR, some new chemical compositions of bromate oscillators have been discovered: oscillations without catalysts [3]; oscillations with mixed substrates [4]; heterogeneous oscillators with organic [5] and inorganic [6] substrates (in that case, elementary bromine is scrubbed from the reaction solution by an inert gas-stream, by liquid-liquid extraction or by reaction with a solid [7]); oscillations in the presence of silver ions that buffer bromide concentrations on a very low level. In [8,9] attempts were made to classify bromate oscillators on the above described chemical phenomenology. The aim of the present paper is to propose a scheme from the kinetic standpoint.

The behaviour of all classes of bromate oscillators can be described by involving nine processes designated (R1)-(R9)

$$2H^+ + Br^{iii} + Br^V + (kRed + lRBr) \rightarrow 2Br^{III} + mBr^- + nO_x + P \qquad (R1)$$

$$pO_x + RBr \rightarrow pRed + gBr^- + P \qquad (R2)$$

$$2Br^{iii} \rightarrow 2Br^V + Br^i \qquad (R3)$$

$$Br^{iii} + Br^i \rightarrow Br^V + Br^- + H^+ \qquad (R4)$$

$$Br^{iii} + Br^- + H^+ \rightarrow 2Br^i \qquad (R5)$$

$$Br^i + Br^- + H^+ \rightarrow Br_2 + H_2O \qquad (R6)$$

$$Br_2 + RH \rightarrow RBr + Br^- + H^+ \qquad (R7)$$

$$Br_2 \rightarrow 2Br^- \qquad (R8)$$

$$Ag^+ + Br^- \rightarrow AgBr \qquad (R9)$$

In these processes Roman numerals refer to oxidation states of bromine such as HOBr, $HBrO_2$ and BrO_3^- for I, III and V, respectively.

The overall reaction may be characterized as follows:

1. The autocatalytic oxidation will be presented by R1.

The mechanism of R1 for autocatalytic oxidation of metal ions is well established and the kinetic constant is known [10]. In metal ion catalyzed oscillations, Red is the reduced form of the catalyst and O_x the oxidized, in which case l and m are zero, and k and n equal two. In uncatalyzed systems, R2 can be neglected and Red is the substrate itself (an aromatic compound) which will be first irreversibly oxidized to O_x and in the later stages of the reaction the first oxidation product can be further oxidized and, in addition, the stoichiometric coefficient m may be changed. P is a complicated mixture of oxidation products that no longer contributes to the reaction (may be, CO_2 or simple aliphatic compounds).

The autocatalytic reaction will be terminated and inhibited by R3-R5. The kinetic constants of these processes, besides R3, are known only approximately. Reactions R2 and R6-R9 control the contractions of the inhibiting species Br^- and HOBr. In uncatalyzed systems R1 contributes to these processes. Kinetic constant of R6 is well-known, R6 is a very fast reaction and may buffer the concentrations of Br^- and Br^i. In the presence of silver ions concentrations of Br^- will be controlled and buffered by R9.

For any given reaction mixture the autocatalytic mechanism will be the same but the mechanism of termination and inhibition of autocatalysis depends on concentrations of inhibitors and these may be controlled by the organic subset and the experimental constraints.

While the mathematical analysis of the model is still in progress, there are the following preliminary results:

- we need a critical BrO_3^- for onset of oscillations;
- at critical concentrations of Br^- or HOBr the autocatalytic oxidation will be switched off or on;
- Br_{crit} is proportional to BrO_3^-;
- $HOBr_{crit}$ is proportional to BrO_3;
- the mechanism of control may be changed depending on concentrations of inhibitors or other constraints.

That means there are three possibilities:

1. Bromide controlled oscillations,
2. HOBr-controlled oscillations,
3. Br^- ions are membered in the inhibiting process.

There are several suggestions that the bromate driven oscillations cannot be explained by a uniform mechanistic scheme in the full parameter space [11]. In a narrow range of parameters only one of the above described three possibilities may be realized.

Below we will give an experimental example for change of control mechanism in the course of the uncatalyzed bromate driven oscillating reaction with chromotropic acid (CA).

The reaction was followed by a bromide sensitive electrode in a stirred tank reactor. It is well-known that bromide concentrations can be measured by a bromide sensitive electrode only when above 10^{-6}M. According to [12], we assume that the response of the electrode in the range of Br^- 10^{-6}M depends on variation in HOBr. In Fig.1 we can distinguish between three types of oscillations. At the beginning, oscillations are on a high level of Br^- ($5\cdot10^{-5}$ - 10^{-6}M). This is followed by oscillations where Br^- is between 10^{-5}M and 10^{-6}M, and in the third stage oscillations are on a level of Br^- 10^{-6}M. To our opinion, this reaction is an example for the changing of control mechanism. At the beginning we observe the bromide-controlled oscilla-

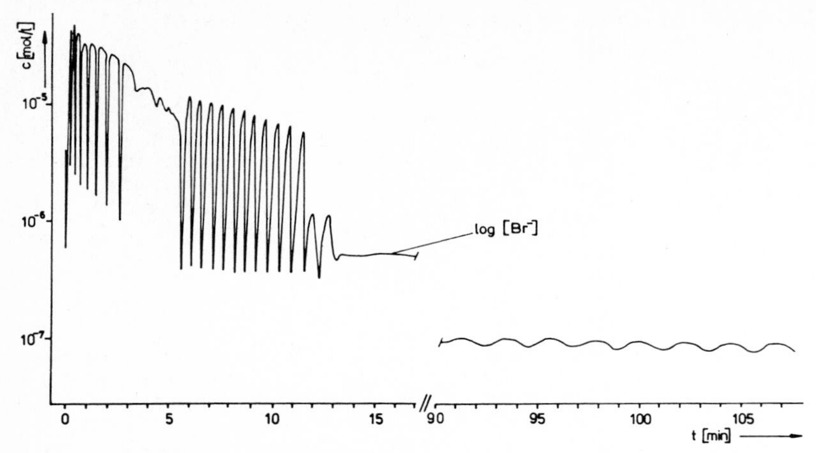

Fig. Potential of a bromide-sensitive electrode during oscillating reaction of chromotropic acid (CA). Concentrations: CA - 8.75×10^{-2}M; BrO_3 - 6.75×10^{-1}M; H_2SO_4 - 1 M, 17°C

tions, then the mechanism where Br^- and HOBr contribute to the inhibiting processes and in the third stage the HOBr-controlled oscillations, By acidic bromate, CA will be oxidized and brominated in subsequently irreversible steps. That means the organic substrate will be changed in the course of reaction. In other words in Rl, kRed and lRBr will be changed and the proportion of bromide production in autocatalysis will be varied.

References

1. B.P.Belousov. Rev. Radiat. Med., 195, 145 (1959)
2. A.M.Zhabotinskii. Concentrational oscillations, Nauka, Moscov, (1974)
3. L.Kuhnert, H.Linde. Z. Chem. 17, 19 (1977)
4. R.P.Rastogi, H.J.Singh, A.K.Singh."Kinetic of Physicochemical Oscillations, Aachen 1979" preprints
5. Z.Noszticzius, J.Bodiss. J. Am. Chem. Soc. 101, 3177 (1979)
6. L.Adamcikova, P.Sevcik. J. Chem. Kinet. 14, 735 (1982)
7. L.Kuhnert. React, Kinetic. Catal. Lett. (submitted)
8. R.M.Noyes. J. Am. Chem. Soc. 102, 4644 (1980)
9. M.Orban, C.Dateo, P. De Kepper, I.R.Epstein. J. Am. Chem. Soc. 104, 5911 (1982)
10. A.B.Rovinsky, A.M.Zhabotinskii. Theor. Exp. Chem. 15, 25 (1979)
11. A.M.Zhabotinskii.Acta Chim. Acad. Sci. Hung. 110, 283 (1982)
12. Z.Noszticzius. Acta Chim. Acad. Sci. Hung. 106, 347 (1981)

Electrical Field Effects on Propagating Pulse and Front Waves

M. Marek and H. Ševčíková

Dept. of Chemical Engineering, Prague Institute of Chemical Technology
166 28 Prague 6, Czechoslovakia

Concentration pulses and fronts propagate in reacting media due to the interaction of autocatalytic reaction steps with diffusion and/or ionic migration. Travelling waves are observed in excitable tissues [1] and in morphogenetic fields [2]. Existing spatial and/or temporal electric field gradients affect the transport of ionic components. We shall summarize here the results of our experimental studies of the effects of a homogeneous electric field on the concentration pulse and front waves.

We constructed a thermostated tubular reactor [3] (100 mm long glass capillary of inner diameter 0.6 mm), placed between two electrolytic cells separated from the reactor by membranes. The propagation of waves generated spontaneously at both ends of the capillary were followed optically. The waves were exposed to the electric field only temporarily; first the wave velocity without the field was measured; then the field was switched on and the wave velocity in the field was determined.

Reaction medium of the Belousov-Zhabotinski (B-Z) type [4] was used in experiments on the pulse wave propagation [3] and iodate oxidation of arsenous acid [5] was used in the study of front waves [6]. The components of the autocatalytic reaction sequences responsible for the wave propagation are Br^- ions and $HBrO_2$ for pulse waves and I^- ions and I_2 for front waves. The dependences of the wave velocities on the imposed electric gradient are shown in Fig.1. The dependence is nonlinear both for the pulse and front waves. The wave velocity in the electric field increases approximately five times for pulse waves and fifteen times for front waves at the potential gradient $E=-40$ V/cm, when the waves travel to the positive electrode. The waves are slowed down when propagating to the negative electrode ($E>0$).

The wave velocities depend on the decrease or increase of the transport rates (ionic migration) of the key ionic components into the reaction medium in front of the wave; changed concentrations of these ionic components then affect the initiation of the autocatalytic reaction step, the form of the wave profiles and their propagation velocities [3,6].

A number of interesting effects were observed in positive electric fields of higher intensities. The propagating fronts can be slowed down by the positive field and they are stopped at values of $E \in (5; 8$ V/cm), Fig.2a. The stopping of the front wave is connected with the change of the front concentration profile; the increased migration causes such changes in the concentrations of migrating components that the reaction proceeds into a different stationary state. When the field is switched off, then the front wave starts to propagate again but leaves behind a stationary zone with different

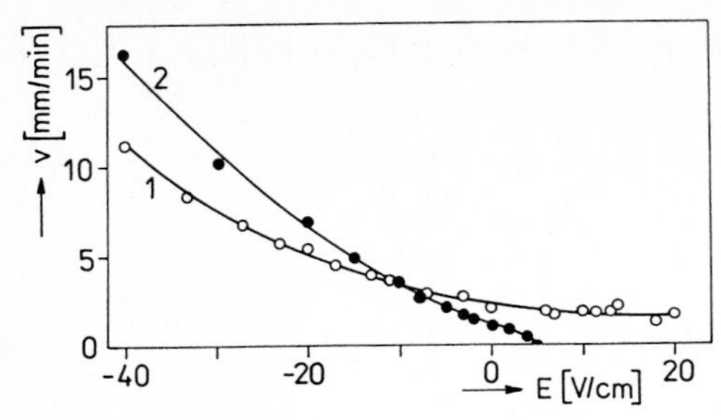

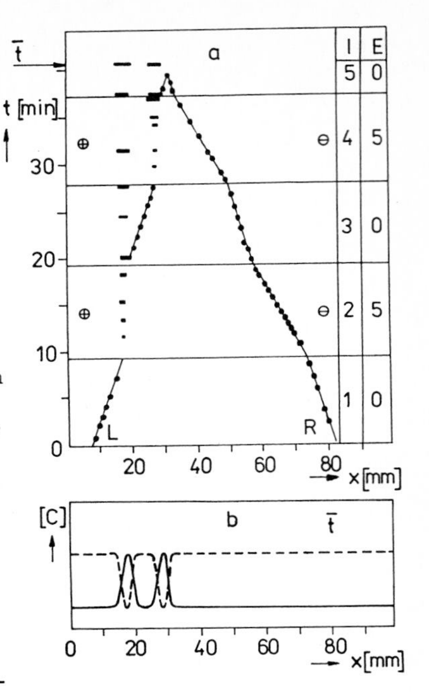

Fig. 1 Dependence of the wave velocity on the imposed electric field at temperature 288.2 K. 1 - pulse wave in the B-Z reaction medium (initial composition: 0.05 M malonic acid, 0.205 M bromic acid, 0.007 M KBr, 0.004 M ferroin); 2 - front wave in the iodate-arsenous acid reaction mixture (initial composition: 0.0155 M arsenous acid, 0.005 M KIO_3, 0.0056 M H^+ and a starch indicator)

Fig. 2 a) Time dependence of the position of the front wave going from the right (R) in the electric field of the intensity E = -5 V/cm; the stopping of the wave going from the left (L) and formation of the stationary structure in the electric field of the intensity E = 5 V/cm; b) Schematic concentration profiles of I^- ions (---) and I_2 (——) in the final inhomogeneous stationary state of the system

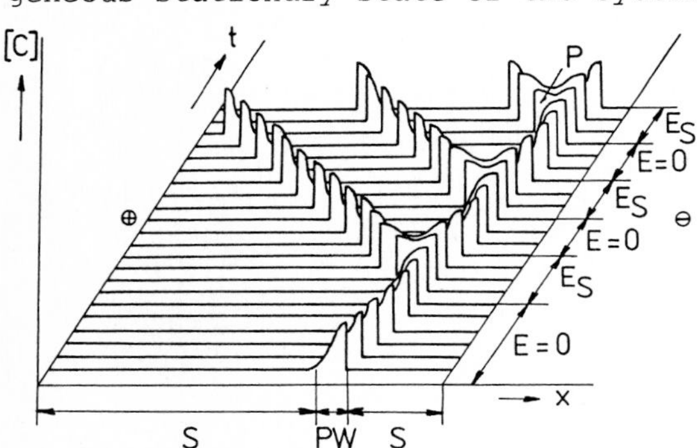

Fig. 3 Controlled generation of new pulse waves by the wave splitting in the B-Z reaction medium. Repeated use of the electric field of the intensity E_S=16.7 V/cm. S - stable stationary state of the system, P - pacemaker, PW - pulse wave

concentrations of reaction products. Simple dissipative structure (Turing structure) is thus formed, Fig.2b.

When the positive field intensities are higher (E>8 V/cm), then the separating effects of the electric field cause such decrease of concentration of I^- ions in the front that the reaction steps are slowed down under the critical limit and the front wave is annihilated. When the electric field is switched on again, the concentration of I ions increases, reactions proceed faster and the front is regenerated.

The pulse waves cannot be stopped; they are irreversibly annihilated at positive field intensities E>20 V/cm. To start another pulse wave, an external concentration perturbation is required. Wave splitting occurs at intermediate values of the field intensities E∈(10; 20 V/cm).

The wave first slows down, then changes its form and a new center of the wave generation (pacemaker) is formed. This pacemaker sends out pulse waves in the direction opposite to the original direction of the wave propagation. When the electric field is repeatedly switched on and off, the generation of new pulses by the wave splitting can be controlled, Fig.3.

A number of the above described phenomena can be simulated by a mathematical model considering diffusion, ionic migration in the applied field and reaction between two characteristic ionic components [7].

References

1. "Cellular Neurophysiology", eds. I. Cooke and M.Lipkin, Jr. (Holt, Rinehart and Winston, New York, 1972)
2. K.J.Whitaker and R.A.Steinhardt. Quart.Rev.of Biophys.15,593 (1982)
3. H.Ševčíková and M.Marek. Physica D (in press)
4. R.Feeney, S.Schmidt and P.Ortoleva. Physica 2D, 536 (1981)
5. A.Hanna, A.Saul and K.Schowalter. J.Am.Chem.Soc. 104, 3838 (1982)
6. H.Ševčíková and M.Marek. Physica D (in preparation)
7. H.Ševčíková, M.Kubicek and M.Marek. "Concentration waves-effects of an electric field", presented at Fourth ICMM, Zürich (1983)

Self-Organization Phenomena and Autowave Processes in Heterogeneous Chemical and Physical Systems

V.V. Barelko

Institute of Chemical Physics of the USSR Academy of Sciences
SU-142432 Chernogolovka Branch, USSR

The studies reviewed in this paper have been carried out at the Institute of Chemical Physics (Chernogolovka Branch) for the last decade. Various systems of essentially different physical nature were studied. These systems include heterogeneous catalytic reactions, boiling on solid heat-generating surfaces, heat transfer between electrically heated filaments and rarefied gas media, chemical solid-state conversions at low and superlow temperatures, and some examples of phase transitions. In spite of these systems being very different, the considered phenomena belong to the same phenomenological class, i.e. to the class of autowave self-organization processes. Another significant feature bringing them together is heterogeneity of processes and their localization on a solid surface.

In [1-11], heterogeneous catalytic systems are treated theoretically and experimentally, the attention being focused on the autowave transitions which occur in the catalytic zone in response to a local external perturbation. The catalytic oxidation of ammonia, hydrogen, and carbon oxide on a platinum wire has been experimentally studied. The specific behaviour of heterogeneous catalytic systems allows one to identify several types of travelling waves with different mechanisms. Thus, the mechanism of ammonium oxidation is characterized by a nonlinear temperature dependence of the rate of heat release and by the conductive heat transfer over the catalyst. CO oxidation has revealed a non-thermal autowave mechanism that was observed in the gas boundary layer at the catalyst surface when the introduced reactant worked as a decelerator due to the diffusion gas phase transfer along the catalytic surface. The fact that a nonlinear mechanism responsible for formation and multiplication of catalytic active centres (of the branched-chain type) has been identified in all these systems, favours to consider one more non-thermal class of catalytic travelling waves controlled by diffusion transfer of active centres just on the surface of a catalyst.

The dissipative structures of the "stopped" wave type, i.e. non-uniform steady states realized in the region of parameters where the travelling wave velocity reduces to zero, have been studied both theoretically and experimentally. The interaction of thermal and branched-chain mechanisms has been shown to enlarge greatly the region of existence of these structures.

The stability and the number of such non-uniform steady states have been shown to depend on the way in which the thermal performance of a catalyst is controlled.

The spatially inhomogeneous structures with anomalously extended and complicated profiles of the temperature fronts have been observed in experiments on CO oxidation (at small CO concentrations). When the same reaction (with a high CO concentration) runs in a diffusive regime, a spontaneous loss of stability of the homogeneous steady state was observed resulting in the production of a single hot highly active domain in a certain range of velocities of the reaction mixture flow.

In the reaction of ethylene oxidation, inhomogeneous steady-states were found to occur with non-damped pulsations of the boundaries separating the zones of high and low activities on the catalyst ("the catalytic heart").

To explain the latter three facts, the interaction of heat transfer and gas-diffusion mechanisms of autowave processes has been employed.

The existence conditions for multiple solutions in the form of a travelling wave have been established using the models describing some heterogeneous catalytic systems.

The studies of autowave processes and dissipative structures in catalysis have laid a theoretical basis for a new branch of the theory of chemical reactors which considers the stability of a reactor to local perturbations, the stability of the uniform modes of reactor performance, conditions for the reactor start-up, by the devices locally initiating the process in the catalytic zone.

In [12-16] the travelling and standing waves in the processes of boiling on the solid surfaces of heat elements have been studied. The "crisis" of boiling which occurs as spontaneous transitions between the stationary regimes of nucleate and film boiling is shown to have a distinct autowave character under practically any real condition. The autowave approach to the boiling "crisis" has made it possible to show that no uniform thermal regime exists in the transition between nucleate and film boiling. Dissipative structures are formed here such that nucleate and film boiling steadily coexist.The obtained theoretical and experimental data provide the basis for calculations of technological conditions of transient processes on heat generating elements and of their dynamic behaviour. Also, some new ways to increase the limiting thermal loads in steam generating apparatus can thereby searched for.

Refs. [17-19] are concerned with autowave processes and inhomogeneous steady structures on electrically heated filaments in rarefied light gases. Two physical mechanisms have been identified for these phenomena. One of them takes place in the iron-hydrogen system and is related to non-linearity in the temperature dependence of the electric resistance of the heating element. The second mechanism identified in the system "platinum wire-helium" is due to non-linearity of heat-exchange between the element and the gas medium. The autowave approach is regarded as a new tool to study sophisticated features of interaction between rarefied gas media and metal surfaces. The results obtained have allowed the development of a thermal theory of current stabilizers with filaments.

Refs. [20-22] describe new phenomena of autowave propagation of a chemical reaction over a frozen mixture of reactants at low and ultra-low temperatures. These papers deal with the autowave mechanism of initiation of solid-state chemical reactions. The hypothesis of a positive feedback between the rate of chemical conversion and mechanical destruction of a sample is used there. The physical essence of the hypothesis is that the reaction is heterogeneous with participation of fresh cracks formed in the destroyed sample,the cracks being in turn the direct result of the reaction. The simplest model for this process has been investigated. It is based on the idea of a thermal mechanism of destruction due to appearance of heat loads in a sample reacting exothermally layer by layer.

The model, unlike the traditional models used in the autowave theory, contains a nonlinear source depending not on a variable, but on its derivative, and presents the multiplicity of solutions in the form of travelling waves.

In conclusion let us emphasize that the application of autowave theory to the description of the decay dynamics of metastable phases in the theory of phase transitions, seems to be very promising. A certain phenomenological similarity between the transfer from a metastable to a stable phase state and the transfer between steady states in open systems has been noted in literature (similarity in the curves of the Van der Waals type and the N,Z-shaped characteristics of autowave systems). However, the modern kinetic theory of metastable transitions has been developed with practically no account for the spatial factor and the possibility of autowave transitions in the system has not been considered. This is illustrated by Ref. [14] which reports the autowave rearrangement of the regime during a spontaneous transition from the free convective heat transfer to boiling of a liquid on a solid heat-generating surface. In this work, a cavitational mechanism for layer-by-layer boiling of a liquid, overheated at the heater surface,has been proposed.

References

1. V.V.Barelko. Kinetika i Kataliz, 14, No 1, p.196 (1973)
2. V.V.Barelko, Yu.E.Volodin. Dokl. AN SSSR, 211, No 6, p.1373 (1973)
3. A.G.Merzhanov, V.V.Barelko, I.I.Kurochka, K.G.Shkadinsky.Dokl. AN SSSR, 221, No 5, p.1114 (1975)
4. V.V.Barelko, Yu.E.Volodin. Dokl. AN SSSR, 223, No 1, p.112 (1975)
5. V.V.Barelko, I.I.Kurochka, A.G.Merzhanov. Dokl. AN SSSR, 229, No 4, p.898 (1976)
6. K.G.Shkadinsky, V.V.Barelko, I.I.Kurochka. Dokl. AN SSSR, 233, No 4, p.639 (1977)
7. V.V.Barelko, I.I.Kurochka, A.G.Merzhanov, V.G.Shkadinsky. Chemical Engineering Science, Vol.33, p.805 (1978)
8. S.A.Zhukov, V.V.Barelko. Dokl. AN SSSR, 238, No 1, p.135 (1978)
9. V.V.Barelko. Yavleniya Begushchikh Voln v Reaktsiyakh Glubokogo okisleniya na Platine. V sb. "Problemy Kinetiki i Kataliza", vol.18, Moscow, "Nauka", (1981) pp.61-80
10. V.V.Barelko, S.A.Zhukov.Khimicheskaya Fizika, No 4, p.516 (1982)
11. Yu.E.Volodin, V.V.Barelko, A.G.Merzhanov. Khimicheskaya Fizika, No 5, p.670 (1982)
12. S.A.Zhukov, V.V.Barelko, A.G.Merzhanov. Dokl. AN SSSR, 242, No 5, p.1064 (1978)
13. S.A.Zhukov, V.V.Barelko, A.G.Merzhanov. Dokl. AN SSSR, 245, No 11, p.94 (1979)
14. S.A.Zhukov, V.V.Barelko, A.G.Merzhanov. Int.J.Heat Mass Transfer, Vol.24, p.47 (1980)
15. S.A.Zhukov, L.F.Bokova, V.V.Barelko. Int. J. Heat Mass Transfer, Vol.26, p.269 (1983)
16. S.A.Zhukov, V.V.Barelko. Int.J.Heat Mass Transfer, Vol.26, p.1121 (1983)
17. V.V.Barelko, V.M.Beybutina, Yu.E.Volodin, Ya.B.Zeldovich. Dokl. AN SSSR, 257, No 2, p.339 (1981)
18. Yu.E.Volodin, V.M.Beybutina, V.V.Barelko, A.G.Merzhanov. Dokl. AN SSSR, 264, No 3, p.604 (1982)
19. V.V.Barelko, V.M.Beybutina, Yu.E.Volodin, Ya.B.Zeldovich. Thermal Waves and Non-Uniform Steady States in a $Fe+H_2$ System. Preprint, Chernogolovka (1980)

20. A.M.Zanin, D.P.Kiryukhin, V.V.Barelko, I.M.Barkalov, V.I.Goldansky. Dokl. AN SSSR, 260, No 6, p.1397 (1981); Dokl. AN SSSR 261, No 6, p.1367 (1981); Khimicheskaya Fizika, 1, No 2, p.265 (1982)
21. V.V.Barelko, I.M.Barkalov, D.A.Vaganov, A.M.Zanin, D.P.Kiryukhin, Dokl. AN SSSR, 264, No 1, p.99 (1982)
22. I.M.Barkalov, V.V.Barelko, V.I.Goldansky, D.P.Kiryukhin, A.M.Zanin. Porogovy Yavleniya i Avtovolnovye Protsessy v Nizkotemperaturnykh Tverdofaznykh Khimicheskikh Reaktsiyakh. Preprint, Chernogolovka (1983)

Threshold Effects and Autowave Processes in Low-Temperature Solid-State Chemical Reactions

I.M. Barkalov

Institute of Chemical Physics, Academy of Sciences of the USSR
SU-142432 Chernogolovka Branch, USSR

It is known that during elastic deformation of a solid the accumulated potential energy concentrates near the surfaces formed by a fracture of the solid [1]. This energy transformation is accompanied by considerable changes in the physical and chemical properties of a substance at the surface [2]. This facilitates chemical conversions in a solid at low temperatures when the molecular mobility is absent. These simple considerations were the starting point for the present investigation.

Experiments were done with the systems which are chemically active only in the liquid state. The following reactions were studied:

1. Chlorination of saturated hydrocarbons [3,4].
2. Hydrobromination of olefines [5-7].
3. Polymerization.

Stabilized free radicals were accumulated in these systems during low-temperature radiolysis. These radicals give rise to a chain of chemical conversions, provided that the required reaction conditions are satisfied. In the absence of such conditions no reaction was observed either during γ-irradiation or after it.

Samples were subjected to brittle rupture by thermoelastic stresses, viz. the rate of temperature change was chosen so high as to provide temperature gradients causing the sample fracture.

In the course of heating or cooling, the temperature of the outer surface of a sample and integral thermal effects were measured (for details of calorimetric technique see [8]).

Test experiments with non-irradiated samples of vitrified $MCH+Cl_2$ [9] have shown that slow heating of a sample from 4.2 to 77°K causes no cracking and no thermal effects (see Fig.1-I and 1-II, solid lines). On the contrary, fast heating results in rupture of the samples. The thermal effect of this process was recorded by the use of calorimeter (Fig.1-III and 1-IV, solid lines). Formation of cracks in the samples was visually and acoustically detected.

When the sample containing stabilized radicals (^{60}Co γ-irradiation) was heated slowly without ruptures, the reaction was not observed (Fig.1-I and 1-II, dashed lines). If brittle rupture occurred during heating, the reaction took place (Fig.1-III and 1-IV, dashed lines).

Brittle rupture of a sample was also possible during fast cooling. Fig.1 (V and VI) presents the data for a sample which was cooled rapidly enough to produce rupture. As would be expected, the failure of the irradiated sample was accompanied by the initiation of a chemical conversion, which was not observed in the non-irradiated samples.

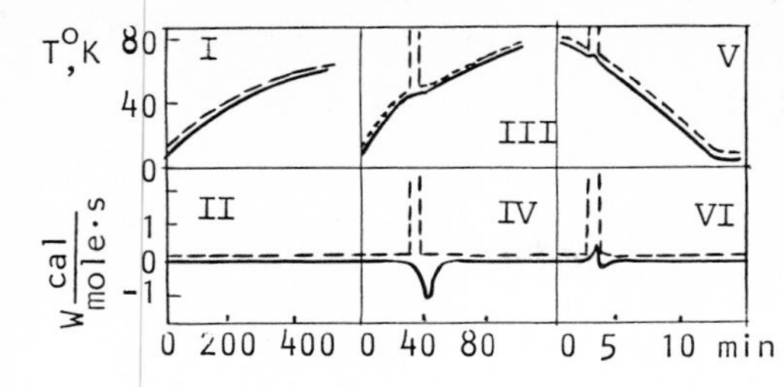

Fig. 1 Time dependence of the sample temperature (Cl_2 + Methylcyclohexane, molar ratio 1:3) - I,III, V and thermal effects - II,IV,VI. Solid lines - non-irradiated samples; dashed lines - samples irradiated by γ-rays of ^{60}Co at 77°K, dose 2.7 Mrad

To verify the decisive role of ruptures in the initiation of the chemical conversion, a direct experiment with mechanical breakage of a sample was performed at a fixed temperature. The sample containing stabilized radicals was broken by turning a frozen-in thin metalic rod at 4.2°K. At the moment of breakage the chemical reaction was switched on and spread over the sample. Such an out-burst of chemical reactions was observed experimentally both in vitreous [9,10] and polycrystalline [11] systems.

It may be concluded that the failure of a sample plays the role of a trigger which switches on a certain positive feed-back mechanism. To ascertain the nature of this mechanism the dynamics of reaction outburst was investigated [10].

Some features of this dynamics evidence against the classical thermal feed-back mechanism being dominant in the phenomenon. Since mechanical breakage plays the decisive role in chemical conversion initiation, the following self-activation mechanism can be proposed. The chemical reaction, which arises as a response to the formation of a new fracture surface, stimulates further development of breakage due to temperature and density gradients and corresponding stresses.

It is important that systems described by this nonlinear model are characterized by autowave processes in samples of sufficient extent. These processes are formally similar to chain flames [12] and waves of catalytic activity [13].

In earlier experiments [14,15] the reaction waves in all the systems involved were initiated by local disturbances (mechanical breakage or thermal fracturing pulse). Since the reaction is accompanied by changes in the sample colour, the propagation of the reaction front can be recorded by cinema technique. The flat front was detected to be rapidly formed and to propagate parallel to itself from layer to layer, the samples being immersed into liquid nitrogen or helium. The reaction front moves with practically constant speed of about 1-2 cm/sec.

To investigate the structure of the reaction wave front, a number of experiments with thermography of autowave process have been carried out. The wave propagation velocity was calculated from the time the wave travels the distance of 3-5 cm between two thermocouples. Fig.2 shows typical profiles of the traveling temperature wave front versus time and X-coordinate at initial temperatures of 77°K and 4.2°K.

Some structural features of the reaction wave front are not typical for classical thermal self-propagation. Those are: 1) the stage of inert pre-outburst heating is absent at all; 2) jump-like switch-on of the reaction (discontinuities in thermograms); 3) the reaction is switched on at temperatures far below that of thermoactivated reaction outset; 4) small change of the wave propagation velocity in passing from 77°K to 4.2°K.

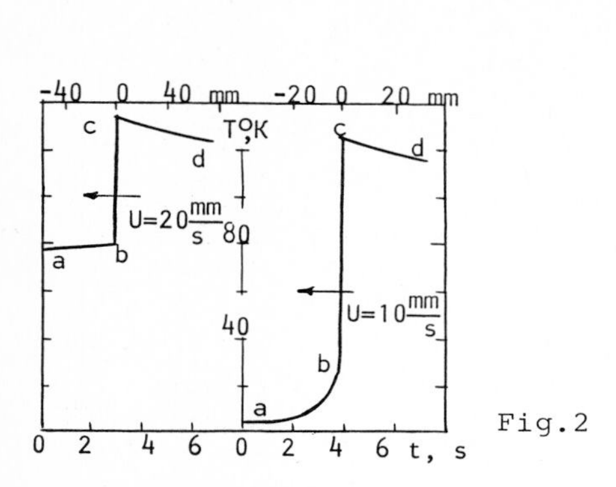

Fig.2

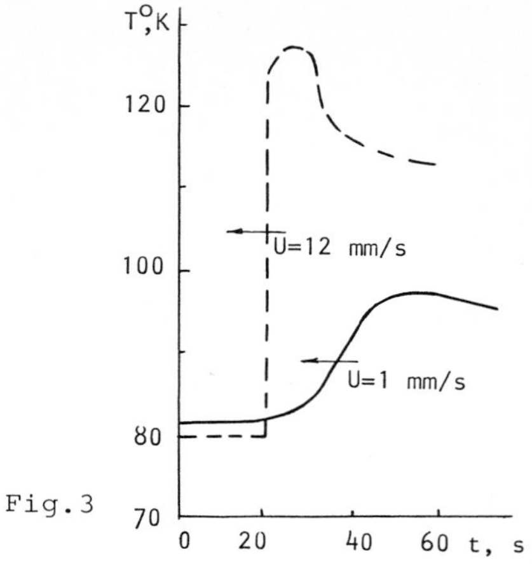

Fig.3

Fig. 2 Temporal and spatial sweeping of typical temperature waves in the reaction $Cl_2 + C_4H_9Cl$ at 77°K and 4.2°K. Dose 2.7 Mrad

Fig. 3 Temporal sweeping of temperature profiles of $Cl_2 + C_2H_4Cl$ reaction waves. The system was placed in the vapours of liquid nitrogen. Initiation of the reaction by slow (solid line) and fast (dashed line) heating (discharge of the capacitor via heater)

Also not typical is the dependence on the sample diameter. The autowave process cannot be suppressed by decreasing the diameter from 10 to 0.5 mm. The reaction wave velocity remains practically constant in this case. This evidences that mechanisms other than the classical one are involved.

The autowave process of copolymerization was observed in the system 2-methylbutadiene-1.3 + SO_2, immersed into liquid helium. In this case almost all of the sample is converted into copolymer due to propagation of a copolymerization wave [16] . The transition to the viscous overcooled liquid at $T>T_g$ does not result in the decay of macroradicals [4].

We think that the rupture, which initiates the reaction, occurs when the temperature gradient reaches a critical value (dt/dx)*. This gradient produces a stress equal to the ultimate strength of the material. The reaction proceeds over a time interval τ, during which the reaction-generated fracture surfaces retain their activity [17]. There are two types of stationary conditions of the reaction, differing in velocity, structure and properties of the wave.

The "slow" reaction wave is similar to the usual combustion wave, viz. the reaction zone is small compared to the heating zone, the temperature in the former being close to the maximum. The "fast" reaction wave has entirely different nature. The temperature at the point of reaction switch-on is practically equal to the initial temperature. In this case the heat transfer only ensures continuous distributions of temperature and temperature gradient. Thus, it is quite natural that the wave velocity is no longer dependent on heat transfer parameters.

In special experiments, when the sample was put into the vapours of boiling liquid nitrogen, we observed both fast waves initiated by a pulse heater or local breakage, and slow waves initiated by slow local heating (see Fig.3). The velocity of the slow wave was approximately by one order of magnitude smaller than that of the fast wave.

It has been ascertained [18] that the chemical reaction propagates pulsewise both in time and space. As can be seen in a microcinegram of the autowave process, the traveling wave leaves a trace in the form of grooves parallel to the wave front and located 50-150 m apart. Such a striped trace verifies that the wave front moves in a series of short jumps.

The fashion of the grooves pointes to the presence of some curvature of the reaction front which looks flat in macrocinegrams [14,15]. This is likely to be caused either by inhomogeneity of mechanical properties of the sample, of its strength in particular, or by the loss of the flat front stability and by the formation of spatial homogeneity of the type of a dissipative structure [19].

It is interesting to note that the striped "imprints" of chemical conversion, observed in our experiments, resemble very much the patterns produced in fatigue failure of metals [21] and polymers [22]. There is also a deeper analogy,viz. the pulsatory character of failures in chemical reactions is due to cyclicity of loading, just as in the case of fatigue failures.

References

1. V.R.Regel, A.J.Slutsker, E.E.Tomashevskii. Kinetic nature of strength of solids, Moscow, Nauka, 1974 (in Russian)
2. N.K.Baramboim. Mechanochemistry of high-molecular compounds. Moscow, Khimiya, 1978 (in Russian)
3. D.P.Kiryukhin, I.M.Barkalov, V.I.Goldanskii. Khim. Vys. Ener., 11, No 6, 438 (1977)
4. I.M.Barkalov. Uspekhi Khimii, 49, No 2, 364 (1980)
5. D.P.Kiryukhin, I.M.Barkalov, V.I.Goldanskii. Doklady Akad.Nauk SSSR, 238, 388 (1978)
6. D.P.Kiryukhin, I.M.Barkalov, V.I.Goldanskii. J.Ch.Phys., 76, 1013 (1979)
7. I.M.Barkalov, D.P.Kiryukhin, V.I.Goldanskii, A.M.Zanin. Chem. Phys. Lett., 73, No 2, 273 (1980)
8. I.M.Barkalov, D.P.Kiryukhin. Vysokom. Soedin., 22A, No 4,723 (1980)
9. A.M.Zanin, D.P.Kiryukhin, I.M.Barkalov, V.I.Goldanskii, ZhETF Pis'ma, 33, No 6, 320 (1981)
10. A.M.Zanin, D.P.Kiryukhin, V.V.Barelko, I.M.Barkalov, V.I.Goldanskii. Doklady Akad.Nauk SSSR, 260, No 6, 1397 (1981)
11. A.M.Zanin, D.P.Kiryukhin, I.M.Barkalov, V.I.Goldanskii. Doklady Akad.Nauk SSSR, 260, No 5, 1171 (1981)
12. N.N.Semenov. On some problems of chemical kinetics and reactivity, Izd. AN SSSR, 1954
13. V.V.Barelko, Yu.E.Volodin. Kinetika i kataliz, 17, 683 (1976)
14. A.M.Zanin, D.P.Kiryukhin, V.V.Barelko, I.M.Barkalov, V.I.Goldanskii. Doklady Akad.Nauk SSSR, 261, No 6, 1367 (1981)
15. A.M.Zanin, D.P.Kiryukhin, V.V.Barelko, I.M.Barkalov, V.I.Goldanskii. Khimicheskaya fizika, 1, No 2, 265 (1982)
16. A.M.Zanin, D.P.Kiryukhin, I.M.Barkalov, V.I.Goldanskii. Vysokom. Soedin., 24B, No 4, 243 (1982)
17. V.V.Barelko, I.M.Barkalov, D.A.Vaganov, A.M.Zanin, D.P.Kiryukhin. Doklady Akad.Nauk SSSR, 264, No 1, 99 (1982)
18. A.M.Zanin, D.P.Kiryukhin, V.V.Barelko, I.M.Barkalov, V.I.Goldanskii. Doklady Akad.Nauk SSSR, 260, No 6 (1981).
19. N.Glansdorf, I.Prigogine. Thermodynamic theory of structure, stability and fluctuations, J.Wiley, London, (1971)
20. A.A.Griffith. Phil. Trans. Roy. Soc., 221, 163 (1921)
21. C.Lair, G.C.Smith. Philosoph. Mag., 7, No 77, 847 (1962)
22. W.Doll. Journ. Materials Sci., 10, 935 (1975)

Spatial Structures and Pattern Selection in Chemical Systems

D. Walgraef

Service de Chimie Physique II Universite Libre de Bruxelles, Campus Plaine, CP 231
B-1050 Bruxelles, Belgium

Spatial structures are actually observed in an evergrowing class of chemically active media and their possible relation to numerous important biological phenomena is often emphasized [1,2] . Hence the problem of their origin and selection is of particular importance and we review here the main results of the theory of spatial chemical patterns based on a stochastic reaction-diffusion description of nonlinear chemical networks.

The problem of the origin of chemical waves in thin layers of oscillating systems is first discussed. These particular spatial patterns which are often presented as a manifestation of the self-organization of nonlinear systems correspond on the contrary to a spontaneous desynchronization of two-dimensional oscillators. Moreover the spiral waves which are associated with singularities of the phase of the oscillations play here the role of the vortices in 2D superfluids or XY models. These topological defects are associated with the impossibility for the system to develop true long range order (corresponding here to the perfect synchronisation of the oscillations over the whole reactor and which can effectively be maintained only by stirring mechanisms) and are the consequence of the spontaneously broken rotational symmetry (here the phase symmetry) [3].

The stochastic analysis of chemical oscillations arising through a Hopf bifurcation leads to a universal behaviour of the phase fluctuations and of their probability distribution. The wave phenomenon is deduced from this analysis and may be summarised as follows [4]: when the stirring, which maintains the homogeneity of a layer of solution where chemical oscillations are present, is interrupted, isolated centers first appear at random in time and space. They radiate target patterns of various wavelengths and their distribution is a functional of the parameters of the oscillations. When time increases further their average number increases but with a decreasing creation rate. These effects are manifestations of the intrinsic character of these waves. In this regime spiral waves would in general not be observed unless initiated. It is only in the asymptotic regime that clusters of spirals of zero total vorticity could spontaneously appear. The presence of these spatiotemporal patterns may be associated with the absence of true long range order in the system and are a manifestation of the partial desynchronization of the oscillations. The complete desynchronization occurs when isolated spiral waves are likely to develop and this only happens when the stiffness constant of the phase fluctuations is sufficiently small (small amplitude of the oscillations or highly anisotropic limit cycle, or low diffusion rate of the species,...)

---

* Chercheur Qualifie au Fonds National de la Recherche Scientifique de Belgique.

Another form of spatial structure which may spontaneously appear in an initially homogeneous chemical system consists of mosaic-like patterns. These structures are essentially stationary and correspond to spatial variations of the concentration of the chemicals with filamental, layered or polygonal shapes. Zhabotinsky and Zaikin [5] were the first to mention the existence of this phenomenon in thin layers, later on, Showalter [6], Orban [7], Boiteux and Hess [8] again mentioned the existence of such structures in different systems. It is only very recently that the number of systems showing this behaviour rapidly increased principally in the field of photochemical reactions after the pioneering work of Mockel where convoluted bands appear near the surface of an irradiated solution [9,10,11].

These structures are however poorly characterized because no truly systematic experimental investigation has been undertaken. Such an investigation is highly desirable in order to check the theoretical assumptions made about their origin and selection mechanisms.

Contrary to the case of chemical waves, mosaic-like structures correspond to a true self organization of the medium. In this case the long ranged phase fluctuations manifest themselves by the presence of topological defects, dislocations, declinations, grain boundaries, which may ultimately be responsible for the melting or disorganization of the patterns. The origin of these structures may be of chemical or hydrodynamical nature and theoretical predictions about their properties such as wavelength, shape, onset times,...may be given in both cases [12,13].

Acknowledgements

The results discussed here were obtained in collaboration with Drs. P.Borckmans and G.Dewel.

References

1. G.Nicolis and I.Prigogine. Self-Organisation in Non Equilibrium Systems (Wiley, New York, 1977)
2. A.T.Winfree. The Geometry of Biological Time (Springer Verlag, Berlin, 1980)
3. V.L.Pokrovsky. Adv. in Phys., 28, 597 (1979)
4. D.Walgraef, G.Dewel and P.Borckmans. J.Chem.Phys.78, 3043 (1983)
5. A.M.Zhabotinsky and A.N.Zaikin. J. Theor. Biol. 40, 45 (1973)
6. K.Showalter. J. Chem. Phys. 73, 3735 (1980)
7. M.Orban. J. Am. Chem. Soc. 102, 4311 (1980)
8. A.Boiteux and B.Hess. Ber. Bunsenges. Phys. Chem. 84, 392 (1980)
9. P.Mockel. Naturwiss. 64, 224 (1977)
10. D.Avnir, M.Kagan and A.Levi. Naturwiss. 70, 144 (1983)
11. J.C.Micheau and M.Gimenez. Naturwiss. 70, 90 (1983)
12. D.Walgraef, G.Dewel and P.Borckmans. Adv. Chem. Phys. 49, 311 (1983)
13. G.Dewel, P.Borckmans and D.Walgraef. Proc. Nat. Acad. Sc., USA, to appear (1983)

Simulation of Self-Organized States in Combustion Processes

L.U. Artyukh, P.G. Itskova, and A.T. Luk'yanov
Kazakh State University, Su-Alma-Ata, USSR

Recently, the problem of self-organization in non-equilibrium systems has intensively been discussed [1,2,3]. The major role in the process of successive generation and formation of more complicated temporal and spatial structures is played by fluctuations [1]. As shown in [4], the form of dissipative structures can be changed due to loss of the initial stable state, when external parameters are varied, and to subsequent dynamical rearrangement.

In the present paper some possible bifurcations in combustion systems are considered, namely the onset of self-sustained oscillations, formation of new temporal and spatial structures. Approximate analytical prediction of the region of bifurcations is based upon the model of a concentrated (zero-dimension) system [5].

We analysed the following system of one-dimensional equations of heat and mass transfer:

$$\rho(\partial C_i/\partial t)+\rho v(\partial C_i/\partial x) = \partial(\rho D_i \partial C_i/\partial x)/\partial x - w_i,$$
$$\rho C_p(\partial T/\partial t) + \rho v C_p(\partial T/\partial x) = \partial(\lambda\cdot\partial T/\partial x)/\partial x + \sum_i h_i^o w_i. \quad (1)$$

1. The process of nonstationary burning in a flat layer of a substance, in a reactor, in a catalyst layer, etc., was investigated for various ratios of the effective coefficients of diffusion and thermal conductivity (Lewis number) in the absence of convective transport of heat and mass. System (1) was solved under the following conditions:

$$C_i=C, \quad w_i=w=k_o C\exp(-E/RT), \quad v=0,$$
$$t=0: \quad 0<x<1: \quad T=T_H, \quad C=C_H; \quad (2)$$
$$t>0, \quad x=0: \quad \partial T/\partial x=0, \quad \partial C/\partial x=0$$
$$x=1: \quad -\lambda(\partial T/\partial x)=\alpha(T-T_\infty), \quad D=(\partial C/\partial x)=\alpha_D(C-C_\infty).$$

The concentrated system was obtained from the distributed one (1), (2), the spatial derivatives being substituted by three-node finite differences including boundary nodes:

$$\partial C/\partial t=n_1(C-C_\infty)-w_i, \quad (3)$$
$$\partial T/\partial t=n_2(T-T_\infty)+\Sigma h_i^o w_i.$$

The analysis of stability of combustion regimes is reduced to analysis of motion stability in the "temperature - concentration" phase plane applying the first Lyapunov's method. It has been found that when some characteristic parameters, e.g. Lewis number, $Le=(D\rho C_p)/\lambda$, are varied, the motion can become unstable, and new types of motion (structures) may result. In the plane of external parameters T_∞, C_∞, the parameter equations of stability have revealed nonsingular stationary states, steady and damped oscillations. Such bifurcations are plotted in Fig.1 by solid, dotted and dot-and-dashed lines. These lines divide the domain of the process parameters into regions differing in type, number and stability of stationary states for system (3) and approximately for system (2).

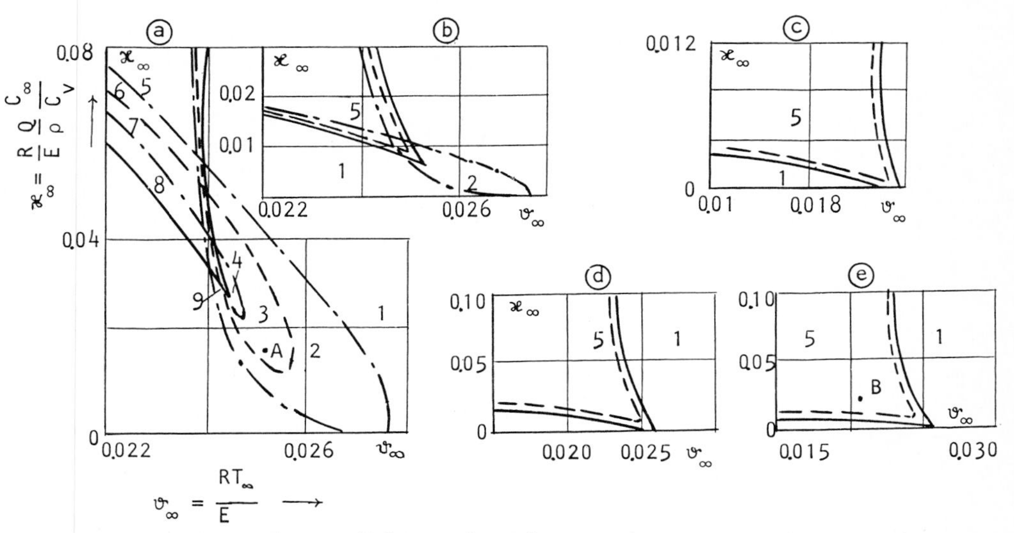

Fig. 1 Domains of possible combustion regimes at various Lewis numbers: a) 0.1; b) 0.5; c) 0.75; d) 1; e) 2. Solid line: nonsingularity boundary (saddles). Dotted line: boundary of steady oscillations. Dot-and-dashed line: focus boundary

Numerical solutions of systems (1),(2) show that the values of stationary functions, the period of oscillations and possible combustion regions, can be predicted from analysis of zero-dimensional system.

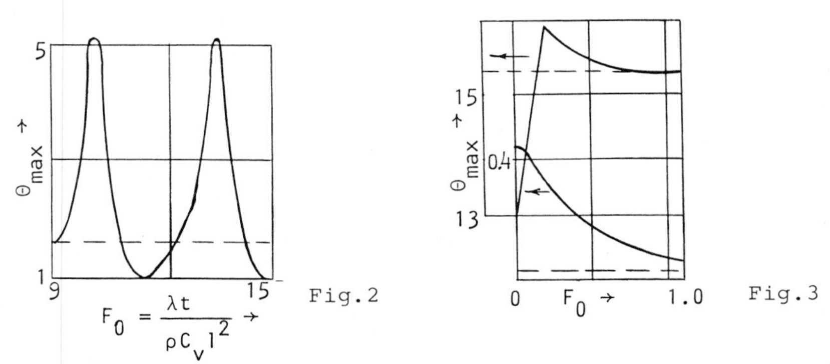

Fig. 2 Auto-oscillatory regime of combustion. Parameters correspond to the point A in Fig. 1a

Fig. 3 Transition to the regime of nonsingular states (point B in Fig. 1e)

2. Heat and mass exchange in chemically reacting systems is considered for the case of exothermic bimolecular oxidation, reaction of bifurcation, and chain cut-off, using the example of combustion of the mixture H_2-O_2-N_2 poor in oxygen. Mathematically, the problem is formulated by the set of diffusion equations for molecular oxygen and atomic hydrogen. The set of equations (1) is solved under the following conditions:

$$w_{O_2} = (A_2/\mu_H)\rho^2 C_{O_2} C_H \exp(-E_2/RT) + (A_p/\mu_H\mu_M)\rho^3 C_{O_2} C_M \exp(E_p/RT),$$

$$w_H = (2A_2/\mu_{O_2})\rho^2 C_{O_2} C_H \exp(-E_2/RT) + (2A_t/\mu_{O_2}\mu_M)\rho^3 C_H C_{O_2} C_M \exp(E_t/RT) +$$
$$+ (2A_7/\mu_H \mu_M)\rho^3 C_H^2 C_M,$$

$$t=0: \qquad C_{O_2} = C_{O_2}(0), \qquad C_H = C_H(0), \qquad T = T_o(0).$$

$$t>0,\ x=x_c: \qquad D_{O_2}(\partial C_{O_2}/\partial x) = v(C_{O_2} - C_{O_{2\infty}}), \qquad D_H(\partial C_H/\partial x) = vC_H, \quad T = T_c;$$

$$x=x_h: \qquad \partial C_{O_2}/\partial x = \partial C_H/\partial x = \partial T/\partial x = 0.$$

The fact that the non-unique stationary states are available depending on the reactor inlet temperature (Fig.4a), was established by the zero-dimensional model and was confirmed by numerical solutions. The hysteresis in the dependence on the gas flow rate (Fig.4b) was obtained numerically.

The theoretically predicted stationary profiles of temperature and concentration for the high temperature combustion regime are consistent with experiment (Fig.5)

The system exhibits damped oscillations in the domain of non-unique regimes for high temperature stationary states.

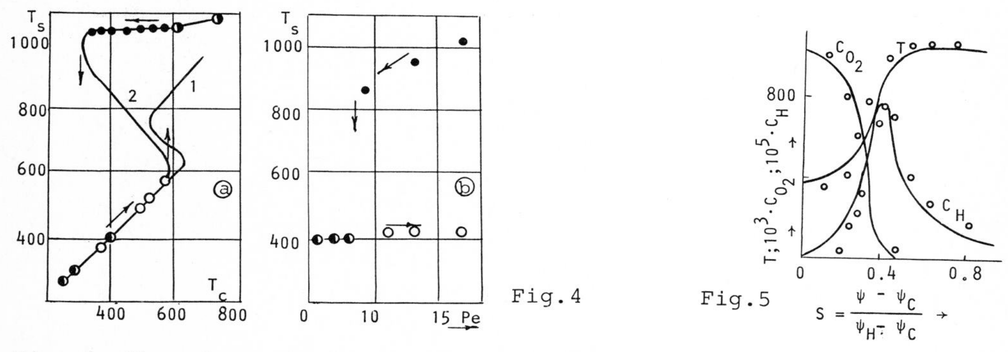

Fig. 4 Nonunique steady states for combustion of the system H_2-O_2-N_2 versus reactor inlet temperature (a), buoyancy rate (b). I- solution of zero-dimensional equation (3), 2- numerical solution of the set (1),(4). $Pe = vC_p(\psi_H - \psi_C)/\chi$, $\psi = \rho dx$

Fig. 5 Stationary profiles of temperature and concentration. Solid lines: numerical solutions. Dotted lines: experiment [7]

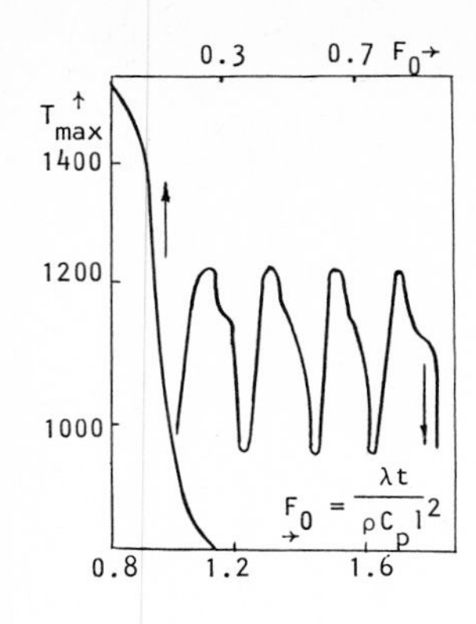

Fig. 6 Transition to stationary states when the rate is disturbed according to the law $Pe = Pe_1(1+\sin\omega F_o)$, $Pe_1=12,78$, $T_c=400^oK$ (1), $T_c=500^oK$ (2)

If the process is periodically disturbed, then both regimes of undamped oscillations and combustion cut-off can occur in the system, depending on the amplitude and frequency of a disturbance, i.e. different temporal and spatial structures are formed and stabilized (Fig.6).

Thus, the approximate analysis based upon a zero-dimensional model and one-dimensional numerical experiments can reveal the phenomena of self-organization in combustion.

References

1. G.Nicolis, I.Prigogine. Self-Organization in Nonequilibrium Systems, Wiley, New York (1977)
2. W.Ebeling. Strukturbildung bei irreversiblen Prozessen. Eine Einführung in die Theorie dissipativer Strukturen, Teubner (1976)
3. H.Haken. Synergetics, Springer-Verlag, Berlin, Heidelberg, New York (1978)
4. B.S.Kerner, V.V.Osipov. Doklady Akad.Nauk SSSR, 269(6), 1366-1370 (1982)
5. A.T.Luk'yanov, L.U.Artyukh, P.G.Itskova. Mathematical Simulation of the Problems of Combustion Theory, Alma-Ata, Nauka (1981)
6. B.V.Volter, I.E.Salnikov. Stability of Chemical Reactors, Moscow, Khimiya (1972)
7. P.L.Stephenson, R.G.Taylor. Combustion and Flame, 20, 231-244 (1973)

Part V

Autowaves in Biological Systems

Leão's Spreading Depression, an Example of Diffusion-Mediated Propagation of Excitation in the Central Nervous System

J. Bureš

Institute of Physiology, Czechoslovak Academy of Sciences, Prague, Czechoslovakia

V.I. Koroleva and N.A. Gorelova

Institute of Higher Nervous Activity and Neurophysiology, Academy of Sciences of the USSR, SU-Moscow, USSR

1. Introduction

Long before the concept of autowaves was formulated and generally recognized, a phenomenon of this class had been described in the vertebrate brain [1,2]. Leão's [2] spreading EEG depression (SD) is characterized by a decrease of electrical activity which propagates from the point of stimulation as a concentric wave with the velocity of 3 mm/min over the cerebral surface. The front of the SD wave is accompanied by a negative slow potential (20-30 mV) and by an increase of extracellular potassium concentration $[K^+]_e$, reaching 50 to 70 mM and lasting 30 to 60 sec [3]. SD can be elicited not only in the cerebral cortex but also in other neural structures. A particularly important preparation for SD studies is the isolated retina of fish, frogs, and birds, where its propagation can be directly observed as a change of light scattering of the retinal tissue [4,5]. SD wave appears as a milky or dark spot spreading in all directions across the retinal surface while the central region of the circular area gradually returns to the pre-depression state. The optical concomitants of SD make it possible to delineate the whole advancing front of SD with an accuracy and completeness which could never be achieved with complex arrays of hundreds of electrodes.

2. Ionic Mechanisms

Although the mechanism of SD is not yet fully understood [6,7,8] it is generally agreed that the K^+ movements between the intra- and extracellular compartments of brain tissue play a key role in the propagation of the process. Excessive stimulation of a circumscribed brain region causes movement of sodium and other extracellular ions into cells and release of an equivalent amount of K^+ ions into the extracellular space. K^+ ions diffusing from the region of high $[K^+]_e$ decrease the membrane potential of adjacent, yet unaffected, neurons until they too become a source of K^+. The regenerative feedback starts when the critical $[K^+]_e$ level of 10 to 12 mM is exceeded and rapidly increases $[K^+]_e$ to the maximum corresponding to equilibration of intra- and extracellular ionic gradients. During the recovery phase, metabolically driven ionic pumps restore the initial composition of the intra- and extracellular fluids and return $[K^+]_e$ to or below the resting level of 3 mM.

3. Mathematical Model

Attempts to describe the above mechanism quantitatively date back to 1959 when A.L.Hodgkin proposed the first mathematical model of SD based on the equations by A.F.Huxley [7,9]. The model assumes that the change of $[K^+]_e$ follows the Fick law for diffusion supplemented by a cubic function describing the rate at which K^+ ions enter or leave the cells. The relationship between the velocity of the wave (u) and amplitude of the $[K^+]_e$ change (y) is described by the equation

$$\frac{D}{u^2}\,\frac{d^2y}{dt^2} = \frac{dy}{dt} + \frac{y(y_o-y)\ (y_a-y)}{Ty_a^2}$$

where D is the diffusion coefficient of K^+ in the brain, T is the time constant of the rapid $[K^+]_e$ increase, y_o and y_a are the threshold and maximum $[K^+]_e$ increments, respectively. When the above equation is solved and the experimentally established values are substituted for D, T, y_o and y_a, the computed velocity u is close to the threshold level. Attempts to establish the critical mass [7] take into account also other factors affecting SD propagation.

4. Critical Mass and Leading Center

It is intuitively obvious that SD cannot be triggered by depolarization of a small group of neurons because K^+ ions released into a large volume of surrounding extracellular fluid cannot raise $[K^+]_e$ to the threshold level. Attempts to establish the critical mass [7] indicate that no SD can be elicited when the volume of the depolarized tissue is less than 0.5 to 1 mm^3. On the other hand, depolarization of a supercritical volume elicited by application of exogenous potassium produces a focus of repetitive SD waves. $[K^+]_e$ decreases from the high suprathreshold value in the center of the focus to the threshold level of 10-12 mM at its periphery, where SD waves are generated. Since each SD is followed by 2-3 min refractoriness, no SD can leave the focus before the adjacent cortex has sufficiently recovered. Conditions at the boundary of the focus were described in detail by Kuznetsova and Koroleva [12].

5. Circular Waves

Concentric waves of SD are not the only form of SD propagation. Theoretical analysis of excitation spread in networks of interconnected excitable cells [13,14] inspired development of cortical [15,16] and retinal [17] preparations, which support SD circulation around anatomical obstacles. An SD wave elicited from a point distant from the obstacle spreads around the lesion, engulfs it and propagates further. Circular SD can be elicited by functional blockade of one branch of the original SD wave when it spreads through a narrow region between the obstacle and tissue boundary. As soon as the SD wave stops, the block is removed so that the other SD branch spreading around the obstacle may pass through this region and start reverberation. The circle wave can also be produced by stimulation of a point in the rear of an SD wave. The new SD wave cannot spread into the refractory region but propagates into the recovered tissue, in the direction opposite to the withdrawing refractoriness. The wave front segment formed in this way in the narrow passage between tissue boundary and the obstacle starts circulation. Under favourable conditions up to 100 cycles of reverberating SD have been recorded around obstacles 3 to 5 mm in diameter with an average cycle time of 4-6 min.

The next step was made possible by the finding [18,19] that SD does not penetrate into neural tissue activated by drugs or by electrical stimulation. Excessive neural activity enhances the K^+ clearing mechanisms so efficiently that the increased $[K^+]_e$ is clamped at a level subthreshold for SD generation (below 10-12 mM). An SD wave stops at the periphery of such a stimulation-induced block, the duration of which can be controlled by adjusting stimulus intensity. The possibility to control SD by closing or opening circumscribed cortical regions to its propagation was used to generate SD reverberation around a stimulated cortical region [20].

6. Spiral Waves

Such experiments raise the question whether gradual decrease of the diameter of the functional obstacle to zero stops SD reverberation. Reverberation in absence of obstacles was theoretically postulated by Selfridge [21] and elaborated by others [22], but experimental demonstration of spiral SD waves is impossible with the limited number of electrodes which can be applied to the rat brain. The phenomenon has been recently demonstrated in the isolated chicken retina [23], which allows direct visualization of SD. Blockade of a retinal region by anodal current caused a gap in the advancing SD wave front. When the current was switched off, the free end of the interrupted SD wave front started to spread sidewards, gradually turned around and continued to propagate into the rear of the preceding SD wave with the free end of the wave front probing for the smallest circumference it could describe in the recovering retina (Fig.). The wave appeared as a spiral extending to the periphery of the retina and limited by its own refractoriness at the center. The latter is not stationary but describes an elliptic loop during each cycle. Furthermore, the center of the spiral gradually migrates across the retina until it approaches retinal margin so much that the next turn is blocked.

7. Conclusion

The evidence reviewed above shows that SD belongs to the family of autowaves, shares their typical properties (critical mass, leading

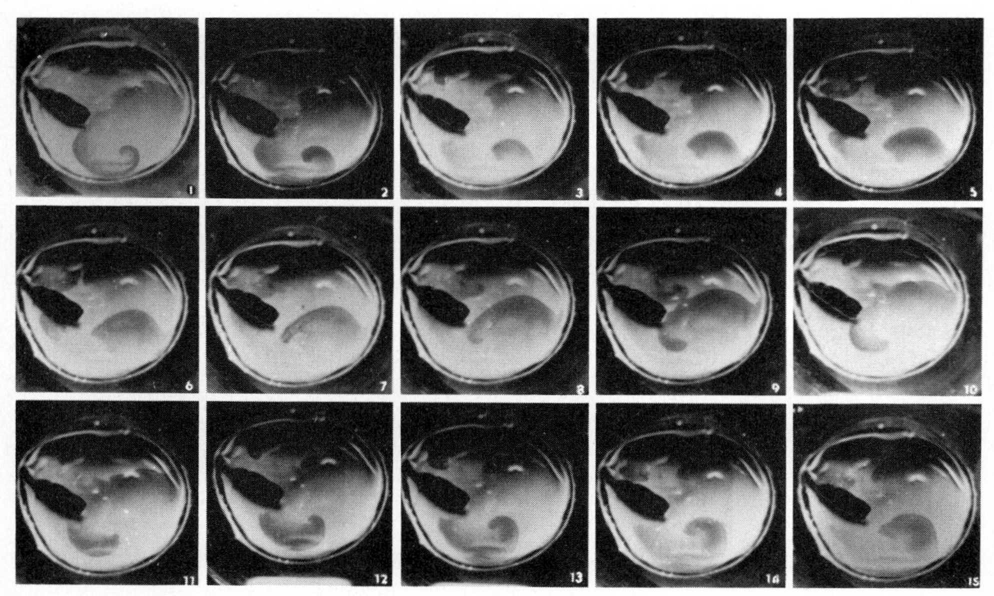

Fig. Photographs of one complete cycle of the spiral SD in the isolated chicken retina. SD appears as a dark area with sharp contours of the advancing wave front. The interval between successive frames is 20 s

center, spiral waves) and can be described by similar formal rules. Its occurrence in the brain indicates that the mass of closely packed neurons can support qualitatively new phenomena, occurring at a more primitive level of cellular communication, independently of the fine synaptic organization of neural networks. The diffuse nonsynaptic interaction mediating reactions of the SD type is an obligatory consequence of the trade-off between the maximal packing density and independent function of individual elements of the system which sets definite limits to maximum activation of synaptic processes and may play an important role in brain pathology.

References

1. K.S.Lashley. Arch. Neurol. Psychiat. 46: 331-339 (1941)
2. A.A.P. Leão. J. Neurophysiol. 7: 352-390 (1944)
3. F.Vyskočil, N.Kříž. J. Bureš, Brain Res. 39: 255-259 (1972)
4. P.Gouras. Am. J. Physiol. 195: 28-32 (1958)
5. H. Martins-Ferreira, G.Oliveira Castro de. J. Neurophysiol. 29: 715-726 (1966)
6. A.A.P.Leão. In: D.P.Purpura, J.K.Penry, D.M.Woodbury and R.D.Walter (eds.), Experimental Models of Epilepsy, Raven Press, New York, (1972), pp.193-196
7. J.Bureš, O.Burešová, J.Křivánek. The Mechanism and Applications of Leão's Spreading Depression of Electroencephalographic Activity. Academic Press, New York-London (1974)
8. C.Nicholson. Neurosci. Res. Progr. Bull. 18: 177-322 (1980)
9. B.Grafstein. In: M.A.B. Brazier (ed.), Brain Function, vol.1, Univ. of California Press, Berkeley (1963), pp.87-116
10. H.C.Tuckwell, R.M.Miura. Biophys. J. 23: 257-276 (1978)
11. H.C.Tuckwell. Int. J. Neurosci. 10: 145-164 (1980)
12. G.D.Kiznetsova, V.I.Koroleva. Foci of Stationary Excitation in the Cerebral Cortex (In Russian), Nauka, Moscow (1978)
13. N.Wiener, A.Rosenblueth. Arch. Inst. Cardiol. Mex. 16: 205-265 (1946)
14. J.S.Balakhovsky. Biofizika 10: 1063-1067 (1965)
15. M.Shibata, J.Bureš. J.Neurophysiol. 35: 381:388 (1972)
16. M.Shibata, J.Bureš. J.Neurobiol. 5: 107-118 (1974)
17. H.Martins-Ferreira, G.Oliveira Castro de, C.J.Struchiner, P.S.Rodriquez. J. Neurophysiol. 37: 773-784 (1974)
18. J.Bureš, I. von Schwarzenfeld, G.Brožek. Epilepsia 16: 111-118 (1975)
19. V.I.Koroleva, J. Bureš. EEG Clin. Neurophysiol. 48: 1-15 (1980)
20. V.I.Koroleva, J.Bureš. Brain Res. 173: 209-215 (1979)
21. O.Selfridge. Arch. Inst. Cardiol. Mex. 18: 177-187 (1948)
22. V.I.Krinsky, A.M.Zhabotinsky. In: M.T.Grekhova (ed.), Autowave Processes in Diffusion Systems (In Russian). Institute of Applied Physics, Acad. Sci. USSR, Gorky (1981) pp.6-32
23. N.A.Gorelova, J.Bureš. J. Neurobiol. Vol.14, N5, 353-363 (1983)

The Autowave Nature of Cardiac Arrhythmias

A.M. Pertsov and A.K. Grenadier

Institute of Biological Physics of the USSR Academy of Sciences, Moscow Region
SU-142292 Pushchino, USSR

This paper briefly reviews the autowave phenomena in heart tissue, the attention being focussed on the rotating waves of excitation (vortices). Vortices are believed to play an important part in the genesis of cardiac arrhythmias, in particular, of fibrillation, which is a major cause of sudden death from myocardial infarction.

1. Autowave Triggers the Contraction of Myocard

Synchronization of the contractions of the heart cells is due to propagation of electrical waves of excitation. Such a wave is a brief (0.2 s) alteration of the potential between the inside and outside of a cell (Fig.1a). Each contraction is preceded by an electrical wave of excitation, which follows a definite route over the heart cells, thus triggering their biochemical contractile apparatus and providing the required sequence of contractions (Fig.1b). The electric current resulting from propagation of the wave causes an alteration of the potential on the surface of the body, which is recorded as ECG and is used in medical diagnostics (Fig.1c).

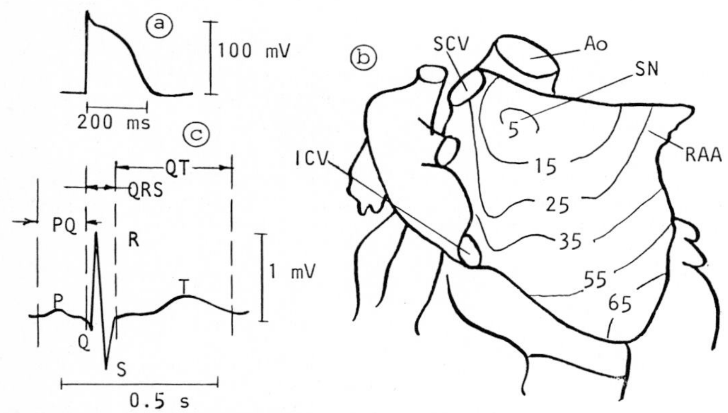

Fig. 1 Electrical impulses and waves in the heart.
a - the change in the myocardial cell potential during propagation of the excitation wave; b - propagation of the excitation wave in the atria (based on [1]). The thin lines show a succession of the wavefront positions, the numbers denoting time in ms. ICV - inferior vena cava, SCV - superior vena cava, Ao-aorta, SN - sinus node, RAA - right atrial appendage; c - ECG (the change in the potential on the surface of the body). The P,Q,R,S,T deflections characterize the rate of the propagation of excitation in the different heart chambers: the P deflection is associated with the contraction of the atria; the QRS complex is caused by the ventricular activity; the QT interval and the T deflection characterize the recovery time and the shape of the back wavefront, respectively

Each cell generates a pulse of fixed amplitude and shape. The consecutive excitation of the cells is due to the local currents at the boundary between the excited and unexcited cells which have different potentials.

The energy required for pulse generation is stored in the heart cells as concentration gradients of Na^+, K^+ and Ca^{++} ions across the cell membrane. These gradients are maintained by the energy derived from nutrients.

In the resting state, the inside of the cell is negative relative to the outside because of the selective membrane permeability to K^+ ions. The resulting potential difference is approximately equal to the Nernst potential for K^+ ions $V=(RT/F)\ln |K_o^+| / |K_i^+|$, where $|K_o|$ and $|K_i|$ are the outside and inside concentrations of K^+ ions, respectively. During wave propagation, a brief increase in membrane permeability to Na^+ and Ca^{++} ions occurs, whose concentration gradients are opposite to that of K^+ ions. This results in a dramatic increase in the membrane potential.

The equations to describe the excitation wave propagation in the heart are the reaction-diffusion equations of the type

$$C \frac{\partial V}{\partial t} = r^{-1} \frac{\partial^2 V}{\partial x^2} + I(V,g)$$

$$\frac{\partial \vec{g}}{\partial t} = \vec{\upsilon}(V,\vec{g}) \tag{1}$$

where C is the specific capacitance of the cell membrane, V the membrane potential, r the longitudinal resistance of the heart fibres, I the total current across the cell membrane; the vector $\vec{g}$ stands for the kinetic variables describing the dynamics of the membrane conductivity for specific ions.

The appropriate kinetic equations (the functions $I(V,\vec{g})$ and $\varphi(V,\vec{g})$) can be derived from voltage clamp experiments. The number of equations for different objects varies from 8 to 11 [2,3]. To describe qualitatively the process of propagation, two equations are, as a rule, sufficient. In this case, all the kinetic variables $\vec{g}$, which describe the relaxation of the membrane potential, are substituted by a single slow variable [4-9]. The equations are similar to the van der Pol equations and have the form*

$$C\dot{V} = f(V)-g$$

$$\dot{g} = (g(V) - g)\cdot/\tau(V) \tag{2}$$

Here $f(V)$ is a nonlinear N-shaped function (it is usually written as a third degree polynomial $f(V)=V(V-a)(V-b)$, $b>a>0$, or is constructed of three linear pieces) and $g(V)$ is a monotonically increasing function.

The function $f(V)$ is commonly derived from the current-voltage membrane characteristics, $g(V)$ and $\tau(V)$ are the established value and the relaxation time constant of the variable g. It is usually assumed that $\tau(V)>>1$. The phase portraits of various second order dynamical systems used in the simulation of excitation in myocard are shown in Fig.2.

---

* System (1) with the simplified kinetic equations (2) is usually called the Fitz-Hugh-Nagumo model after the authors who first demonstrated the applicability of the model to biological excitable membranes

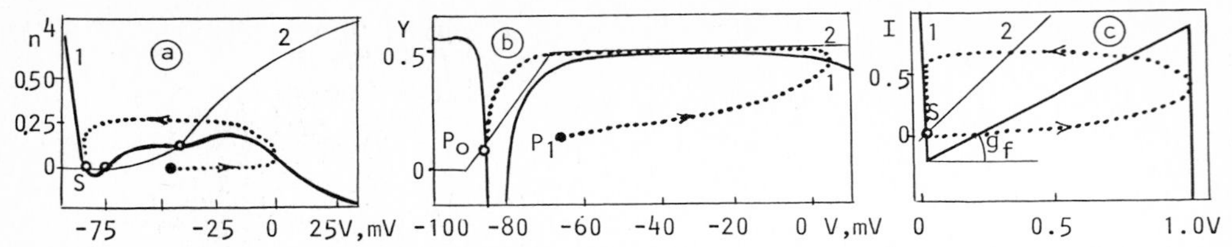

Fig. 2 Phase portraits of the kinetic equations describing the change in the myocardial cell potential in various models.
a - model [6] obtained from Noble's equations by Tikhonov's limit transition. Curve 1, the nullcline $\dot{V}=0$; curve 2, the nullcline of the slow variable $\dot{n}=0$. There is one stable stationary state S. The dashed line shows a phase trajectory corresponding to a pulse of excitation. b-c - simpler phenomenological models proposed in [7] and [8], respectively. The latter is a version of the Fitz-Hugh-Nagumo model.

The wave of excitation propagating over the heart is a typical autowave [10]. It shows no damping when propagating, it retains its shape, it does not reflect from the boudaries, and it annihilates in collision with a wave emitted by a different source. There is no interference of autowaves in the heart.

2. Wave Sources in Myocard. Rotating Waves

The waves of excitation in the heart are generated by the specific cells of the sinus node. They exhibit an auto-oscillatory nature,the period of oscillations of the membrane potential being about 1 s under normal conditions. On the contrary, the cells of the working myocard have a single stable stationary state, corresponding to the resting potential. They respond to externally applied current with single pulses. However, excitation sources which are not associated with the transition of the cells to the auto-oscillatory mode may also occur in the working heart under some pathological conditions.These sources are due to the generation of rotating waves, or "reverberators" ("leading cycles" in the physiological context).

Reverberators were predicted by Balakhovsky in 1965 [11], but it was not until 1973 that they were observed experimentally by Alessie et al. [12] in isolated rabbit atria. The reason for such a gap lies not only in the fact that the work [11] escaped the experimentalists' notice, but also in a number of experimental difficulties. The major of them involves visualization of excitation waves, in particular, of vortices, in the heart. To follow the route of an excitation wave, it is necessary to register the moments of arrival of the wavefront at a great number of points on the myocard surface and then to reconstruct the isochronal map. The complexity of the problem is in the necessity to have a great number of sensors (up to 400) to obtain a high spatial resolution and to provide simultaneous multichannel recording and special complicated processing programs. The technique of mapping is now available only in a limited number of scientific laboratories in the world.

Fig.3 shows a reverberator as recorded from an isolated rabbit atrium by the multielectrode mapping technique. The rotation period is about 100 ms, that is, it is one order of magnitude less than the auto-oscillation period of the sinus node cells. It is determined by the relaxation time of the membrane potential, or,in other words, by the refractory period. In the centre of the reverberator, there is a region of about 2-3 mm in size, where the cells are unexcited. This is the vortex core.

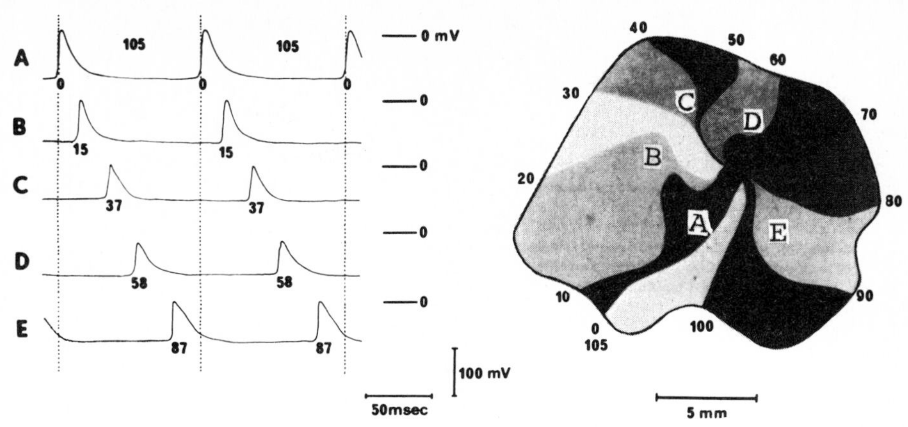

Fig. 3 A reverberator produced by the rotating wave of excitation in isolated rabbit atria (from [13]). Right, an isochronal map of the successive wavefront positions (the numbers indicate time in ms). The rotation is clockwise, the period being 105 ms. Left, pulses of the membrane potential at the points A, B, C, D and E.

Once appeared, such a vortex governs the rhythm of the sinus node cells, this resulting in a sharp increase in the heart rate - paroxysmal tachycardia, a disease which ranks, along with fibrillation, among the most severe heart disorders.

3. Theoretical Problems

Though there are some adequate phenomenological models to describe the propagation of excitation in myocard, no analytical approach to solve the problem is, as yet, available. Practically, only one-dimensional cases lend themselves to analysis. The classical work of Balakhovsky, which contained first hints at the reverberator, employed a radically different mathematical approach, the so called axiomatic model of excitable media. This type of model was first developed by Wiener and Rosenbluth (1946) to describe the wave processes in myocard [14]. Myocardial tissue was simulated by a network of automata with a finite set of states and a prescribed behaviour. These simplified models proved to be very useful. They allow the prediction of new properties of reverberators and the formulation of the first theory of fibrillation (Krinsky, 1965 [15]).

The best approach to date for studying reverberators in more detailed dynamical models (in the form of differential equations of type (1)-(2)) is computational experiment.

Fig.4a shows the dependence of the reverberator period and the refractory period on the parameter g_f as obtained numerically using the model of [8]. The parameter g_f determines the slope of the falling portion of the function f(v) (Fig.2c, curve 1) and the value of Na^+ ion current during excitation. The reverberator period and the refractory period are seen to lengthen sharply as g_f decreases. The effect is nontrivial, because this parameter is not contained in the equation for the membrane potential relaxation and is not expected to affect significantly these characteristics (for qualitative explanation, see [9]).

Fig. 4b gives the same dependences measured experimentally with varying concentrations of Na^+ ions in the environment. It is seen

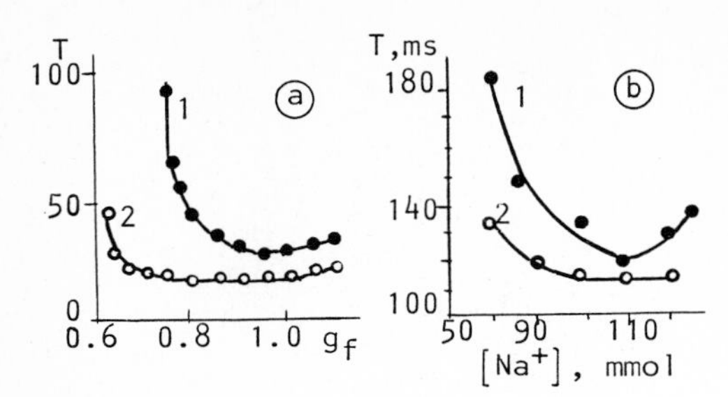

Fig. 4 Dependence of the reverberator rotation period on the parameters governing the sodium membrane current. a - the reverberator period (curve 1) and the refractory period (curve 2) as functions of the parameter g_f (numerical experiments using the model [8]). b - the corresponding functions obtained with isolated rabbit atria at different extracellular concentrations of Na^+

that the predictions of the model are in good agreement with the experimental results. The above plots not only illustrate the predictive power of the model but also allow an explanation of the antiarrhythmic effect of the whole class of the empirically found compounds - inhibitors of Na^+-current.

4. Mechanism of Generation of Rotating Waves

There are presently known several mechanisms of generation of reverberators [10,15-18]. For a reverberator to appear, spatial parameter gradients and high frequences of wave generation are, as a rule, necessary. The appearance of a reverberator is confined to a narrow region of frequences close to a critical one, which is determined by the refractory period.

The mechanism of the generation of reverberators in a medium inhomogeneous with respect to the refractory period* is illustrated by Fig.5. The reverberators are born at the boundaries of domains with large refractory period. In these domains, the propagating wave leaves behind itself short-lasting inexcitable islets (Fig.5a). If the time interval between two successive waves is small enough, a wavebreak occurs at an islet, whose edges pass round the temporarily inexcitable region (Fig.5b) If this region is large enough and the difference between the wave period and the refractory period is not too large, the region may become excitable again and the islet will disappear before the edges of the wavebreak have had time to merge (Fig.5c). Then a second wave rushes to the "window" which has appeared and two oppositely rotating reverberators result.

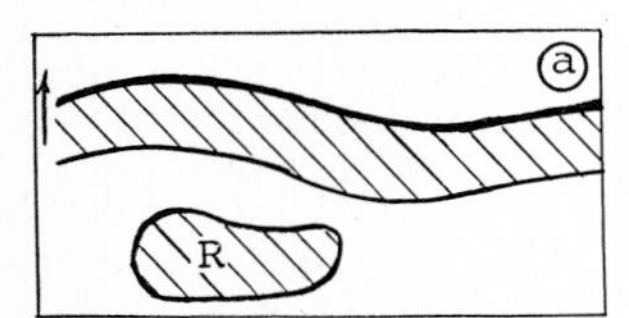

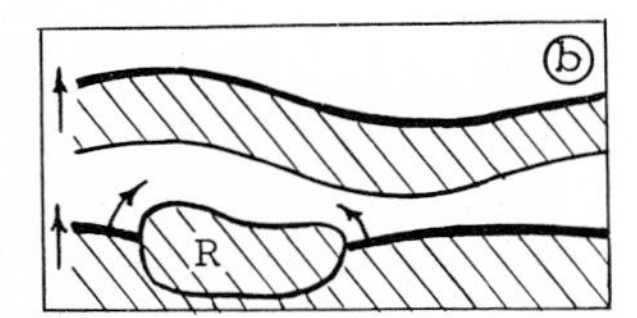

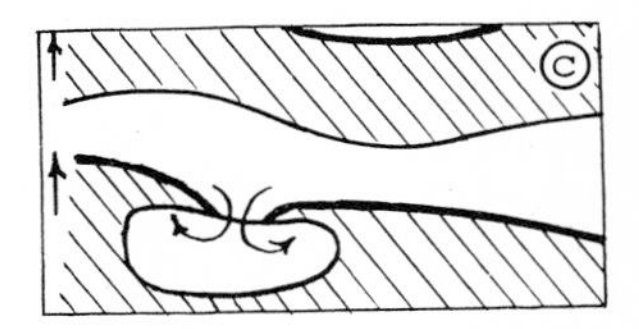

Fig. 5 A scheme of the mechanism of reverberator generation in media with refractory period gradients. Dashed are the regions in the refractory state. The solid line shows the fore wavefronts,the arrows indicate the direction of propagation. a - formation of a refractory islet. b - appearance of wavebreaks at a small interval between the waves. c - dissappearance of the refractory islet and the occurrence of two reverberators

* The mechanism was proposed by V.I.Krinsky and was analysed in more detail in [17].

5. Rotating Waves in Three Dimensions

Cardiac tissue is essentially three-dimensional (the left ventricule wall thickness exceeds 1 cm, the value being considerably greater than the length of the waves emitted by reverberators). However, little is known of the structure of reverberators in the thick of the heart wall.

While in two dimensions the wave rotates around a point, in three dimensions such points form a filament around which a scroll-like wave rotates. The thread may have a variable shape, in particular, it may be closed, and then a scroll ring occurs. The experiments on the three-dimensional scroll in myocard are described in [19,20].

6. Fibrillation and Rotating Waves

Fibrillation means chaos. However, no detailed theory of chaos in the system described by equations (1)-(2) is, as yet, available. Wiener and Rosenblueth [14] were first to attempt modelling of fibrillation, but the success was due to G.Moe [21] who modified the model of the previous authors by introducing into it the refractory period gradients and could thus observe in computational experiments what was very much like fibrillation. To initiate fibrillation it was sufficient to apply a high-frequency wave train that triggered a persistent, self-sustained chaotic activity. Based on the axiomatic model, Krinsky [15] proposed a theory of fibrillation. According to the theory, fibrillation is a result of the reproduction of vortices due to gradients in the refractory period. The theory shows that the reverberators in inhomogeneous media are unstable and have a finite lifetime determined by parameter gradients. However, they emit high-frequency trains of waves which give birth to new reverberators by the above described mechanism. At a certain critical relationship between the medium parameters, the number of the nascent reverberators may exceed the number of those which died away. This results in a sustained chaotic wave activity, i.e. fibrillation.

The reverberator theory of fibrillation explains qualitatively the transition from paroxysmal tachycardia to fibrillation, the high-frequency (close to f_{cr}) excitation of the cells during fibrillation, the critical mass phenomena (the requirement of the minimum heart tissue extension for fibrillation to occur).

Direct experimental evidence to support the theory is difficult to obtain, mainly because of the myocard being three-dimensional. In a three-dimensional medium, the reverberator can be observed only in certain cross-sections intersecting its filament [20]. Thus identification of vortices in myocard involves serious complications. Nevertheless, the recent experimental results obtained by multielectrode mapping of fibrillation support the main points of the theory. In particular, the reproduction of reverberators in the passage from paroxysmal tachycardia to fibrillation has been well established [22].

Possible Applications

Two clear-cut directions of study may presently find a wide application in clinical medicine concerned with the diagnostics and treatment of cardiac arrhythmia.

The first one rests upon finding the key parameters that control the birth and decay of reverberators in myocard and upon searching for ways to affect these parameters by various drugs (see Fig.4).

The second one is connected with elaboration of experimental procedure of identification and localization of the pathological wave sources in myocard and with refinement of diagnostical methods. This direction may be fruitful in surgery when treating arrhythmias in patients resistent to pharmacological drugs.

Conclusion

In this work we have only touched upon one class of problems related to vortical processes. The real range of problems is significantly broader. In particular, we have by-passed the question of "latent pacemakers" and "trigger automaticity" (local autowave sources of the type of leading centres occurring, along with reverberators, in the working myocard and causing its disfunction [7]). Here there are even more blanks than in our knowledge of vortices.

The aim of the work was to attract the attention of experimentalists and specialists in the theory of nonlinear waves to this problem. We believe that the synergistic approach will be fruitful in cardiology and it will also enrich the modern autowave theory.

References

1. D.Durrer, R.T. van Dam, G.E.Freud, F.L.Meijler, R.C.Arzbaecher. Circul. 41, 899 (1970)
2. G.Beeler, H.Reuter, J.Physiol. 268, 177 (1977)
3. R.E.Allister, D.Noble, R.N.Tsien. J.Physiol. 251, 1 (1975)
4. R.Fitz-Hugh, Biophys. J. 1, 445 (1961)
5. J.Nagumo, S.Yoshizava, S.Arimoto, IEE Trans. Commun. Technol. 12, 400 (1965)
6. V.I.Krinsky, Yu.M.Kokoz, Biofizika 18, 1067 (1973)
7. F.J.L.Capelle, D.Durrer, Circ. Res. 47, 454 (1980)
8. A.M.Pertsov, A.V.Panfilov, in: Autowaves in Reaction-Diffusion Systems, Gorky. 1981, 77
9. A.M.Pertsov, A.V.Panfilov, R.I.Khramov, Biofizika, 26, 1077 (1981)
10. V.I.Krinsky, this volume
11. I.S.Balakhovsky, Biofizika 10, 1063 (1965)
12. M.A.Allessie, F.I.M.Bonke, F.J.G.Shopman, Circ. Res. 39, 54 (1973)
13. M.A.Allessie, F.I.M.Bonke, F.J.G.Shopman, Circ. Res. 39, 168 (1976)
14. N.Wiener, A.Rosenblueth, Arch. Inst. Cardiol. Mex. 16, 205 (1946)
15. V.I.Krinsky, Biofizika 11, 676 (1966)
16. B.Ya.Kogan, V.S.Zykow, A.A.Petrov, in: Simulation of Systems, IMAGS Congress 1979, ed. Dekker L. et al., North-Holland Publ. Co., 1980, 693
17. A.N.Reshetilov, A.M.Pertsov, V.I.Krinsky, Biofizika 24, 129 (1979)
18. A.V.Panfilov, A.M.Pertsov, Biofizika, 27, 886 (1982)
19. M.J.Janse, F.J.L. van Capelle, H.Morsink, A.G.Kleber, F.Wiems-Shopman, R.Cardiual, C.N. d'Alnocourt, D.Durrer, Circ.Res. 47, 151 (1980)
20. A.B.Medvinsky, A.V.Panfilov, A.M.Pertsov, this volume
21. G.K.Moe, W.C.Rheinboldt, T.A.Abildskov, Amer. Heart J., 67, 200 (1964)
22. M.A.Allessie, W.Lammers, J.Smeets, J.Hollen. in: Atrial Fibrillation (eds. H.E.Kulbertus, Olsson S.B., M.Schlepper), Mölndal, Sweden, Hässle (1981), 44

Cardiac Arrhythmias During Acute Myocardial Ischemia

F.J.L. van Capelle and M.J. Janse

Department of Cardiology and Clinical Physiology, and the Interuniversity Cardiology Institute, University Hospital, Wilhelmina Gasthuis
Eerste Helmersstraat 104, NL-1054 EG Amsterdam, The Netherlands

There are two basic mechanisms of cardiac arrhythmias; they are abnormal impulse formation and re-entry. In the first case, there is DE NOVO impulse generation. In the second case, the impulse travels through a closed loop, re-entering a zone that was activated earlier by the same wavefront.

In this paper, we shall concentrate upon early arrhythmias during acute regional myocardial ischemia caused by occlusion of a major coronary artery. We have performed mapping experiments in the isolated dog and pig hearts under these circumstances (1) and have used a computer model to elucidate the mechanisms underlying our experimental findings. The model was based on interaction between a large number of excitable elements with different characteristics. Interconnecting various types of elements,using various network configurations, we tried to simulate some of our own experimental results, and so to learn something about the mechanisms that could help explain them. A more detailed description of the model can be found elsewhere (2).

Early arrhythmias in acute ischemia

In fig 1, the activation sequences during a basic beat propagated from the atrium, and a spontaneous ventricular premature beat are shown. The recordings are from the subepicardium of a dog heart after 4 minutes of occlusion of the circumflex artery. The activation maps represent the rectangular area shown in the schematic drawing of the heart. The earliest premature activity occurred in healthy tissue, close to the ischemic border. This was a consistent finding: whenever we could localize the earliest ectopic activity (this was not always possible, since the electrodes covered only part of the heart), the origin of the premature activation was found at the non-ischemic side of the border. Selected DC electrograms showing the premature complex and the preceding basic beat are also shown. Tracing A comes from an electrode that was located in the ischemic region, as is apparent from the TQ depression of the baseline; tracing B was also recorded from ischemic tissue, but closer to the border; and tracing C is from non-ischemic myocardium at the normal side of the border. During the basic beat, the tissue under electrode A was activated very late. The deep negative "T" wave in the electrogram is caused by the corresponding delay of the repolarization of the tissue. For an account of the relation between the extracellular DC potentials and the local membrane action potentials, see Kléber et al (3). Trace B shows monophasic complexes, indicating that the tissue underneath was inexcitable. A zone of conduction block (shaded area in the activation map) thus separated the zone of delayed activity in the ischemic myocardium from the healthy side of the ischemic border, where the earliest ectopic activity emerged. Since the pattern depicted in fig 1 was found quite frequently, we were led to assume a causal relationship between the occurrence of late activity in the central ischemic zone and the occurrence of premature activity at the healthy side of the border.

What mechanism can be responsible for this state of affairs? Several arguments plead against re-entry as the initiating mechanism. The time

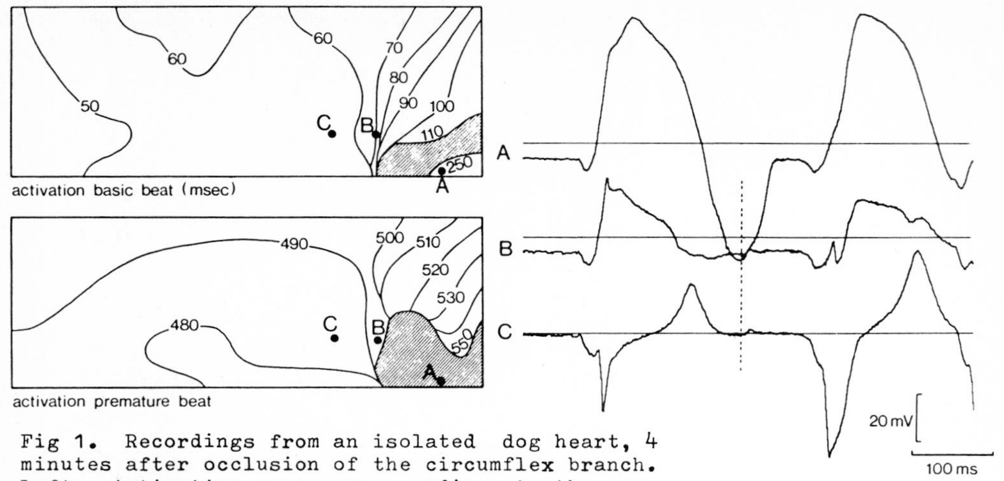

Fig 1. Recordings from an isolated dog heart, 4 minutes after occlusion of the circumflex branch. Left: Activation maps corresponding to the rectangular area shown in the schematic drawing of the heart. Right: Tracings of a basic beat propagating from the atrium, and a ventricular premature beat. Tracings A, B and C correspond to locations shown in the maps. Note the deep negative "T" wave following the basic beat in the ischemic area (A). Isochronal lines separate areas activated within the same 10 msec interval in the activation maps. Time t=0 corresponds to the P-wave of the basic beat. Zones of conduction block are shaded in the activation maps. Reproduced by permission of the American Heart Assoc., Inc., from Janse and van Capelle (1982), Circ. Res. 50: 527-537.

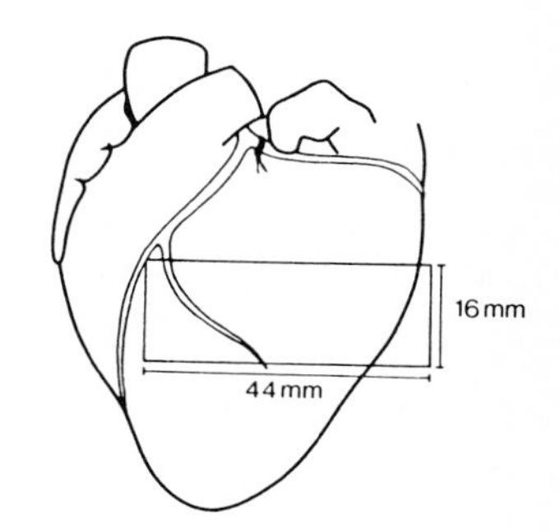

interval between the last propagated activity of the basic beat and the earliest ectopic activity is about 200 msec, and yet no activity bridging this time gap could be found. In contrast, re-entry could be easily demonstrated during both ventricular tachycardia and during fibrillation (1). The pattern was constant in these cases: in the initial beats, early activity emerged from the normal myocardium, close to the border; in subsequent beats, conduction within the ischemic area became slower and islands of conduction block appeared; and, finally, circular movement of the activation around an area of conduction block developed. This strongly suggests that the mechanism underlying the initiation of the tachycardia is different from the one responsible for its maintenance later on.

Computer simulation studies

An important concept, emerging from earlier work with the computer model, is suppressed automaticity (2). As explained before in greater detail, the automatic properties of pacemaker elements can be modified when they are connected to non-pacemaker elements. It is intuitively clear that a small pacemaker cell cannot be expected to maintain its automaticity when it is coupled tightly enough to a very large non-automatic cell, because the second cell, if large enough, would impose a primitive voltage clamp upon the pacemaker, and the ensemble would be silent. Inversely, a large pacemaker would not be stopped by a tiny non-automatic cell, no matter how tight the coupling. An interesting intermediate situation, where the pacemaker element is incompletely suppressed, is illustrated in fig 2. Two

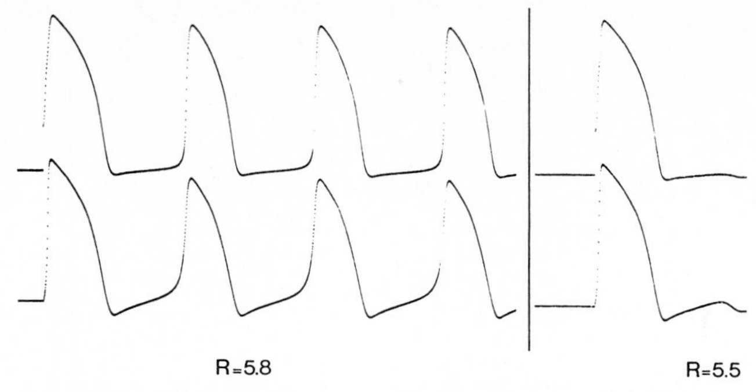

Fig 2. Interaction between a pacemaker element (lower trace) and a non-pacemaker element (upper trace) at critical values of the coupling resistance R. By permission of the American Heart Assoc.Inc., from van Capelle and Durrer (1980), Circ. Res. 47: 454-466.

excitable elements were connected by a variable resistance. One element (upper trace) was a non-automatic element, the other (lower trace) was a pacemaker. Cell coupling was tight, and the non-pacemaker element was large enough to inhibit overt pacemaker activity, as can be seen from the initially stable baseline in both panels of fig 2. The ensemble could be stimulated, however, and upon stimulation of the non-pacemaker element, sustained rhythmic activity ensued (left panel), which could be stopped again by an appropriately timed stimulus (not shown). When the coupling between the elements was made a little tighter (or when the size of the non-pacemaker was increased a little), sustained automatic activity no longer followed the stimulated response, but threshold afteroscillations

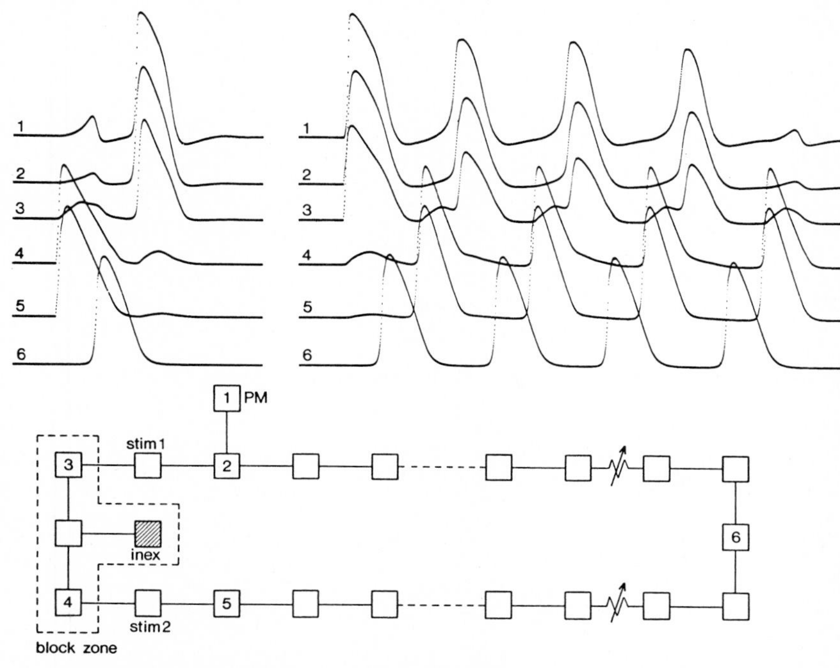

Fig 3. See text. Reproduced by permission of the American Heart Assoc., Inc., from Janse and van Capelle (1982), Circ. Res. 50: 527-537.

appeared instead (right panel). There is a striking resemblance with the phenomenon of "triggered activity" as documented by Wit and Cranefield (4).

It is of interest that suppressed automaticity can promote, in yet a different way, the occurrence of extrasystoles in an area that is adjacent to a block zone. In the configuration shown in fig 3, a zone with bidirectional conduction block separates the two ends of a long cable, consisting of non-automatic excitable elements. In the present case, the conduction block was due to the inclusion of a passive and depolarized element in the block zone. This element depolarized its neighbours, thereby inactivating the element in the middle of the block zone, and also, to a lesser degree, the adjacent elements (3 and 4). Therefore, the four elements included within the broken line in fig 3, together act as a zone of conduction block. At one side of the block zone, suppressed pacemaker activity is present, due to the automatic element 1, which is tightly coupled to its non-automatic neighbour (2). Upon stimulation of the element labeled "stim 2", the excitation wave fails to cross the block zone (note local responses in elements 3 and 2 in the upper left panel of fig 3). The pacemaker element (1) shows an enhanced local response, but it does not fire. In the meantime, the impulse is traveling through the cable, activating elements 5 and 6 on its way, to reach the opposite side of the block zone somewhat later. Full-blown activation of elements 1, 2 and 3 therefore occurs, but again the impulse fails to cross the bidirectional block zone, and no further activity ensues. Note, however, the subthreshold afterdepolarization of element 1. In the upper right-hand panel of fig 3, the stimulus is applied to the element labeled "stim 1". Again the impulse fails to cross the block zone, yielding only a local response in elements 4 and 5. Traveling through the cable,it subsequently reaches the other side of the block area and initially fails to re-excite the elements at that side of the block zone. This time, however, the local response in elements 3, 2 and 1 occurs at the same time as the afterdepolarization of element 1, due to its original activation. The two subthreshold responses are summated in element 1, enhancing each other, and as a result a new action potential is generated. The extrasystole is conducted along the cable, reaching the opposite side of the block area, and subsequently gives rise to a second premature beat originating from element 1 by the same mechanism. In this way, a short run of premature impulses results. After the third premature beat of element 1, it can be seen that the synchrony of the afterpotential and the local response, propagated through the block zone, is lost, and as a consequence the run is terminated. The occurrence of the premature impulses depends on the conduction time through the loop. The latter can be varied by changes in one or two of the coupling resistances in the cable, as indicated in the figure. The length of the run of premature impulses that are evoked can therefore be modulated easily by slight changes of the conduction time through the cable.

References

1. Janse MJ, van Capelle FJL, et al. (1980) Circ Res 47:151-165
2. van Capelle FJL, Durrer D (1980) Circ Res 47:454-466
3. Kléber AG, Janse MJ, et al. (1978) Circ Res 42:603-613
4. Wit AL, Cranefield PF (1976) Circ Res 38:85-98

Properties of Rotating Waves in Three Dimensions. Scroll Rings in Myocard

A.B. Medvinsky, A.V. Panfilov, and A.M. Pertsov

Institute of Biological Physics of the USSR Academy of Sciences, Moscow Region
SU-142292 Pushchino, USSR

Introduction

The discovery of rotating spiral concentrational waves in the B-Z reaction and in some biological objects, such as retina, colonies of microorganisms, cardiac tissue, has provoked a great number of studies of this interesting phenomenon [1]. The problem of structure and properties of such waves in three dimensions remains among the least studied. However, it is of particular interest in the case of myocard because the rotating waves are of the key importance in the genesis of most dangerous heart diseases [2]. Myocard is essentially three-dimensional, the heart wall thickness being comparable with the wavelength.

Topological analysis reveals several possible three-dimensional structures of rotating waves [3], one of which, a scroll ring (or a vortex ring) has been observed experimentally in the B-Z reaction [4,5]. The scroll ring produces in three dimensions a wave pattern which is a family of enclosed toroidal surfaces, resembling the confocal structures of liquid crystals [6]. However, the topological properties of three-dimensional rotating waves cannot bring to any definite conclusions about the feasibility of a given structure and its dynamical properties. Therefore, the only reliable approach at present is to study the problem by both computational and laboratory experiments.

Using numerical methods, we have obtained a solution of reaction-diffusion equations in the form of a scroll ring. The properties of this wave structure have been studied, and experimental evidence has been presented for the occurrence of three-dimensional scroll rings in myocard.

Scroll Ring as a Solution of Reaction-Diffusion Equations

We used a model of Fitz Hugh - Nagumo type with excitable kinetics*

$$U_t = \Delta U + f(U) + V,$$
$$V_t = \varepsilon(U-V) \qquad (1)$$

Here Δ is a three-dimensional Laplacian operator, $f(U)$ a nonlinear N-shaped function, and ε a small parameter. The system was studied under no-flux (Neumann) boundary conditions. The solution was obtained

* This type of model can adequately describe biological excitable systems (nerve and muscle fibres) [7]. The original model [8] was used to describe nerve pulse propagation.

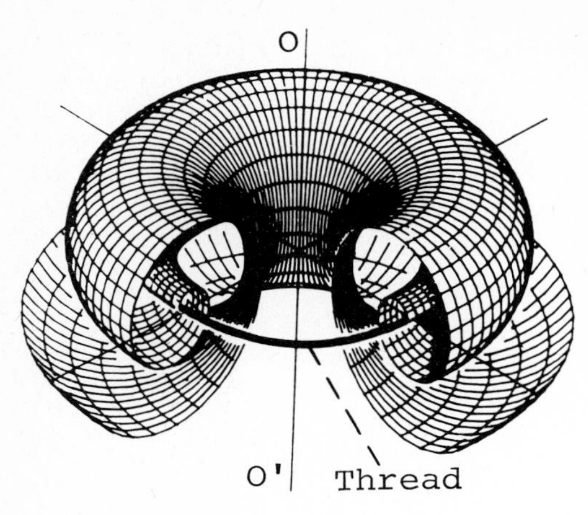

Fig. 1 The isophase surfaces of a reconstructed toroidal scroll. The section normal to the scroll thread is an Archimedean spiral

for the piecewise linear approximation of f(U):

$$f(U) = \begin{cases} 4U \\ -g_f(U-0.02), \\ 15U \end{cases} \qquad (2)$$

g_f being an adjustable parameter. Calculations were done in the cartesian and cylindrical coordinate systems (mesh size 30x30x30 segments) using the simple Euler method.

Fig.1 shows a reconstruction in three dimensions of a toroidal scroll obtained by numerical integration of system (1)-(2). The scroll is unstationary and has the following properties:

1. The position of its thread is not stabilized in space. The thread drifts along the torus axis in the direction determined by the vector $\vec{n} = \frac{|\vec{r}\ \vec{V}|}{|\vec{r}||\vec{V}|}$, where $\vec{r}$ is the radius-vector of the distance of the rotation axis to any arbitrarily chosen point on the thread. $\vec{\omega}$ is the vector of the angular velocity of scroll rotation (Fig.2a).

2. The thread length, l, gradually decreases and upon reaching a certain critical value, l_c, the scroll disappears*.

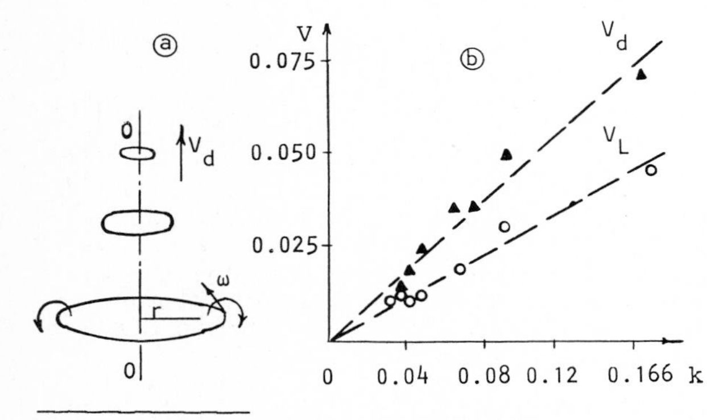

Fig. 2 Evolution of the thread (scheme) (a). The rates of drift (V_d) and shortening (V_c) of the thread as functions of its curvature, g_f=0.9 (b)

*Properties 1 and 2 resemble the properties of the quantized vortex rings.

3. The velocities of the thread drift, V_d, and of decrease in its length, V_l, are functions of the thread curvature ($k = \frac{2\pi}{l}$). As seen from Fig.2b, these functions are well approximated by straight lines. V_d and V_l are very small, their maximum values (at $l \approx l_c$) being two orders of magnitude less than the velocity of wave propagation. Note that the plot of V_l versus k being linear, the life time of a toroidal scroll is a quadratic function of the initial thread length l_o. Actually,

$$T = \frac{1}{2\pi}\int_{l_c}^{l_o} \frac{dl}{V} \sim l_o^2 . \qquad (3)$$

4. The rotation period of a toroidal scroll is only slightly different from that of a two-dimensional scroll with the same parameters of the medium. It remains unaltered (within the experimental error) during the evolution of the thread.

It is remarkable that the scroll in three dimensions proves to be unstationary, whereas the two-dimensional ones are stable and stationary, the parameter values being equal. Non-stationarity of the scroll rings is due to the fundamental property of active media, namely the wave propagation velocity being dependent on the wave front curvature. Because of the thread geometry, the sign of the curvature is reversed during rotation, which leads to non-stationarity and, as a result, to a drift of the thread and to its shortening.

The results obtained suggest that non-stationarity of scrolls with closed threads is a universal property of active media. It seems to underlie the experimentally observed phenomena of annihilation and breakdown of scrolls [4].

Scroll Rings in Myocard

In cardiac tissue, the three-dimensional rotating waves arise from waves of electrical excitation, synchronizing the contractions of myocard. To study the scrolls in the thick of the heart wall is a very complicated experimental problem. Therefore, only circumstantial evidences of their three-dimensional structure can, as a rule, be derived.

We have obtained some experimental evidence of the existence of scroll rings in myocard, namely:

1. A wave pattern typical of scroll rings [6] was discovered on the surface of the heart wall (retracting circular waves).

2. The wave period was close to that of simple vortices in myocard.

3. As a rule, the life time of the source of retracting waves was short.

Fig.3a shows schematically a mechanism of wave generation on the myocard surface, following the occurrence of a scroll ring with its thread lying in the plane parallel to the surface. The real picture, as recorded from the surface of rabbit heart atrium, is shown in Fig.3b,c,d. The thread plane is somewhat tilted with respect to the surface, and a horseshoe shape of the wave initially results.

The experiment was carried out using a model of tachycardia initiated by specifically phased electric pulses.The technique of multielectrode mapping was employed [7].

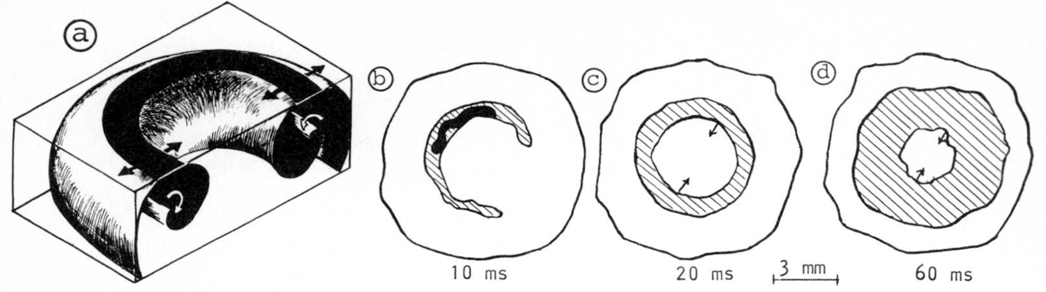

Fig. 3 A scroll ring in the heart wall. a - a scheme of retracting wave pattern formation during the scroll rotation, the inner boundary of the ring of excitation progressively shrinks to the centre. b, c, d - a record of the wave pattern observed experimentally on the surface of myocard during arrythmias. The region of the initial excitation is blackened. Dashed is the region involved in excitation

A hypothetical mechanism for the generation of a scroll ring in myocard is shown in Fig.4. In principle, it does not differ from the mechanisms of vortex generation in two dimensions [1,2,7]. For a scroll wave to appear it is sufficient that an extended region with an increased refractory period R be present in the thick of the myocard wall. The wave which gives birth to a scroll ring starts propagating after the time period $T<R$. Because of this, it cannot penetrate the region of a large R (Fig.4a,b). The scroll ring appears if the situation shown in Fig.4c takes place when the region recovers from the refractoriness. The shape of the wavebreak at this instant of time determines the thread geometry. Fig.4d illustrates further evolution of the scroll wave.

The occurrence of extended inhomogeneities in the refractory period in the thick of the heart wall is typical for some stages of myocardial infarction. This means that the attacks of paroxysmal tachycardia, a dramatic increase in the heart rate, may be connected with generation of scroll rings.

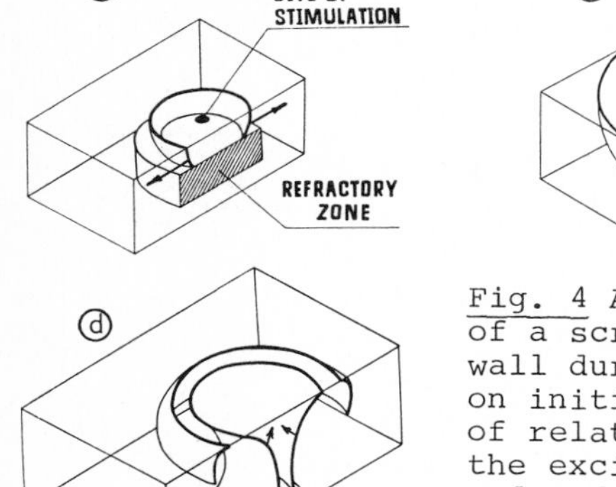

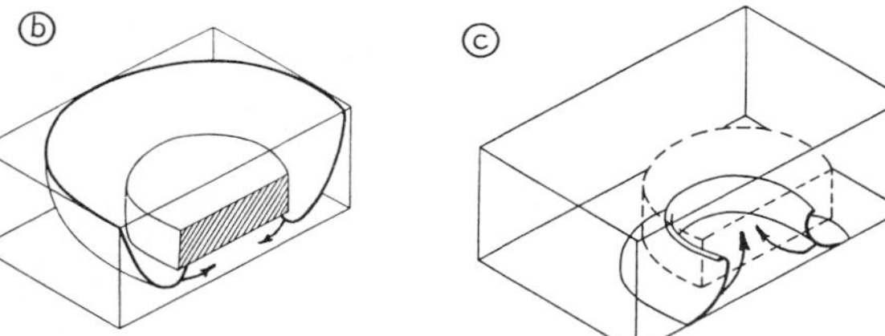

Fig. 4 A hypothetical mechanism of generation of a scroll ring in the thick of the heart wall during arrhythmias. a, b - wave propagation initiated by a stimulus applied at the phase of relative refractoriness; c - invasion of the excitation wave into the region of increased refractoriness following the recovery of excitability (the birth of a scroll ring). The shape of the ring thread is determined by the boundary of the cross section of the region of increased refractoriness and is close to a circumference. d - emergence on the myocard surface of a wave generated by the scroll ring and formation of retracting and outward-spreading waves

References

1. V.I.Krinsky (the article in this book).
2. A.M.Pertsov, A.K.Grenadier (the article in this book)
3. A.T.Winfree, S.H.Strogatz. Physica, 8D, 35 (1983)
4. A.T.Winfree. Science, 181, 937 (1983)
5. B.J.Welsh, J.Comatam, A.E.Burgess. Nature, 304, 611 (1983)
6. P.G. de Gennes. The Physics of Liquid Crystals. Clarendon Press, Oxford (1974)
7. V.I.Krinsky, A.M.Zhabotinsky. In: Autowave Processes in Systems with Diffusion. Gorky (1981) p.6
8. R.Fitz Hugh. Biophys. J., 2, 11 (1962)
9. M.A.Allessie,F.I.M.Bonke,F.J.G.Schopman. Circ.Res. 33, 54 (1974)

Waves and Structures in Space: Ecology and Epidemiology

Yu.M. Svirezhev and V.N. Razzhevaikin

Computer Center, Academy of Sciences of the USSR, SU-Moscow, USSR

The problems, concerning the spatial dynamics of natural systems, have recently become a matter of great interest to researchers in ecology, medicine (mainly epidemiology) and mathematics. Such phenomena as "patchness" of species number distribution in an ecosystem, or "life waves" (or epidemic waves) spreading over the space, cannot be simulated without application of modern mathematical methods such as those used by the theory of partial differential equations.

The point is that the indicated phenomena can be adequately described by the simplest, in a certain sense, equations of this type, namely, by systems of parabolic type equations. The easiest way to write such systems is to add the terms, describing spatial displacements of individuals, to the right hand side of the proper point system. These terms can describe both stationary currents in a medium (streams, winds) and stochastic displacements of individuals (e.g. in the case of an active search for food). Restricting ourselves to the latter case, we can write the initial system in the form;

$$\frac{\partial N_i}{\partial t} = D_i \Delta N_i + f_i(N_1, \dots, N_m). \tag{1}$$

$i=1,\dots m$, $N_i(x,t)$ is the density of the i-th species at a point X of the region $\Omega \subset R^n$ at time t, $\Delta = \sum_{j=1}^{n} \frac{\partial^2}{\partial x_j^2}$ is the Laplace operator in the region. If this region has a boundary, it is also necessary to take into account boundary conditions, e.g. neutral ones:

$$\left.\frac{\partial N_i}{\partial \vec{n}}\right|_{\partial\Omega} = 0 \tag{2}$$

($\vec{n}$ is an external normal vector).

It is well-known that in the case of m=1 the system (1)-(2) has no solutions of the type of spatial structures (SS), i.e. no solutions which are stationary, stable and spatially inhomogeneous. For m=2, we can get these structures in a "predator-pray" system with the right hand sides

$$f_i(N_1, N_2) = N_1\left(\frac{\varepsilon N_1}{1+N_1} - N_2\right), \; f_2 = N_2(N_1 - \gamma N_2).$$

Here $N_{1,2}$ are the biomass densities of the prey and predator respectively. The pray growth rate is considered to be small, if it's density is small, e.g. when it is difficult to find a sex partner. When the region Ω has the form of a circle, the proper choice of coefficients results in spatial structures of a hat-like shape.

For a wide class of systems, it is possible to write rather concrete conditions for the existence of SS. For instance, when using the diffusion coefficient ratio as a single parameter $\lambda=D_2/D_1$, the conditions take the form

1) det $A>0$, 2) tr $A<0$, 3) $a_{11}<0$ for $a_{22}>0$ or $a_{11}>|\varkappa_1|$,

where $A=||a_{ij}||$ is the matrix of the initial point system, linearized near the stable equilibrium point, $\varkappa_i$, $(i=0,1,\ldots)$ are Laplace operator eigenvalues for conditions (2) [1].

Interesting results can be obtained by changing Neumann boundary conditions to Dirichlet ones. From the biological viewpoint it corresponds to embedding an ecosystem into a large reservoir in a stationary regime. Mathematical interest is caused by the possibility to get such effects as single branch bifurcations for a one-dimensional area, which can be observed at $n>1$ under conditions (2).

In the case, when the number of parameters is greater than one, it is convenient to distinguish parametric areas corresponding to different conditions of SS existence. For systems, close to those of a Volterra type, one can introduce parameters characterizing this closeness. Thus, for the well-known systems, such as Gierer-Meinhardt systems or "predator-pray" systems, we can write explicite conditions, under which the corresponding families in parameter space can fall within the regions of SS appearance. For details see [2].

Processes of space propagation can also be simulated by system (1). Since natural waves move as fronts, it is here enough to take $n=1$, $\Omega=R$, the spatial coordinate being considered normal to the front surface. The classical work by Kolmogorov, Petrovskii, Piskunov allows the problem on the character and velocity of the wave of transition from unstable into stable equilibrium for $m=1$ to be exhaustively solved. If some conditions formulated there are violated the character of solutions may be changed. Nevertheless, it appears that the properties and character of wave propagation can be studied in some detail and, moreover, some interesting peculiarities can be discovered. It can be done by analysis of the proper self-similar systems.

Consider an example. Let the dynamics of an epidemic process (e.g. gonorrhea or any other venereal disease) be described by system (1) with $m=2$, $f=f_1=-f_2$, which corresponds to conservation of the total number of sick (N_2) and healthy (N_1) persons. If $D_1=D_2$ and $f=-v+wN_2$, where v and w are the contagion and recovery rates, then for

$v(N_1,N_2) = \frac{\alpha N_1 N_2}{N_1+N_2}$; $w(p) = \min\{w^*, \frac{c}{p}\}$ $p = N_2/(N_1+N_2)$ in the case of

adequate choice of coefficients two consecutive epidemic waves are obtained, which correspond to transfer from one equilibrium state to a second, and then to a third one [3]. It is possible to calculate the velocities of the waves and to define the character of propagation for the case, when one wave overtakes the other. In fact, the problem can be reduced to the case $m=1$. More complicated problems arise in "resource-consumer" type systems with $m=2$ and $D=0$ (N_2 corresponds to a fixed, e.g., plant resource). It turns out that in this case the front propagates just as described above, the front velocity being defined by the diffusion coefficient D_1, corresponding to consumer mobility. The problems of practical interest can also be solved here. One of them is to find the velocity of wave propagation through a "dead zone", i.e. through the zone devoid of resource. Investigation of bifurcations of self-similar solutions also allows periodic traveling waves, which can be interpreted as "waves of life", to be revealed. These results are confirmed by numerical experiments [4]. Let us point out the most typical of those systems:

$$\begin{cases} \frac{\partial R}{\partial t} = Q - V(R)N \\ \frac{\partial N}{\partial t} = D\Delta N + (kV(R)-m)N. \end{cases} \tag{3}$$

Here $R(x,t)$ and $N(x,t)$ are the resource and consumer densities, X is a one-dimensional spatial coordinate. Q is the resource biomass growth rate. $V(R)$ is a trophic function, characterizing the rate of resource consumption by unit consumer biomass and depending on resource density. This function is usually considered to increase monotonously $V(0)=0$, $V(+\infty) = V^* < +\infty$. k is an economical coefficient, which shows the efficiency of resource biomass into consumer biomass digestion, m is the mortality rate of the consumer. If Q=const, the system (3) has a stable stationary solution

$$N^* = kQ/m, \; R^* = V^{-1}(m/k).$$

In the case of a local outbreak of the consumer population, this distribution establishes following a wave of "eating out" which has the velocity $v = 2\sqrt{D[kV^*-m]}$. The dead zones, where $Q=0$ and $R(x,o)=0$, are overcome with constant velocity which does not depend on the zone size

$$v_z = (kV^* - m)/\sqrt{D/kV^*}.$$

If $Q=0$ (i.e. the resource is unrenewable) and the initial resource density is R_o, the outbreak of consumer population generates a single wave, propagating with the velocity

$$v = 2\sqrt{D[kV(R_o)-m]}.$$

The case of a self-restoring resource is of special interest, e.g. for $Q=\alpha(R)R$, $\alpha(o)>o$, $\alpha'(R)<o$, $\alpha(\hat{R})=o$. When the parameters are properly chosen, the self-similar system with $\xi = x + vt$, $v = 2\sqrt{D(kV(\hat{R})-m)}$ has periodic solutions, corresponding to a sequence of waves, propagating with the above velocity. It means that a local outbreak of consumer population, arising at a certain moment of time, starts generating waves, which propagate over consumer-occupied, ecologically active medium. For more details see [5].

In conclusion we should try to show, how the attempts to solve some ecological problems leads to a new, non-traditional formulation of mathematical problems. Let $N(x,t)$ and $R(x,t)$ describe, as above, the consumer and resource distributions in a one-dimensional area $-\infty<x<+\infty$. We assume that the uptake rate of the resource located at ξ by the unit consumer biomass located at x is

$$P(|x-\xi|) \; V(R(\xi,t)).$$

Here $P(|x-\xi|)$ can be presented by the density of the normal distribution centered at ξ with dispersion σ^2. Then the equation for this model takes the form

$$\begin{cases} \frac{\partial R(x,t)}{\partial t} = Q - \int_{-\infty}^{+\infty} P(|x-\xi|)\cdot V\cdot(R(x,t))N(\xi,t)d\xi \\ \frac{\partial N(x,t)}{\partial t} = k\int_{-\infty}^{+\infty} P(|x-\xi|) \; V(R(\xi,t))N(x,t)d\xi - mN(x,t). \end{cases} \tag{4}$$

Here Q is the input flow of the resource, k is the efficiency, m is the coefficient of natural mortality. Let L be the specific spatial scale of the problem. If $\sigma << L$, that is the effective radius of consumer-to-resource interaction is small, system (4) comes to its asymptotic analog

$$\begin{cases} \frac{\partial R}{\partial t} = Q - V(R)\left(N + \frac{\sigma^2}{2}\frac{\partial^2 N}{\partial x^2}\right) \\ \frac{\partial N}{\partial t} = kN\left(V(R) + \frac{\sigma^2}{2}\left(V''\left(\frac{\partial R}{\partial x}\right)^2 + V'\frac{\partial^2 R}{\partial x^2}\right)\right) - mN. \end{cases} \tag{5}$$

This system is interesting as it is, but for better understanding of its properties we take $V(R)=\alpha R$ and linearize it in the vicinity of the spatially homogeneous stationary solution $\{R^*, N^*\}$, where $N^*=kQ/m$, $R^*=m/k\alpha$. Denoting $z_1=R-R^*$ and $z_2=N-N^*$, and taking Q,m,k to be constant, we get

$$\frac{\partial \vec{z}}{\partial t} = A\vec{z} + D\frac{\partial^2 \vec{z}}{\partial x^2}, \quad \vec{z} = \{z_1, z_2\}, \tag{6}$$

where

$$A = \begin{Vmatrix} -\frac{\alpha kQ}{m} & -\frac{m}{k} \\ \frac{\alpha k^2 Q}{m} & 0 \end{Vmatrix}, \qquad D = \begin{Vmatrix} 0 & -\frac{m\sigma^2}{2k} \\ \frac{\alpha k^2 Q\sigma^2}{2m} & 0 \end{Vmatrix}.$$

Since the eigenvalues of matrix D are purely imaginary, the system (6) is not parabolic according to Petrovskii. Similar equations arise in quantum mechanics and therefore we refer to systems (4)-(6) as Schrödinger type systems. No substantial theory of such quasi-linear systems has as yet been offered. Meanwhile, their solutions, if any, may possess rather interesting properties. In particular, there may exist spatially periodic discontinuous solutions, belonging to the type of finite functions. Such solutions describe the so-called self-rarefaction processes in plant communities, when a stable discrete structure arises from the initially everywhere compact biomass distribution. However, this is a mere hypothesis and these new mathematical objects, Schrödinger systems, call for intensive study.

References

1. V.N.Razzhevaikin. - Zh. Vych. Matem. i Matem. Fiz., 20, 1328, (1980)
2. N.V.Belotelov, D.A.Sarancha. - Biofizika, 28, No 6, (1983)
3. V.N.Razzhevaikin, Yu.M.Svirezhev. - In: Proceedings of the Intern. Conference "Applied Mathematics in Immunology and Medicine", IFIP, Moscow, 1982, Published by North-Holland ed., 1983
4. V.N.Razzhevaikin. - In: "Mathematical Models in Ecology and Genetics", Moscow, Nauka, 1981, pp. 36-51 (in Russian)
5. S.Yu.Gavrilets, A.A.Gigauri, V.N.Razzhevaikin, Yu.M.Svirezhev - In: Proc. of All-Union Scientific Research Institute of System Investigations, Moscow, 1983 (in Russian)

Synergetics and Biological Morphogenesis

L.V. Belousov

Moscow State University, Faculty of Biology, SU-Moscow, USSR

The interest which biologists, particularly developmental biologists, presently show in synergetics, is not due to a transient scientific fashion. Indeed, the fundamental biological problems of organization and wholeness [1-3] closely correspond to those referred to in synergetics as cooperativity and coherence. In this paper we discuss three interconnected synergetic approaches to biological morphogenesis: (i) qualitative analysis of morphogenesis using the concepts of instability, stability and parametrical regulation; (ii) examination of the morphogenetical role of auto-oscillations and autowaves; (iii) morphogenetical imitating models.

(i) This approach has been initiated by Waddington [4], who has created the allegory of "epigenetic landscape", further elaborated in [5,6]. One of the main tasks of this approach is an experimental study of developmental instabilities. From the general viewpoint [7], the only possible way to reduce symmetry of a developing system and to create a non-preformed pattern is to bring the system through certain instabilities, this being the fundamental property of ontogenesis. As experimental criteria of instability we have taken significant temporal growth of variability of certain properties of a developing system. Those properties are: intensities of synthetic processes, a set of differentiation pathways, arrangement of rudiments.

The experiments have been carried out with embryos of sea urchins and amphibians.

For sea urchin embryos [8] the variability of protein synthesis (the ratio of standard deviation to the arithmetical average intensity of ^{3}H-leucin incorporation, Fig.1-A) has been compared with the dynamics of energy metabolism (concentration of free radicals, Fig.1-B) and with morphological (beginning of gastrulation) and metabolic differentiation, the latter being estimated by the value of regional differences of ^{3}H-leucin incorporation intensity, Fig.1-C.

It is seen that both morphological and metabolic differentiation are immediately preceded by a sharp increase of metabolic variability. At the same time, the qualitative set of synthesized proteins increases and then falls [9]. The observed phenomena may point to the loss of synthetic stability just before the irreversible differentiation. The temporal coincidence of the stability loss and the activation of energy metabolism may be interpreted in terms of a trigger model [10,11], which considers energy metabolism as a non-specific parameter controlling the transition from a non-differentiated to a differentiated state via instability.

Two series of experiments have been made with amphibians [12]. In the first series, the sets of differentiation potencies of non-induced, temporarily induced and completely induced explants of presumpti-

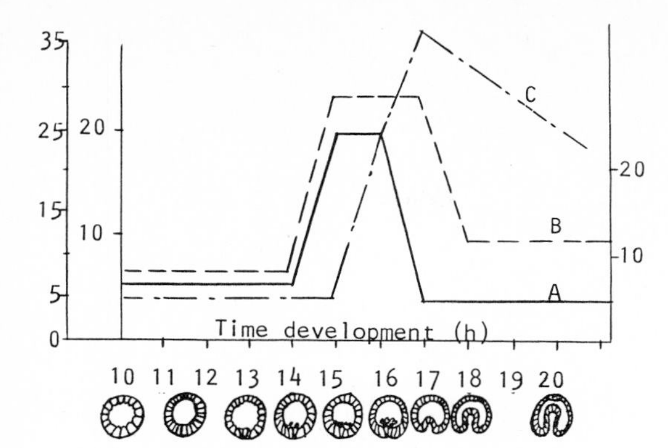

Fig. 1 Metabolic patterns of the successive stages of development of a sea urchin embryo (*Strongylocentrotus intermedius*). Ordinates are numbers of silver grains per 25 μm^2 (per unit area of radioautograph). Experimental data are averaged. Other notation is given in the text

ve neuroectoderm have been compared (Fig.2). Though both non-induced (Fig.2-A) and completely induced (Fig.2-C) explants reveal only a small number of potencies, just after induction (Fig.2-B) a real "explosion of potencies" takes place. Namely, notochord, hypochord and somites, which never arise normally from this tissue, develop in highly variable arrangement.

In the second series a sharp increase of morphological variability of the dorsal tissue explants was detected (Fig.3-B,C) just during the transformation of the initial "blastula-like" morphogenetic field Fig.3-A) to the "axial" field (Fig.3-D).

According to the presented data, it is probable that both morphogenesis and cytodifferentiation at certain crucial moments actually pass through instabilities. Hence, the problem is to reveal the parameters controlling the ontogenetical behavior at these moments. Earlier we have proposed [14] the following multilevel hierarchy of ontogenetical parameters: genetic parameters, biochemical parameters (intensity of energy metabolism, concentrations of gene-regulating and inducing substances), and morphogenetic field (structurally stable patterns of cellular motions, e.g. those corresponding to Fig.3-A,D). The characteristic time scales of these parameters decrease successively down to a few hours (morphological field). The time scales of the dynamical variables of morphogenesis (cytoskeletal and membrane reconstructions) are, however, even shorter (minutes). It is of special importance that at certain crucial moments of strong developmental instabilities those processes are tightly coupled with the initial stages of cytodifferentiation [15].

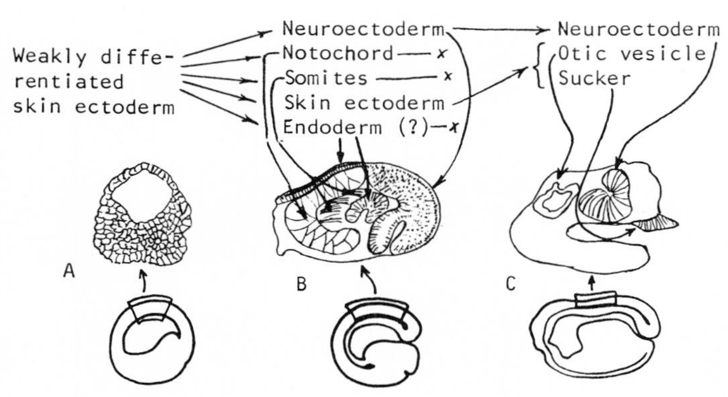

Fig. 2 Changes in the set of differentiation potencies during primary embryonic induction. Schemes of explantations are in the lower part, experimental results in the upper part. For other explanations see text

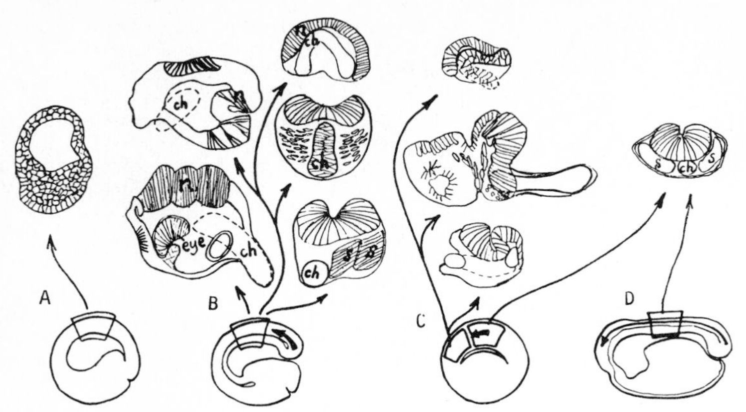

Fig. 3 Changes of morphological variability in the explants from successive stages of amphibian embryo development. Solid lines in the lower part indicate the wave of the formation of an "axial" morphogenetic field [13,28]

In our opinion, further elaboration of this approach might lead to replacement of the traditional model of development, which considers development as a complicated network of highly specific causal relations, by the concepts of potential relief, parametric regulations and interlevel relations.

(ii) In the recent literature, autowaves in developing systems are related, as a rule, to slime moulds [16]. Table 1 shows that the autowave phenomena are much more widespread. In addition to the presented data, linear growth oscillations, which are possibly coupled with autowaves, have been detected in brown and red algae and in fish embryos [26]. Remarkable is the similarity of the wave periods in objects 1.-9., listed in the Table, and of the rates of all waves, but for object 9.. However, little is known about the nature of waves in objects 2.-11. Contact cell polarization is connected with cytoskeletal reconstructions [25] and with activation of translation [27], whereas Hydrozoa autowaves seem to be based upon osmotical processes and also on cytoskeletal transformations. One of the main functions of autowaves, at least in objects 2., 10. and 11., is the generation of mechanical stresses of pressure (2.) or tension (11.) leading to polarized geometrical and/or topological reconstructions of embryonic tissue [17,18,28]. Contact cell polarization can also transmit certain types of cytodifferentiation [12]. Linear oscillations in Hydrozoa are affected by ionic shifts, electrical currents, cyclic nucleotides and biogenic amines. On the other hand, oscillations are not influenced by exogeneous cAMP and therefore, are not identical with slime mould autowaves [26].

(iii) Morphogenesis-imitating models may be divided into chemical (reaction-diffusional - RD) and mechanical ones. At present, RD-models are mostly developed [29]. However, there is no direct evidence of the correspondence between certain morphological structures and concentrational peaks, which are postulated in these models. Under these conditions more attention should be paid to mechanical models which principally can reproduce the complicated shapes even in chemically homogeneous or slightly inhomogeneous systems [17,18,30].

These models are also attractive because they indicate a clear connection between successive geometrical shapes of developing rudiments. Naturally, this does not mean that RD-models are to be completely rejected. We suggest, however, that at least during formation of the main embryonic rudiments and cell types, the running autowaves, rather than stationary dissipative structures, play a leading role. It is desirable that the models describing polarized propagations of autowaves together with possible chemical and mechanical effects should be elaborated.

Table Autowaves in developing systems

No	Developing object	Type of wave	Rate, μm/min	Period min	References
1.	Slime mould *(Dictyostelium discoideum)*	Chemotactical relay wave	43	5	16
2.	Hydrozoa *(Thecaphora)*	Proximo-distal wave of cell reorientation	30-70	5-16	17,18
3.	Hydrozoa *(Hydra attenuata)*	Peristaltic annulations	10-100	5	19
4.-8.	Eggs:				
4.	*Potonema vulgaris* (Nematoda)	Cortical peristaltic waves	15-20	12-16	20
5.	*Pollicipes* (Crustacea)	Same, during meiosis and segregation	14-20	2	21
6.	*Limnaea stagnalis* (Gastropoda)	Subcortical motion of yolk granules	12	30	20
7.	Amphibians	Postfertilization and precleavage waves of cortical contraction	60	30	21,22
8.	Teleostei	Circular waves of ooplasmic invagination	3800-300	1-22	23
9.	Blastoderm of chick embryo	Waves of cell contraction	200*	2-3	24
10.	Neuroectodermal explants from gastrula stage, amphibians	Contact cell polarization (CCP)	5-10	- *	12
11.	Ectodermal Ectomesodermal explants from tail-bud stage, amphibians	(CCP)	60-100	- *	25

* CCP is an irreversible wave of transformation of a cell from a morphologically unpolarized state to polarized one. Objects 1.-7. and 10., 11. were observed at 15-22°C, object 9. at 37°C, object 8. in the range 1-30°C.

The author is greatly indebted to Drs. D.S.Chernavskii, G.I.Solyanik, and Yu.A.Labas for helpful discussions.

References

1. H.Driesch. Philosophie des Organischen. Leipzig, Engelmann (1921)
2. A.G.Gurvich. Selected works. Moscow, Meditsina (1977) (in Russian)
3. I.I.Shmalgauzen. Selected works, Moscow, Nauka (1982) (in Russian)
4. C.H.Waddington. Organizers and Genes. Cambridge Univ. Press (1940)

5. R.Thom. Stabilite structurelle et morphogenese. Renjamin, N.Y. (1972)
6. E.C.Zeeman. Lect. Math. Life Sci., (1974), 7, 69-161
7. M.V.Volkenstein, D.S.Chernavskii. J. Social Biol. Struct., (1978), 1, 95-108
8. T.V.Ostroumova, L.V.Belousov, E.G.Mikhailova. Ontogenez, (1977) 8, 323-334
9. P.A.Bodard, B.P.Brandhorst. Devel. Biol., (1983) 96, 74-83
10. Yu.M.Romanovskii, N.V.Stepanova, D.S.Chernavskii. Mathematical modelling in biophysics. Moscow, Nauka (1975) (in Russian)
11. D.S.Chernavskii, G.I.Solyanik, L.V.Belousov. Biol. Cybern., (1980) 37, 9-18
12. L.V.Belousov, K.V.Petrov. Ontogenez* (1983) 14, 21-29
13. L.V.Belousov. Ontogenez* (1979), 10, 120-129
14. L.V.Belousov. Zh.Obshch.Biol. (1983), 44, 23-30 (in Russian)
15. M.J.Bissel, H.G.Hall, G.Parry. J. Theor. Biol., (1982) 99, 31-68
16. G.Gerisch. Biol. Cellulaire (1978) 32, 61-68
17. L.V.Belousov. Publ. Seto Mar. Biol. Lab. (1973) 20, 315-366
18. L.V.Belousov, Ya.G.Dorfman. Amer. Zool (1974) 14, 719-734
19. R.D.Campbell. In: Devel. and Cell Biol., Eds. Coelentorates P. and Tardent R., Elsevier/ North-Holland Biomedical Press (1980) 421-428
20. V.N.Meshcheryakov, L.V.Belousov. Spatial organization of cleavage, VINITI, Moscow (1978) (in Russian)
21. V.D.Vacquier. Devel. Biol. (1981) 84, 1-26
22. M.Yoneda, Y.Tobayakawa, H.Y.Kubota, M.Sakai. J. Cell Sci. (1982) 54, 35-46
23. P.N.Reznichenko. Transformation and change of the functional mechanisms in the ontogenesis of lower vertebrates, Moscow, Nauka (1982) (in Russian)
24. C.D.Stern, B.C.Goodwin. J. Embr. Exp. Morphol. (1977) 41, 15-22
25. L.V.Belousov, N.N.Luchinskaya. Tsitologiya (1983) 25, 939-944
26. Yu.A.Labas, L.V.Belousov, L.A.Badenko, V.N.Letunov. Doklady Akad. Nauk SSSR,(1981) 257, 1247-1250
27. L.E.Zavalishina, L.V.Belousov, T.V.Ostroumova. Ontogenez*, (1980) 11 491-500
28. L.V.Belousov. Sov.Sci.Rev.,Sect.D (Biology)(1982) 3, 343-370
29. H.Meinhardt. Models of Biological Patterns Formation, Acad.Press, L. (1982)
30. L.A.Martynov. In: Mathematical Biology of Development, Eds.Zotin A.I.,Presnov E.V., Moscow, Nauka (1982) 135-154 (in Russian)

Collective Phenomena in the Multicellular Development of *Dictyostelium Discoideum*

B.N. Belintsev

Institute of Molecular Biology of the USSR Academy of Sciences, SU-Moscow, USSR

The problem of modelling of embryogenesis will be simplified if we initially restrict ourselves to phenomenological description of the observed activities of individual cells. Then the key problem is: how are these activities transformed into the global behaviour of cellular ensembles?

In this work we deal with the multicellular development of the slime mold *Dictyostelium discoideum* (Dd). It seems to be the most suitable experimental model because the number of arbitrary assumptions in the description of its development can be reduced to a minimum. The slime mold cells show the simplest and reliably reproducible behaviour pattern during the multicellular morphogenetic phase [1]. In addition, they possess an intercellular signalling system which is among the best studied [2].

Fig.1 shows the scheme of the successive phases of the *Dictyostelium* life cycle. The coherent properties of the population of Dd amoebae is the response to food depletion. At the early morphogenetic stage (interphase) the coherence is manifested by the spatial organization and co-ordination of aggregative cell movements [3] (Fig.1c).

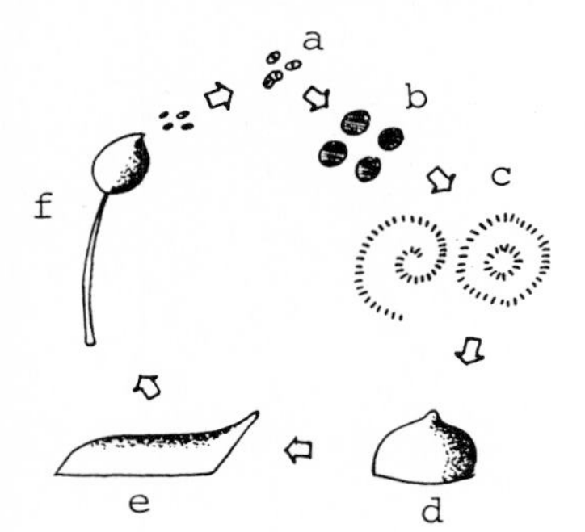

Fig. 1 The scheme of Dictyostelium multicellular development

The system of chemical signalization underlying the co-ordinated cell movement remains the same throughout the development from separate amoebae to migrating slug [2]. The 3-5 cyclic AMP (cAMP) is the signal substance - an attractant [4]. The cells respond chemotactically to cAMP spatial gradient, and they are able to produce and release cAMP.

The following succession of cAMP secretion regimes is reliably reproduced in the course of natural development as well as in the experiments with suspension of Dd cells [5].

i. The continuous weak emission of cAMP.

ii. The relaying competence exhibited by the majority of cells (the cells secrete cAMP pulse in response to an external pulse of cAMP) alongside with autonomous signalling by a small portion of cells. The latter are located in the centre of every aggregation field.

iii. The intense continuous cAMP release by the cells at the aggregate tip.

These changes in the regimes of individual secretion can be detected by alteration of the collective behaviour of the cell population. A small-scale cell clusterization is usually observed at the early interphase and is followed by the pulsatile large-scale cell aggregation as the properties (ii) develop.

The theoretical problem is to deduce the observed modes of collective behaviour from the known or assumed properties of individual cells.

1. Formulation of the Mathematical Model

It is assumed that the whole sequence of the dynamic modes i to iii is governed by the same structural scheme of biochemical reactions [6]. Then a parameter should exist which influences the dynamics of cAMP production and release, on the one hand, and is determined by the duration of cell starvation (developmental age), on the other. The net protein content can play the role of such a parameter.

Protein content is reported to decrease during the development of starving Dd cells [7]. Owing to this the Dd cells can "sense" their developmental age [8]. On the other hand, a low molecular product of protein degradation (such as ammonia) can inhibit the synthesis of cAMP [7,8].

The kinetic parameters of the dynamical model under investigation are the concentrations of cAMP, C, and the inhibitor, h, the reserve of the inhibitor precursor, R, and the cell surface density, b. The equations simulating the signal secretion account for the externally induced intracellular cAMP production and its release into the extracellular space (expressed phenomenologically as a nonlinear autocatalysis with saturation); cAMP diffusion and linear decay, the latter being the effect of phosphodiesterase; the inhibition of the autocatalytic step.

A possible (though not unique) form of the appropriate mathematical model can be written as follows:

$$\partial c/\partial t = bf(c,h) - qc + D_c \nabla^2 c \qquad (1a)$$

$$\partial h/\partial t = mRc - ph \qquad (1b)$$

$$\partial R/\partial t = -\varepsilon Rc \qquad (1c)$$

$$\partial b/\partial t = -n(b-b_o) - \mu\nabla(b\nabla c) + D_b \nabla^2 b. \qquad (1d)$$

Here the evolution equation for the cell density distribution (1d) incorporates the chemotactic response to an external signal ∇c [9] (the drift term) and the cell diffusion.

The gradual change of parameter R according to Eq. (1c) determines the slowest time scale in the system (hours). This change is supposed to control the switching of individual cAMP secretion regimes from i to iii [8].

2. Cell Clustering

At the beginning of starvation the R value is supposed to be great. This implies the large quantity of the inhibitor h (Eq. (1b)). As the plot in Fig.2 shows, the rate of cAMP production f is little dependent on both C and h values in this case. Therefore, considering f(C,h) in Eq. (1a) as a constant parameter f_o (which is determined from experiment), one gets the regime i.

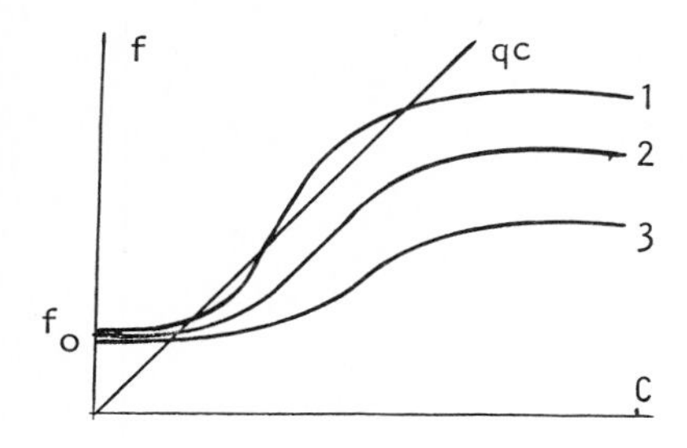

Fig. 2 The rate of cAMP production as a function of cAMP and inhibitor concentrations. Curves 1 to 3 correspond to the values of $h: h_1 < h_2 < h_3$

Now Eqs. (1a) and (1b) form a closed system and can be studied separately. The chemotactic term in Eq. (1d) can, in general, promote the local cell accumulation in spite of diffusion. For cell clustering to occur, the nonuniform attractant distribution is necessary. The question is whether the exfoliation can spontaneously originate in the entirely homogeneous system ($b \equiv b_o$; $C \equiv C_o$). The kinetic equations do allow such possibility. The exfoliation and cell clustering in the population of amoebae occur as a result of breaking the stability of homogeneous steady state distribution $b \equiv b_o$, $C \equiv C_o$ [10]. Small amplitude modes of random perturbations with a definite wave-number grow due to this instability, the nonlinear term in Eq. (1d) being sufficient for its stabilization at a certain macroscopic level [10]. The instability threshold is reached with the increase of parameters b_o, μ, and f_o. This actually takes place at the early multicellular development.

3. Wave Aggregation

The next developmental stage follows a decrease in R. The new features of the dynamics of cAMP production (ii) result from the alteration of the phase portrait of subsystem (1a),(1b), following the decrease in R. The simultaneous presence of two activities - relaying and signalling - in the originally uniform cell population is believed to be due to preclusterization of the cells [11]. Indeed, during cell accumulation the cAMP concentration is increased, and, accordingly, the R value falls more significantly here than in the surrounding space (see Eq. (1c)). As a result, the following situation may occur: the cluster cells have overcome the threshold of oscillatory instability (nullcline III in Fig.3), while the surrounding cells have not yet (nullcline II in Fig.3). In other words, the pacemaker area turns out to be encircled by the excitable medium. The resultant global behaviour presents a problem to be examined.

At a sufficiently large cell density ($>7 \cdot 10^4$ cells/cm^2), the clusterization is only the first step in the experimental aggregation process. It is followed by the stage of cell accumulation organized via propagation of attractant waves, which are generated by the autonomously signalling cells. Every aggregation centre controls cell motion at a range of 1-2 cm, which exceeds significantly the range of attractant diffusion (~100 μ).

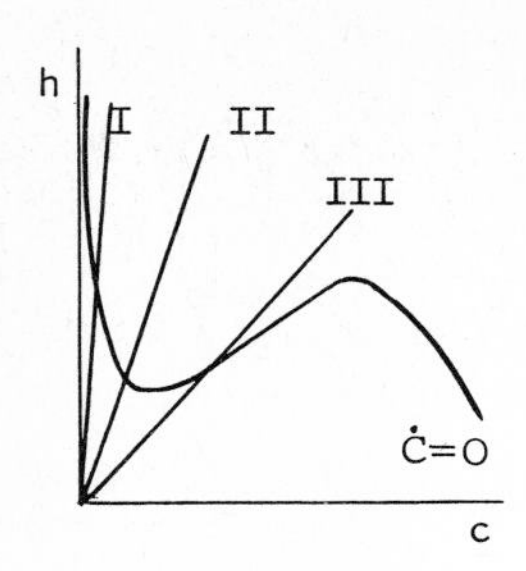

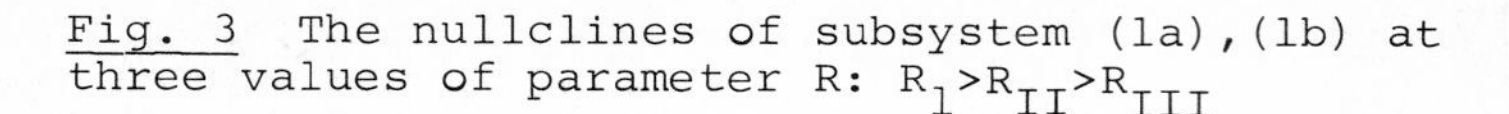
Fig. 3 The nullclines of subsystem (1a),(1b) at three values of parameter R: $R_{I}>R_{II}>R_{III}$

Two kinds of wavefront geometry can be observed in the same aggregation field: spiral and concentric rings (3). The question is whether they are generated by the same pulsing activity of signalling cells.

On the time scale of the process in question (10 min) the cell rearrangement has no time to occur and, therefore, is neglected. Thus, only Eqs. (1a),(1b) describing the signal production are to be considered.

For the total field of aggregating amoebae, the same Eqs. (1a), (1b) can be used. However, the values of parameter R in Eq. (1b) must be taken different in the pacemaker area and in the surrounding field of relaying cells (nullclines III and II in Fig.3, respectively).

As a first approximation, one can consider separately the behaviour in these two regions. In the vicinity of the oscillatory instability point (emergence of the pulsing activity), the approximate analytical investigation for the inside circular domain was carried out [11]. The result can be represented by the bifurcation diagram shown in Fig.4.

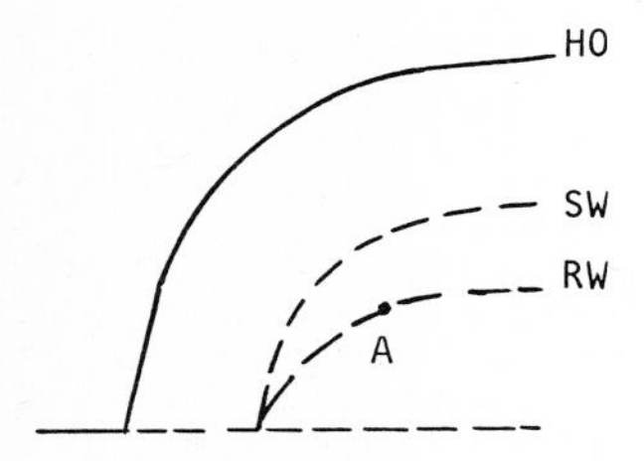

Fig. 4 Branching of the oscillatory regimes in the circular domain. Solid lines - stable regimes, dashed lines - unstable regimes

It is seen that the homogeneous oscillations are the only stable solution near the first bifurcation point (branch HO). The alternative stable regime appears as a result of the secondary bifurcation (at a point A). This new solution represents nonhomogeneous oscillations, viz. a propagating circular wave (branch RW). The interval between the primary and secondary bifurcation points decreases with the increase of the oscillating area (for details, see [11]). There is also a solution like a standing wave (branch SW), but it is unstable.

Now the global behaviour in the aggregation field can be described qualitatively by the oscillation modes outlined. The space surrounding the pacemaker area reveals all the properties inherent in the axiomatic excitable medium. Such a medium is known to be able to transmit the signal with a definite velocity. Hence the circular pacemaker region apparently generates expanding concentric rings when the oscillations in that region are spatially uniform (branch HO). In the alternative regime of propagating circular waves, a pacemaker loop is formed on the boundary of the oscillating region and therefore the rotating spiral fronts must be excited in the surrounding area [11]. This corresponds qualitatively to the observed large scale behaviour in the Dd aggregation field.

References

1. W.F.Loomis. Dictyostelium discoideum. A Developmental System N.-Y. Acad. Press (1980)
2. R.H.Kessin (1981) Cell 27, 241
3. F.Alcantera, M.Monk (1974) J. gen. Microbiol. 85, 321
4. T.Konijn, D.Barkley, Y.Chang, J.T.Bonner (1968) Am. Nat. 102, 225
5. A.D.J.Robertson, J.F.Grutsch (1981) Cell 24, 603
6. M.S.Cohen (1977) J. theor. Biol. 69, 57
7. M.Sussman, J.Schindler, H.Kim, in: Dev. and Dif. in the Cellular Slime Moulds, eds. Cappuccinelli and Ashworth (1977) Elsevier/North-Holland
8. P.S.Hagan, M.S.Cohen (1981) J. theor. Biol. 93, 881
9. E.F.Keller, L.A.Segel (1970) J. theor. Biol. 26, 399
10. B.N.Belintsev (to appear in Izv.Acad.Sci. of USSR, Ser. Biol.)
11. B.N.Belintsev, I.D.Judin (to appear in Izv. Acad. Sci. USSR, Ser Biol.)

Study of "Target Patterns" in a Phage-Bacterium System

G.R. Ivanitsky, A.S. Kunisky, and M.A. Tzyganov

Institute of Biological Physics, USSR Academy Sci., Moscow Region
SU-142292 Pushchino, USSR

A phage-bacterium system can be considered as a population model and as one of the simplest models of epidemic spreading.

In the present work the ring-shaped structures produced by phage-infected bacteria on agar have been studied.

It is known that interaction between the phage and bacterium sensitive to the phage action results in the bacterium lysis accompanied by liberation of newly formed phage virions into the medium. Phage action on bacteria (at certain concentrations) on agar leads to formation of local lysis areas which we term plaques. The formation of a plaque is initiated due to infection of a bacterium by a phage at the early stage followed by interaction of the phage progeny and the bacteria from the bacterial lawn. The size of the plaque depends on that of the phage virion (the smaller the phage, the higher its diffusion coefficient), on the rate of phage adsorption on the surface of uninfected cells, and on the period of intracellular phage development. Morphology of the plaque depends, as a rule, on the type of bacteriophage [1,2]. Certain phages produce plaques surrounded by a typical halo. A possible mechanism of the halo formation was proposed by Adams and Park [3]. They showed, that phage-infected bacteria of a type 2 strain of *Klebsiella pneumonia* liberate an enzyme capable of hydrolyzing the capsular polysaccharide of this organism. The action of the enzyme on the host cell increases its resistance to the phage infection and thus leads to the formation of haloes about the plaques.

A mathematical model of reaction-diffusion type was studied under the assumption that phagolysis, possibly, causes liberation of inhibitory enzymes, which, at certain concentrations, increase the adsorption resistance of bacteria to phages [4]. A computer analysis of the model has demonstrated a possibility to create the conditions providing formation of the ring-shaped structures with characteristic alternation of zones of bacterial lysis and secondary growth. The kinetics of formation of the ring-shaped structures was supposed to be essentially determined by the rate of bacterial growth, which, in turn, depends on the composition of the culture medium. The above assumptions were confirmed by the experiments with bacteriophage T7, whose plaques possess haloes. In particular, the results obtained show, that essential increase of pepton content in the bottom agar composition results in formation of the plaques with multiring structures. Fig.1 demonstrates a plaque produced by bacteriophage T7 in the medium with the following concentration of the components (in g/l), top layer agar: NaCl-5, agar Difco-7, pepton-10; bottom agar: NaCl-8, agar Difco-12, pepton-42 (i.E. 4-fold pepton concentration).

Similar experiments were performed with bacteriophages T4, T2 whose plaques form no haloes. The bacteriophages studied are known to be markedly larger in size than T7, and essentially different

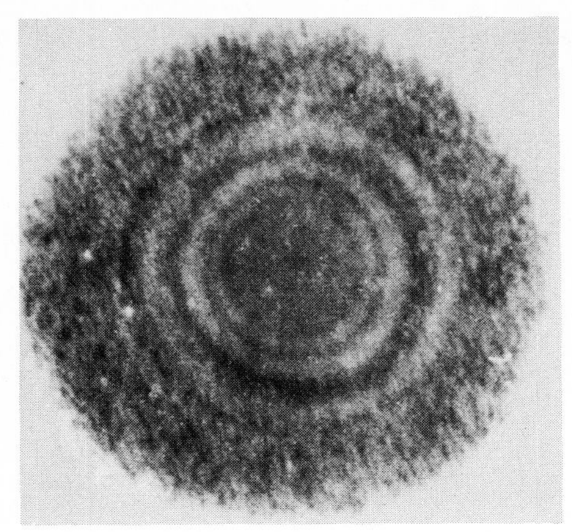

Fig. 1 The plaques produced by phage T7 at a 4-fold pepton concentration

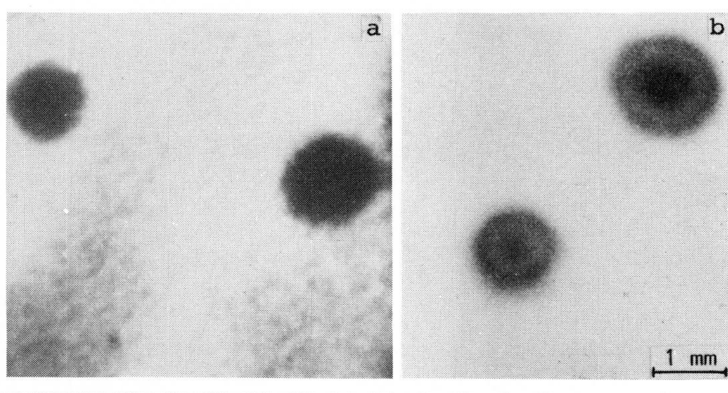

Fig. 2 The plaques produced by phage T4 at a standard (a) and 7-fold pepton concentration (b)

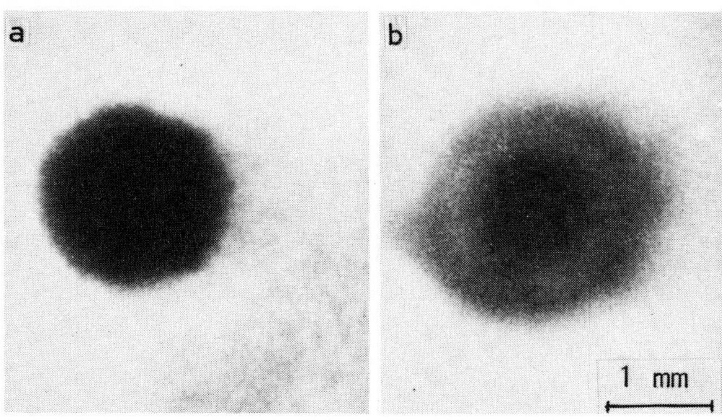

Fig. 3 The plaques produced by phage T2 at a standard (a) and 7-fold pepton concentration (b)

from it in the latent period and yield of phages upon bacterial lysis. The plaques produced by phages T4 and T2 are much less in diameter than those produced by T7. Fig.2 illustrates the plaques produced by T4 24 hr after the inoculation at both a standard and 7-fold pepton concentration. Similar results for phage T2 are shown in Fig.3.

Thus, the experiments show that the plaques with multiring structures can be obtained for different bacteriophages. Varying the composition of the culture medium, we produce, along with changes in diffusion coefficients due to agar viscosity, also a change in the rate of bacterial growth, which is, seemingly, the main cause of enhanced secondary growth of the bacteria.

The influence of the agar thickness and cultivation conditions (temperature, moisture) on the parameters of the ring-shaped structures was not analyzed in detail in the present study. Nevertheless we found that the plaques obtained are structurally identical,though their sizes vary to a great extent, since the onsets of infectioning for different bacteria do not coincide.

Annihilation of the rings (Fig.4) takes place upon interaction of the ring-shaped structures with each other.

The behaviour of the plaques near an obstacle is shown in Fig.5.

The plaques at different stages of their development were studied using the method of densitometry. It was shown that after propagation

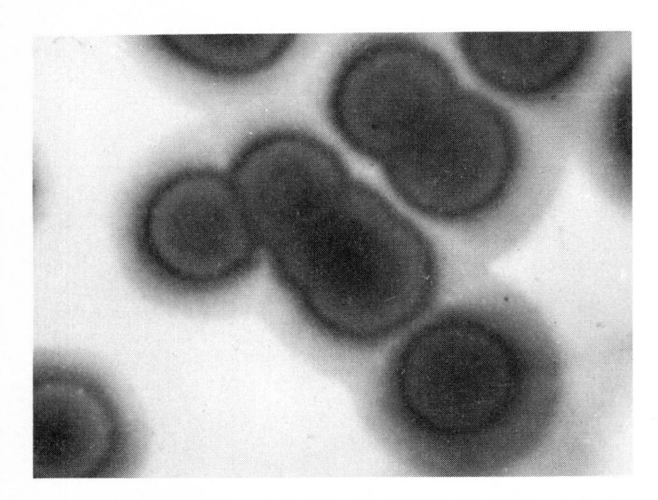

Fig. 4 The result of interaction between the plaques (phage T7)

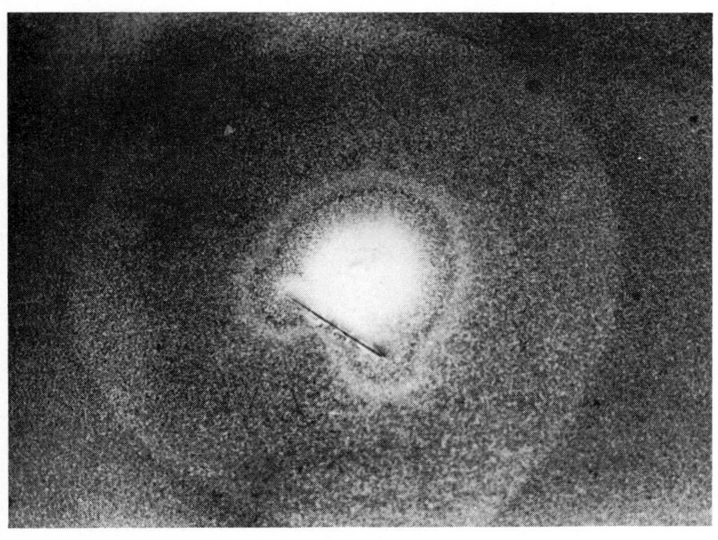

Fig. 5 The plaque (phage T7) near an obstacle

of a lysis wave due to the secondary growth, the haloes with stationary ring-shaped structures are formed. However it should be noted that we failed to observe the initial stage of the halo formation using densitometry.

References

1. M.H.Adams (1959) Bacteriophages. N.-Y.-London. Interscience Publ.
2. S.E.Luria, J.E.Darnell, Jr., D.Baltimore, A.Campbell (1981) General Virology. Moscow, Publ. House "Mir"
3. M.H.Adams, B.H.Park (1965) Virology, 2, 719-736
4. G.R.Ivanitsky, O.D.Veprintseva, A.A.Deev, A.S.Kunisky, A.A.Khusainov, M.A.Tzyganov (1982) Proc. I-st All-Union Biophysical Congress. Moscow, p.34

Plasmodium of the Myxomycete *Physarum Polycephalum* as an Autowave Self-Organizing System

S.I. Beilina, N.B. Matveeva, A.V. Priezzhev, Yu.M. Romanenko, A.P. Sukhorukov, and V.A. Teplov

Institute of Biophysics, Academy of Sciences of the USSR, SU-142292 Pushchino, USSR

and

Lomonosov State University, Faculty of Physics, SU-117234 Moscow, USSR

The plasmodium of the acellular slime mold *Physarum Polycephalum*, when migrating over a substrate, looks like a protoplasmic sheet with a network of channels, which transfer into separate protoplasmic strands or veins in the posterior. Within these channels and strands there is an intensive reciprocating endoplasmic streaming with the period of about 1 min. The streaming is caused by nonstationary gradients of intracellular pressure due to quasiperiodical contractile activity in the relatively stationary ectoplasm. The movement of plasmodium is driven by the advancing of the endoplasm in each cycle. There is an additional periodicity of about 30 min, connected with the frontal zone extending and the tail region tearing down. As a rule, the two processes are out of phase (Fig.1).

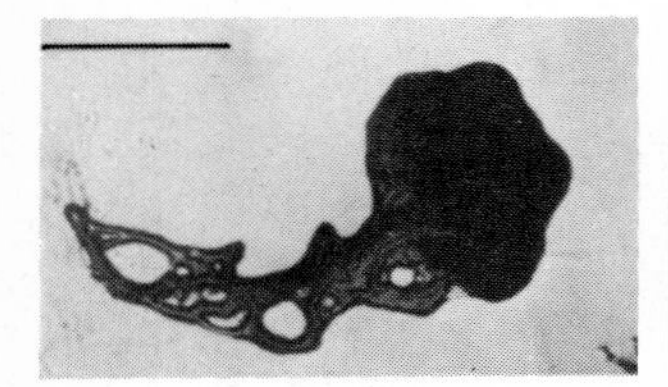
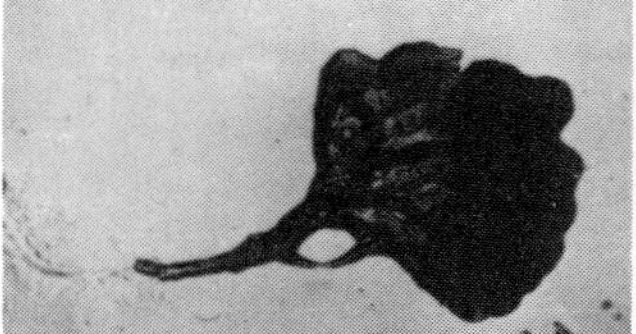
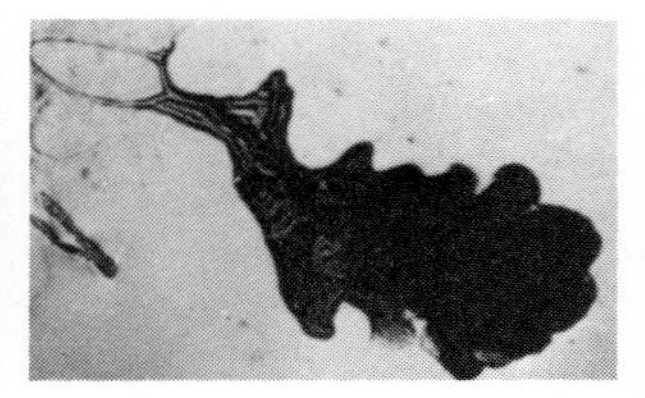

Fig. 1 Cyclical changes of the plasmodial form in the course of migration. 40 min between the frames. Bar 1 mm

The basic proteins of the contractile apparatus (as in muscle) are actin and myosin, but in the plasmodium the contractile machine is being periodically destroyed and reconstructed during migration [1]. Though this is a very interesting molecular aspect of self-organization, in this paper we restrict ourselves to macroscopic phenomena of motility, namely, to autowaves (AW). The following processes have been experimentally found to take place: 1) wave-like longitudinal and circular propagation of contraction-relaxation zones within the frontal region [2,3]; 2) quasistochastical oscillations of thickness in local sites of the frontal region [3];3) nearly synphase radial pulsations of the strands [4] ; 4) radial pulsations of the strands in the form of propagating or standing waves [5,6]; 5) periodical contractions and torsion oscillations of isolated fragments of the strands [7].

AW phenomena in migrating plasmodium are shown in Fig.2 as a phase diagram. The diagram was constructed on the basis of cinema pictures of a large plasmodium, migrating on non-nutritive agar gel along the track bordered by paraffin strips. Pulsations of thickness were recorded photometrically at the points situated along the direction of migration. In Fig.3 another essential manifestation of the AW properties of plasmodium is illustrated. It is the ability of mutual

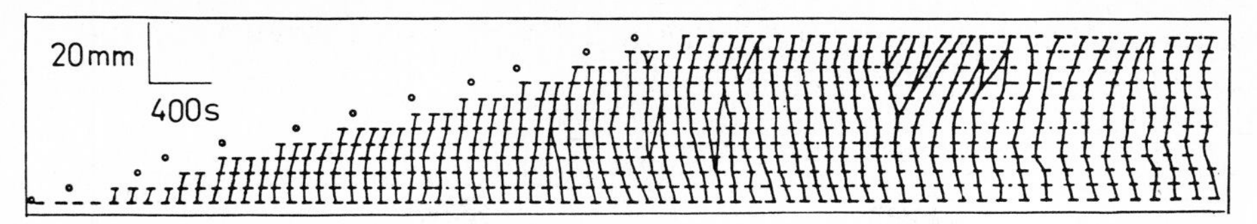

Fig. 2 Phase diagram of contractile activity in 12 points along the longitudinal axis of migrating plasmodium. The dark bars parallel to the time axis represent the contraction periods. The lines connect the points with the same phase. Circles correspond to the advancing front positions

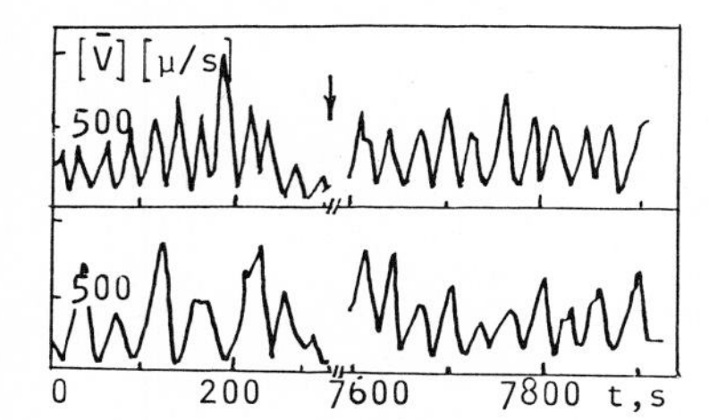

Fig. 3 Synchronization of the stream velocity oscillations in two interconnected protoplasmic strands. The arrow indicates the moment of connection

synchronization of two strand fragments, which are initially independent and have different oscillation frequencies [8,9]. Shown are the oscillograms of protoplasmic flow velocity modulus in such fragments before and after their being joint by a third one. Modulus of the average flow velocity has been registered by a laser Doppler anemometer on-line with a computer [10]. The resulting single frequency is seen to be established some time after the fragments being reconnected. It has been also shown that when period-changing stimuli are applied the resulting synchronous frequency is defined by the more powerful portion of the plasmodium [11,12].

Let us consider one of the possible mathematical models of the AW contraction-relaxation process in a thin long strand of radius R located along the Z-axis. According to this model, the pressure in a given strand cross-section, $P(t,z)$, is created by small alterations of the radius, $x<<R$, and is the sum of the active, P_a, and passive, P_p, components. The passive component can be easily defined via Young modulus, E, and the viscosity coefficient, χ, of the strand wall:

$$P_p = EhR^{-2}x + \varkappa hR^{-2} \cdot \frac{\partial x}{\partial t} \cdot \tag{1}$$

Here h is the wall thickness, $h<<R$. The active pressure is assumed to be

$$P_a = \alpha[1-\beta x^2(t-\tau)] \frac{\partial x(t-\tau)}{\partial t}, \tag{2}$$

where α characterizes the magnitude of the active force, β is a nonlinear limitation factor [13]. P_a is time-delayed against the radius deviation, $x(t,z)$ by a time interval τ, which is less than the system period [14]. An alternative form to define the active pressure is the equation

$$\frac{\partial P_a}{\partial t} = \tau^{-1}[-P_a + f(x, \frac{\partial x}{\partial t}], \tag{3}$$

where $f(x,\partial x/\partial t)$ defines the existence of a negative friction in the system. The axisymmetric stream of the endoplasm with a parabolic velocity profile is described by the following hydrodynamic equation for the velocity averaged over the strand cross-section, $\bar{V}_z$,

$$\rho \frac{\partial V_z}{\partial t} = - \frac{\partial P}{\partial z} - 8\mu R^{-2} \bar{V}_z \,, \tag{4}$$

provided that the Reinolds number is very small: Re~10^{-3}. Here ρ and μ are density and viscosity of endoplasm, respectively. Relationship between $\partial x/\partial t$ and $\bar{V}_z$ is derived from the boundary conditions and the continuity equation

$$\frac{\partial x}{\partial t} = V_r \Big|_{r=R} = -0.5\ R \frac{\partial \bar{V}_z}{\partial z} \,. \tag{5}$$

Using (1)-(5) we obtain the equation describing the AW motion in a strand

$$\frac{\partial^2 x(t+\tau)}{\partial t^2} + \frac{1}{\rho R}\left[\frac{8\mu}{R}\frac{\partial x(t+\tau)}{\partial t} - \frac{\varkappa h}{2}\cdot\frac{\partial^3 x(t+\tau)}{\partial t \partial z^2}\right] - \frac{Eh}{2\rho R}\frac{\partial^2 x(t+\tau)}{\partial z^2} +$$
$$+ \frac{R}{2\rho}\frac{\partial^2 P_a(t,z)}{\partial z^2} = 0. \tag{6}$$

If τ is small compared to the relaxation time of the system, this time being of the order of the autooscillation (AO) period, then $x(t+\tau)=x(t)+\tau\frac{\partial x}{\partial t}$ holds. Thus, the solution of (6) can be found in the form of stationary waves: $x_n(t,z) = x_{on}(t)\cdot\cos(\frac{\pi n z}{L})$, where L is the length of the strand. For the amplitude of the n-th mode we get the Van-der-Pol type equation, n=1 corresponding to the first mode of AO in a "short" strand fragment. The AO frequency weakly depends on n and is estimated by the equation

$$\Omega_n = k\sqrt{\frac{Eh}{2\rho R}\left(\frac{\pi n}{L}\right)^2 \Big/ \left[1 + \frac{\varkappa h \tau}{2\rho R}\left(\frac{\pi n}{L}\right)^2\right]} \approx k\sqrt{\frac{E}{\varkappa \tau}} \,, \tag{7}$$

where k depends on the relaxation time of the system. Thus, the AO period in the quasiharmonic regime is independent on both viscosity and density of endoplasm as well as on strand geometry. At $E = 10^5 din/cm^2$, $\varkappa=10^6 P$, R = 1 mm, h = 0.1R, τ=1-4 s we obtain $T=2\pi/\Omega \sim 20\sqrt{\tau}$ =20-40 s and $\bar{V}_z \sim 0.1 - 0.5$ mm/s. Therefore, the model is self-consistent, i.e. the actual values of mechanical parameters do not contradict the velocity and period values and the condition $\tau \ll T$.

When simulating the processes of the plasmodium form organization and migration, the following experimental evidences must be taken into account:

1) the AW period is controlled by both internal and external parameters, in particular, by externally applied stimuli;
2) all types of plasmodium motile behavior (spreading, taxis, and migration) can be described by a piecewise linear dependence of temporal variations of the plasmodium area [15];
3) the increase of the plasmodium area extension time correlates with the decrease in AW frequency and vice versa (see Fig.4 and [12] for the case of chemotaxis), in some cases such events occur step-wise;
4) permanent water flow through a sheet of filter paper, put under an isotropically spreading plasmodium, makes the plasmodium take a polar form with the frontal edge counter-directed to the flow (the higher the flow rate, the lower is the migration velocity);
5) plasmodium avoids substrates containing the products of its excretion [11].

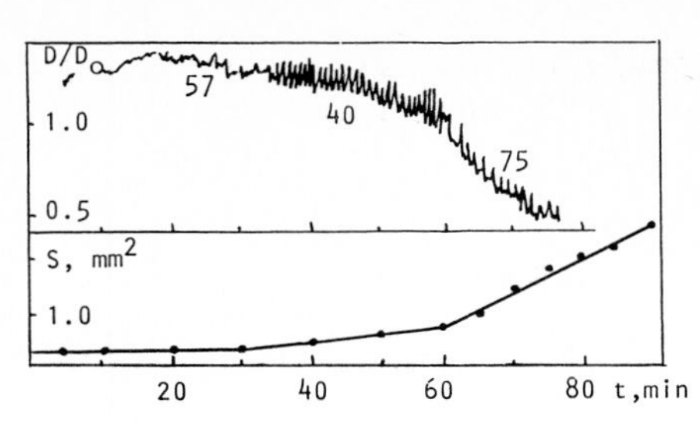

Fig. 4 The correlation between the rate of area growth and the frequency of contractile activity during spreading of the plasmodium regenerated from a protoplasmic drop

These findings allow one to suggest that migration is controlled via the plasmalemma (as in the case of chemotaxis) and is induced by extracellular regulator(s), the products of plasmodium itself, on reaching a threshold concentration. Thereby the coexistence of such opposing processes as mixing of the plasmodium intracellular contents and genesis of the polar form can be understood.

References

1. G.Isenberg, K.E.Wohlfarth-Bottermann. Cell Tissue Res., 173, 495 (1976)
2. P.A.Stewart. In: "Primitive Mobile Systems in Cell Biology", Acad. Press, New-York, 1964, p.69-78
3. Z.Baranowski. Acta protozool., 17, 377 (1978)
4. A.Grebecki, M.Cieslawska. Cytobiologie, 17, 335 (1978)
5. Z.Baranowski, K.E.Wohlfarth-Bottermann. Eur. J. Cell Biol., 27, 1 (1982)
6. V.B.Ermakov, A.P.Priezzhev. Biofizica, 6, (1983) (in Russian)
7. N.Kamiya, Y.Yoshimoto. In: "Aspects of Cellular and Molecular Physiology", Univ. Tokyo Press, Tokyo, 1973, p.167-189
8. Z.Hejnowicz, K.E.Wohlfarth-Bottermann. Planta, 150, 144 (1980)
9. U.Achenbach, K.E.Wohlfarth-Bottermann. Planta, 150, 180 (1980)
10. M.V.Evdokimov, A.V.Priezzhev, Yu.M.Romanovskii. Avtometriya, 3, 61 (1982) (in Russian)
11. V.A.Teplov, S.I.Beilina, N.B.Matveeva, D.B.Layrand, V.V.Lednev In: "Progress in Protozoology", Abstracts VI Int. Congr. Protozool., Warszawa, 1981, p.363
12. N.B.Matveeva, S.I.Beilina, V.A.Teplov, D.B.Layrand, V.V.Lednev. VINITI, No 2112-81, 1982 (in Russian)
13. Yu.M.Romanovskii, E.B.Chernyaeva, V.G.Kolin'ko, N.P.Horst. In: "Autowaves Processes in Systems with Diffusion", Gorkii, 1981, p.202-219 (in Russian)
14. V.G.Kolin'ko. To be published in "Dynamics of Cells Populations", Gorkii (in Russian)
15. S.I.Beilina, A.A.Budnitskii, D.B.Layrand, N.B.Matveeva, V.A.Teplov. In: "Progress in Protozoology", Abstracts IV Int. Congr. Protoz., Warszawa, 1981, p.28

Part VI

Evolution and Self-Organization

Violation of Symmetry and Self-Organization in Prebiological Evolution

L.L. Morozov and V.I. Goldanskii

Institute of Chemical Physics, Academy of Sciences of the USSR, SU-Moscow, USSR

1. Introduction

Profound ideas of synergetics play an important role in physical theories from high-energy physics to cosmology. They are also of great interest for chemists and biologists. In this communication we should like to outline the role of these ideas in the scientific picture of prebiological evolution, which is one of the most interesting and so far unsolved problems of the natural sciences.

Nowadays, studies of the origin of life are being developed mainly in two directions. In the first trend started by the experiment of Miller and Urey, attempts have been made to synthesize the basic biomolecules. The second trend,which was called "chemistry on a sheet of paper", consists in analyzing hypothetical chains of chemical conversions, capable to select the key biopolymers for primary organisms. In fact, the bulk of accumulated scientific data is just a collection of facts and hypotheses about possible conditions and ways to form elements of living matter. Though many of these hypotheses ("If-hypotheses" as termed by T.Ulbricht) are backed up by experiments and high-level models, any of them can be chosen arbitrarily. The common opinion is that life may originate in an organic medium as soon as necessary physical conditions are met, but, in fact, nobody knows what conditions are necessary, but for those of the present-day terrestial life. Must the organic medium be condensed and "hot" (something like a "warm, thick and salty" primordial bouillon [1]) or rarefied and cold [2] as in dark clouds? Must the organic matter be subjected to corpuscular and electromagnetic radiation or must it be protected against them? Answers to these and many other questions concerning the physical conditions for the origination of life are absent. Moreover, the principles such answers can be based on, are not clear.

In this communication we suggest the third way to solve the problem. We try to show that a physicochemical theory, equipped with synergetical ideas and methods, can help in understanding the state of prebiological matter, which had provided the conditions for the formation of the main properties of the bioorganic world.

Here we take only one property of the biosphere as a criterion: violation of mirror symmetry in biomolecules. Long ago, Pasteur regarded this violation as the most important material attribute of life. The case is that in bioorganic matter only L-aminoacids and only D-sugars but not their mirror isomers are employed. The biosphere is a self-reproducing, almost absolutely chirally-pure state of a tremendous number of molecules. (The term "chiral" was introduced by Lord Kelvin to designate objects non-identical to their mirror images [3]. We use the term "chiral purity" to describe a state where only one of two chiral forms is utilized).

To describe this phenomenon we shall use a parameter of chiral polarization [4,8]

$$\eta = (L-D)/(L+D), \tag{1}$$

where L and D are the quantities of the left and right mirror isomers of chiral molecules. In the sense of statistical physics $\tilde{\eta}=\langle\eta(\alpha);\ \eta(\alpha')\rangle$ is the parameter of ordering, α being the time or space coordinate. $\eta=0$ corresponds to the racemic state of the medium, in which equal quantities of L- and D-molecules are chaotically mixed; $\eta=\pm1$ corresponds to chirally pure states. In these terms, the origination of bioorganic world may be considered as a transition from the state $\eta=0$ to the state $|\eta|=1$, i.e., a transition with strong violation of chiral (mirror) symmetry. It is for more than a century that attempts have been made to reveal the reason for breaking the symmetry of biomolecules. The case is that there is no understanding of the general requirements the process of life origination should meet in order to break the mirror symmetry of a substance and to create an organic world with a high degree of chiral purity.

These questions are considered below.

2. Primary Factors of Symmetry Violation

a) Fluctuation Factor (FF)

Natural statistical fluctuations are the first cause to produce asymmetry in the initially symmetric mixture of mirror isomers. They can give rise to states which have some degree of chiral polarization. Chiral polarization of the initial state, conditioned by the fluctuation factor η_{of}, is a random value selected from the distribution with the mean value $\bar{\eta}=0$ and variance σ. If the number of chiral particles in the initial state is N_C and the variance $\sigma_N \sim N_C^{1/2}$, then the value $\sigma_\eta \sim N_C^{-1/2}$ may be taken to estimate the measure of FF.

b) Advantage Factor (AF)

The second possible reason for symmetry violation is that mirror isomers are not equivalent when engaged in chemical reactions. The only global factor of such non-equivalence is violation of purity in weak interactions. In principle, there are two mechanisms which might be responsible for this effect being manifested in molecular transformations. One of them is associated with neutral currents, i.e. with the fact that electrons can interact with nuclei not only via electromagnetic, but also via weak forces [9]. The second mechanism is based on longitudinal polarization of β-decay electrons [10,11], which, when interacting with mirror isomers, may, in principle, destroy one of them. A quantitative measure for AF may be written as

$$\varepsilon = (k_L-k_D)/(k_L+k_D), \tag{2}$$

where $k_{L,D}$ are rate constants of a reaction of optical isomers. The estimated values of AF are extremely small. Non-equivalence of mirror--isomeric molecules due to neutral currents is $\varepsilon_{int} < 10^{-17}-10^{-20}$ [12]. The estimate for radiation-chemical transformations conditioned by emission of naturally polarized α-decay electrons is, seemingly, not far from this value. As reported in [13], it amounts to $\varepsilon_{p.r.} \ll 10^{-10}-10^{-11}$. When analyzing the above schemes the question arises about the mechanisms capable of amplifying the microscopic asymmetry, produced by either FF or AF, and of transforming this asymmetry into the chiral purity of the final state. In fact, the question is what laws had governed the transformations of organic matter in the origination of the early biosphere; how the individual transformations had been organized and how they had interacted; by what dynamic laws the development of chiral polarization in the early biosphere had been controlled.

We use the following method of analysis [14].Consider a molecular system, consisting of two subsystems C and S. Subsystem C contains mirror-antipodal molecules L and D, subsystem S contains a non-chiral substance with concentrations of chiral molecules [L], [D] and [S],respectively. The dynamics of transitions of the substance between the two subsystems may be described by a system of differential equations:

$$
\begin{aligned}
d[L]/dt &= f_1([L], [D], [S]) \\
d[D]/dt &= f_2([L], [D], [S]).
\end{aligned}
\tag{3}
$$

The transformation $([L], |D|) \rightarrow (\eta, \Theta)$, where η is chiral polarization and Θ is the overall fraction of chiral molecules in the system $(\Theta = ([L] + [D])/([L] + [D] + [S]))$, leads to the equations:

$$
\begin{aligned}
d\eta/dt &= \varphi_1(\eta, \Theta) \\
d\Theta/dt &= \varphi_2(\eta, \Theta)
\end{aligned}
\tag{4}
$$

in the phase space with axes η and Θ. A set of points $\eta(\Theta, \eta_o, \Theta_o)$ makes up a trajectory of the evolutionary process in the space (η, Θ). The ensemble of trajectories is called a phase portrait of the molecular chirality evolution. In this space we shall consider violations of mirror symmetry in various scenarios of prebiological evolution.

3. Violation of Mirror Symmetry in the Early Biosphere. Evolution or Catastrophe

All versions of prebiological evolution available in modern literature can be subdivided in two classes.

Evolution. Origination of life is considered as a result of evolutionary improvements of organic matter. The biosphere arises as a new state of the initially lifeless matter in a continuous process having a stable dynamic trajectory in the space of proper variables. Today, the most popular version of origination of life is the evolution without singularities and discontinuities, based upon evolutionary perfection [1] of organic matter in a chain of consecutive events. In particular the well-known model of genesis of molecular ordering in prebiological evolution given by M.Eigen [5] is in this line.

Catastrophe. In the last decades an alternative to the consecutive evolution has been formulated. It considers evolution as bifurcation, catastrophe, phase transition, i.e. as such a process, in which the former phase trajectory loses stability. In this scenario of life origination there is a break of the chain of consecutive evolution, when the initial lifeless state becomes unstable. Models of prebiological evolution with phase transitions, with formation of dissipative structures, are connected with the work by I.R.Prigogine and his school [6]. Recently a model of this kind has been suggested by F.Dyson [7].

It is not clear *a priori* which scenario is to be favoured. To make a choice, it is necessary to study the possibilities of mirror symmetry violation in dynamic systems of the two types under real physical conditions.

Scenario of evolution. According to the general theory of dynamic systems [15], the process of stable continuous evolution is characterized by a quite definite topology of dynamic equations. For the dynamics of chiral polarization, as well as for any other parameter of order [16], such topology is featured by the equation $\dot{\eta}=f(\eta)$, where only the first three terms of the expansion $f(\eta)=\alpha_o + \sum_i \alpha_i \eta^i$ are taken into account [4,8,14,17]. In the case of mirror symmetry $\alpha_o=\alpha_2=0$. In the presence of AF all the terms in the equation for $\dot{\eta}$ must be kept. This kind of dynamics may be realized in many different processes:

i. Destruction of chiral material: $L(D) \xrightarrow{k_{L(D)}} S'$ ($k_{L(D)}$ are isomer destruction rate constants, see, e.g., [18]).

ii. Autocatalysis: $S + L(D) \xrightarrow{k_{L(D)}} 2L\ (2D)$ ($k_{L(D)}$ are rate constants of autocatalytic reproduction of the chiral substance). The first model of this type was suggested by P.Jordan [19].

iii. Autocatalysis + destruction of chiral particles in interaction:
$S + L(D) \xrightarrow{k_{L(D)}} 2L(2D)$; $L + L(L+D,\ D+D) \xrightarrow{k_a} S$ (k_a is a rate constant of collisional destruction of chiral molecules [20])

iv. The Eigen type of selection. Norden [18] used the Eigen [5] scheme of autocatalysis for selection of molecules of mirror isomers. The Eigen type equation for mirror isomers is

$$dx_{L(D)}/dt = (a_{L(D)}g_{L(D)} - d_{L(D)})x_{L(D)} - \phi_{L(D)}x_{L(D)} + \psi_m x_{L(D)}. \quad (5)$$

Here a_i and ϕ_i are the rate constants of the reproduction of isomers and of their dilution, g_i is the fraction of exactly reproducible particles, ψ_m is the rate constant of erroneous reproduction, $(a_i g_i - d_i)$ are selective values introduced by Eigen, they correspond to the reproduction rate constants, in our notation $k_{L(D)}$.

For every scheme mentioned above, the dynamic equation for chiral polarization is

$$d\eta/d\tau = \varepsilon(1-\eta^2), \quad (6)$$

where ε is the advantage factor, $\tau = 1/2(k_L+k_D)tS$. Naturally, the possibilities of chiral polarization development are the same in all processes of evolutionary type, since in a symmetrical world (AF=0) $d\eta/d\tau=0$ and η preserves the initial value, and in a non-symmetrical world (AF>0) $d\eta/d\tau>0$ and the non-zero advantage factor takes the system to the final chirally pure state. The sign of this pure state is predetermined by the sign of AF, no matter how small the values of ε may be. Thus, it may seem that we have an ideal model of amplification, which produces complete vanishing of mirror symmetry in an asymptotically attainable final state. However, such a model actually corresponds to ideal conditions, non-existing in nature. In terms of evolution of molecular chirality, this means that factors always exist, which cause relaxation of chiral polarization down to zero, i.e. to the state, corresponding to the racemic composition of mirror isomers mixture.

a) True racemization (see [27]). In this case the racemization term $k_r\eta$ appears in the right part of the dynamic equation (6), k_r being the racemization rate constant.
b) Replication errors. Interaction of molecules cannot be absolutely selective in autocatalytic reactions (as well as elsewhere). In schemes of self-replication, along with the correct reactions S+L(D)→2L(2D), the erroneous reactions S+L(D)→L+D arise. In (6) there appears the relaxation term $k_m\eta$, where k_m is the relative rate constant of erroneous replication [17].
c) Fluctuation of dynamic trajectory. Using the method developed in [28] it can be shown that because of chaotic fluctuations of mirror isomers concentrations an additional relaxation term σ_η arises in the deterministic equation (6) [29]. (σ is the fluctuation variance). Thus the dynamic equation for η in a model of true evolution is

$$d\eta/d\tau = \varepsilon(1-\eta^2) - \xi_R\eta, \tag{7}$$

where ξ_R is the rate constant of relaxation (racemization) of η. The last term corresponds to the third major factor of the process of symmetry violation, to the Racemization Factor (RF), ξ_R is the measure of RF.

Thereby the conclusion about the possibility of evolutionary violation of symmetry changes radically. That is in a symmetrical case of $\varepsilon=0$, the system relaxes from the initially asymmetric state into the only possible state $\eta=0$. In a non-symmetrical case, there is a point attracting phase trajectories $\eta_\infty<1$. This point is defined by the expression

$$\eta_\infty = \frac{\varepsilon}{\xi_R}\left(1+\sqrt{1+\frac{\varepsilon^2}{\xi_R^2}}\right)^{-1},$$

i.e., in fact, by the AF to RF ratio. For a chirally pure state to arise, the advantage factor must be much higher than the racemic factor.

Analysis of these factors is beyond the scope of this communication (see [21]). Nevertheless, let us make some estimations. Theoretical estimates for AF caused by neutral currents give ε_{int} 10^{-17}-10^{-20}. Assume that errors in the reproduction of molecules chirality are the only source of racemization. (Naturally, taking other factors into account only worsens the estimate). The best estimate corresponding to minimal errors in biomolecular reproduction of nucleic acids in contemporary organisms is $\xi_R \sim 10^{-8}$. Thereby, the chiral polarization which may originate in the process of organic medium evolution is $\eta_\infty \sim 10^{-9}$-10^{-12}. If AF is caused by polarization, then $\varepsilon_{p.r.} < 10^{-10} - 10^{-11}$. An experimental estimate of RF which is due to radiation racemization [22] is $\xi_R \sim 10^{-1}$-10^{-2}. In this case $\eta_\infty \sim 10^{-8} - 10^{-10}$.

Thus, for the continuous consecutive evolution to produce the existing strong violation of mirror symmetry, it is necessary that either the AF should be by many orders of magnitude greater, or the RF smaller, than in the surrounding world of today. Thus, until the required physical state of the medium has been proved to be possible, such processes cannot be considered adequate to explain the origination of the early biosphere (unless some very particular If-hypothesis has appeared). It is highly improbable that the chiral purity has originated in a continuous evolution of organic matter under real physical conditions of our world.

Scenario of Catastrophe

As is known, for bifurcation processes to appear the dynamic equation for a parameter of order must contain at least a cubical nonlinearity [16]. In this case, the dynamic equation for the chiral polarization is [4]

$$d\eta/dt = \alpha_o + \alpha_1\eta + \alpha_2\eta^2 + \alpha_3\eta^3. \quad (8)$$

In particular, the simplest mirror-symmetrical ($\alpha_o = \alpha_2 = 0$) system ($\alpha_1 = \alpha(\lambda-\lambda_c)$, $\alpha_3 = -\beta$) behaves in the following way. If the system parameters are in the subcritical area $\lambda<\lambda_c$, then the only stable state is racemic, $\eta_\infty = 0$. When the critical conditions are reached ($\lambda>\lambda_c$), the racemic state loses stability but two new stable states arise

$$\eta_{1/2} = \pm\left[-\frac{\alpha}{\beta}(\lambda-\lambda_c)\right]^{1/2}$$

and the system spontaneously comes to one of such polarized states.

As an example of the physical realization of a catastrophe-type process (see [23-25,17]) we shall discuss a scheme of transformation of mirror isomers taking into account all FF, AF and RF of the chiral polarization dynamics. These factors comprise the following reactions:

a) autocatalysis $S + L \xrightarrow{k_L} L + L$, $S + D \xrightarrow{k_D} D + D$

b) destruction of mirror isomers during interactions (annihilation) $L + D \xrightarrow{k_a} S'$

c) racemization $L \underset{}{\overset{k_r}{\rightleftarrows}} D$ as the simplest form of RF.

The dynamic equation for the chiral polarization in this scheme contains three terms:

$$\dot{\eta} = \psi(AF) + \psi(FF) + \psi(RF), \quad (9)$$

where $\dot{\eta} = \partial\eta/\partial\tau$; $\tau = 1/2(k_L+k_D)tS$; $\psi(AF) = \varepsilon(1-\eta^2)$;

$\varepsilon = (k_L-k_D)/(k_L+k_D)$ is the advantage factor, $\psi(FF) = [k_a/(k_L+k_D)] \times$ $\times\theta\eta(1-\eta^2)$ is the term describing spontaneous violation of symmetry due to the fluctuation factor $\psi(RF) = -k_R\eta$; $k_R = k_r/(k_L+k_D)$.

The principal difference between this model and that for the continuous evolution is that when critical conditions are met the noises connected with the racemization factor are suppressed and the system tends to the chirally pure final state $\eta_\infty = \pm 1$.

Thus, the origination of the chiral purity of the early biosphere is the result of a process of a quite definite class. Most probably, the genesis of this property is associated not with some kind of stable evolution but with a catastrophe in the definition by R.Thom. This concept means that a developing medium attains a critical point and then loses stability of the initial symmetrical state.

This is a strong assertion with far-reaching consequences. If we were perfectly unaware of the physical conditions in real chemical processes, we could arbitrarily take either evolution or catastrophe as a possible explanation. But the values of constants we have in our world make us choose the processes of a catastrophe type, viz. the birth of the early biosphere is the process of destabilization of a lifeless chaotic symmetrical medium, a Biological Big Bang (BBB).

4. Primary Prebiological State

The conclusion made in the previous section takes the problem to the field of concrete physical analysis. Since the intrinsic properties of the biosphere, such as chiral purity and, seemingly, some others, must be generated in the course of a phase transition, it is necessary to introduce a concept of the under-critical prebiological state of organic matter and to find out necessary and sufficient conditions this state must for a "strong" symmetry violation to occur. Theoretical physics can specify the objects in which the chiral catastrophe of an organic medium could take place. Let us discuss briefly some approaches to estimate these conditions.

The critical parameter of a dynamic system, responsible for the chiral catastrophe, λ_c, is a function of the parameters, describing individual processes. Those parameters are: the rate constant of autocatalytic processes k_a^i, the advantage factor ε, the stock of chiral material θ, selectivity of interactions of chiral molecules w^i, constants determining the racemization factor k_R.

$$\lambda_c = f(k_a^i,\ \varepsilon,\ \theta,\ w^i,\ k_R). \tag{10}$$

All these parameters are associated with the characteristics of the state of a substance (the stock of a chiral substance, its density, temperature, fluxes of substances and energy). Theory and experiment must estimate the required set of state characteristics for λ to reach a critical level of symmetry violation. In addition to analytical methods of dynamic systems stability, a great deal of information is furnished by computer-movies [25], which can reveal rather subtle characteristics of symmetry violation.

Let us discuss in more detail the question of fluctuations in the initial prebiological state of a substance. Consider the following mental experiment: suppose that at a certain moment $t(\lambda=\lambda_c)$ the necessary critical level is reached in the course of evolution. Then, after an individual fluctuation, the chirally pure final state can, in principle, be formed as a result of the development of chiral polarization. The real pathway of the development of the process depends on the distribution of individual fluctuations in space and time. Two examples can be given. If fluctuations are rare and their distribution is sufficiently narrow, i.e. the probability is small that an arbitrarily strong fluctuation of chiral polarization arises in a unit time, then the state generated by the first symmetry-violating fluctuation, colonizes all the accessible medium.

If fluctuations are frequent and their distribution is broad, i.e. the probability that a fluctuation is soon followed by another one, capable of inverting the development of asymmetry, is not small, then the process of chiral polarization will take the form of random walk about the symmetrical state during a long period of time τ_{exp}, symmetry violation expectation time. The parameters determining τ_{exp} are: the frequency of fluctuations, their distribution, rate of condensation, all these parameters being functions of the state of the medium.

Assuming that in the initial state there are fluctuations of chiral polarization with frequency τ_o^{-1} and variance σ, and that individual modes of symmetry violations, corresponding to different values of chiral polarization, have increment α^{-1}, we have obtained [26] the following expression for τ_{exp}.

$$\tau_{exp} = \tau_o \exp (\sigma/\alpha\tau_o). \qquad (11)$$

Analysis of concrete dependences which determine τ_{exp} is beyond the scope of this communication (see [26]). Here we only note that even the most simple estimations of τ_{exp} point rather exactly to the time and other characteristics of the origin of life.

Thus, τ_{exp} is also a function of the state of a substance, just as the critical characteristics of symmetry violation.

Since τ_{exp} evolves as the Universe evolves, one can estimate the probability of formation of the early biosphere with the violated symmetry in a cosmological object by comparing τ_{exp} with the existence time of this object. Therefore, the estimates of critical conditions and τ_{exp} may point to the time and place of the origin of life in the Universe [30].

Conclusion

We tried to demonstrate that the ideas of self-organization and symmetry violation not only outline a general approach to the problem of life origin, but they may turn out to be the very tool to understand the pathways and conditions of the origination of biosphere and to characterize the objects in the history of the world, which are responsible for the origination of life.

All the above conclusions ensue from the only criterion, the necessity to form chiral purity in the early biosphere. Naturally, when considering other criteria we could learn more about the problem. But this is the subject matter of further investigations in theoretical biology.

In conclusion we should make some remarks about application of chemical synergetics to the problem of life origination. At present there exist detailed and refined theories of various kinds of cooperative phenomena in chemical systems, associated with dissipative structures, autowave processes, and other synergetic effects in chemistry. However, they operate mainly with hypothetical chemical reactions, at least as concerns non-biological systems. Conventional routine chemistry rarely deals with systems requiring such kind of analysis. That is why chemical synergetic systems which can be studied in laboratory, such as Belousov-Zhabotinskii reaction,are so rare.

It is quite plausible that when considering real physical problems we shall understand that the best laboratory of chemical synergetics is the prebiological history of the early biosphere.

References

1. A.I.Oparin. "Origin of life on the Earth", Publishing House of the USSR Acad. Sci., Moscow, 1957 (in Russian)
2. V.I.Goldanskii. Uspekhi Khimii, 44, 1019 (1975)
   V.I.Goldanskii. Nature, 279, 109 (1979)
3. Lord Kelvin."Baltimore Lectures, 1884, 1893", London, 1904
4. L.L.Morozov. Origins of Life, 9, 187 (1978)
5. M.Eigen. Naturwiss., 58, No 10, 465 (1971)
6. I.Prigogine, G.Nicolis. "Self-organization in Nonequilibrium Systems", J.Willey and sons, N.Y., 1977

7. F.J.Dyson. J. Molec. Evol., 18, 314 (1982)
8. L.L.Morozov. Doklady Acad. Sci. USSR, 241, 481 (1978)
9. Ya.B.Zel'dovich. ZhETF, 36, 964 (1959)
10. T.L.V.Ulbricht, F.Vester. Tetrahedron, 18, 522 (1962)
11. F.Vester, T.L.V.Ulbricht, H.Krauch. Naturwiss., 46, 68 (1959)
12. D.Rein. In:"Origins of Optical Activity in Nature", D.C.Walker ed., Elsevier, Amst.-Oxf.-N.Y., 1979, p.21
13. D.W.Gidley et al. Nature, 297, 639 (1982)
R.A.Hegstrom. Nature, 297, 643 (1982)
14. L.L.Morozov, V.V.Kuz'min, V.I.Goldanskii. In: Skulachev ed., Soviet Scient. Review Ser." Physico-Chemical Biology", v.5, Overseas Publ. Ass., Amst.-N.Y., 1984 (in press)
15. A.A.Andronov, A.A.Vitt, S.E.Khaikin. "Theory of Vibrations", Fizmatgiz, Moscow, 1959 (in Russian)
16. L.D.Landau, E.M.Lifshits. "Statistical Physics", Nauka, Moscow, 1976 (in Russian)
17. L.L.Morozov, V.V.Kuz'min, V.I.Goldanskii. Origins of life (1983) (in press)
18. B.Norden. J. Mol. Evol., 11, 313 (1978)
19. P.Jorden. Naturwiss., 32, 309 (1944)
20. Cs.Fajszi. J. Theor. Biol., 99, 295 (1982)
21. L.L.Morozov, V.I.Goldanskii. Uspekhi Fiz. Nauk (in press)
22. W.A.Bonner, R.M.Lemmon, J. Molec. Evol., 11,95 (1978)
W.A.Bonner, R.M.Lemmon. Bioorganic Chem., 7, 145 (1978)
23. F.Frank. Biochem. et Biophys. Acta, 11, 459 (1953)
24. P.Decker. J. Molec. Evol., 4, 49 (1974)
M.I.Kabachnik, L.L.Morozov, E.I.Fedin. Doklady Akad. Nauk SSSR, 230,1263 (1979)
25. L.L.Morozov, V.E.Kulesh. Doklady Akad.Nauk SSSR, 230, 1135 (1976)
26. L.L.Morozov, V.V.Kuz'min, V.I.Goldanskii. Doklady Akad. Nauk SSSR, (1983) (in press)
27. Cs.Fajszi, J.Czege. Origins of Life, 8, 277 (1977)
L.Keszthelyi, J.Czege, Cs.Fajszi, V.I.Goldanskii. In: D.Walker (ed.) "Origins of Optical Activity in Nature", Elsevier, Amst.-N.Y., 1979, p.229
28. R.L.Stratonovich. "Topics in the Theory of Random Noise", Gordon and Breach, N.Y.London, 1969
29. S.A.Anikin, S.A.Avetisov, V.I.Goldanskii, L.L.Morozov. Phys. Rev. Lett. (in press)
30. L.L.Morozov, V.I.Goldanskii. Nature (in press)

Physical Models of Evolution Processes

Werner Ebeling and Rainer Feistel
Sektion Physik der Humboldt-Universität, DDR-1040 Berlin, German Democratic Republic

1. Introduction

The world we are living in can be understood only as the result of a long evolution process. Big progress in understanding evolution has been achieved in the last 10-20 years due to the development of a theory of self-organizing processes |1|. Evolution can be defined physically as a historical irreversible process consisting of an unlimited series of self-organization steps |2|:

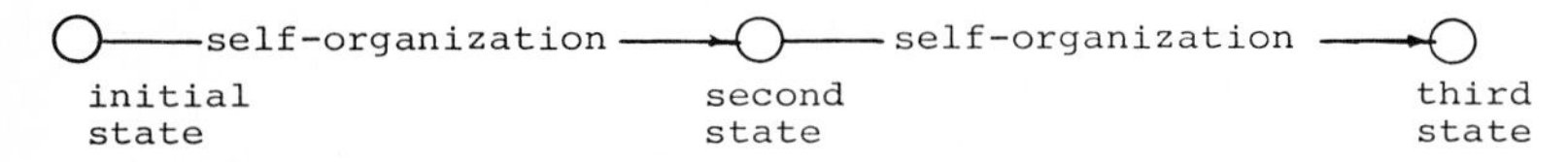

Each step is triggered by a critical fluctuation (e.g. an advantageous mutation) leading to an instability of the original semi-stable state. This fluctuation ignites a self-organization process, which rearranges the elements of the system and leads to a new semi-stable state, where the whole game may repeat on a "higher" level. Considering states of the evolving systems as certain occupation distributions of points of an appropriate space (phase space), evolution may be viewed as moving, switching or hopping of the occupation field through the phase space, depending on the discrete or continuous character of the phase and/or occupation number space (Fig.1). We shall come back to these variants of description in the following chapter. Evolution, in our sight, is restricted to several necessary physical conditions |2|:

(i) The system has to be open to exchange energy and matter and to export entropy, its state must be kept sufficiently far from equilibrium, that is beyond the linear, thermodynamic branch.

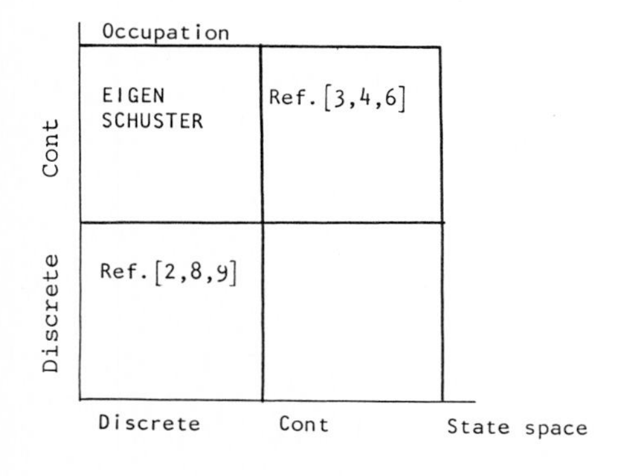

Fig. 1 Possible types of description

(ii) The system should have a sufficient structural reserve to be able to make unlimited chains of self-organization processes. Self-reproduction seems to be the process satisfying this condition in an ideal way.

(iii) Higher evolution processes require the ability to produce, to store and to use information by means of multistability, self-reproduction, finite living time and selection of its subsystems.

(iv) The system is subject to fluctuations which vary the stored information and lead via trial and error to an accumulation of information.

Essential properties of such evolution processes are:
- optimization and adaptation of subsystems
- historicity of the evolutionary pathway
- integration of subsystems to entities of higher complexity
- pattern formation with a tendency to increasing complexity
- hierarchical structure in space, time and function
- evolution accelerates by feedback mechanisms
- differentiation and specialization of subsystems
- branching character of actual and potential evolution pathways
- network character of the relations between the subsystems
- appearance of kinetic transitions (qualitative jumps, symmetry breakings)
- game character, mutual action of stochastic and deterministic elements

An interesting and often discussed question is whether evolution is governed by general principles such as a "generalized second law". In other words, the question is whether a scalar quantity exists which monotonously increases with time [2-6]. It is our opinion that such a well defined quantity, valid for all evolution processes on earth, and exclusively increasing as time goes on, does not exist (except the time itself). Nevertheless such quantities may exist for very special processes (as e.g. the Fisher-Eigen model of the evolution of a single species in a fixed environment), but in general the differential Pfaffian form (Glansdorff-Prigogine principle)

$$df = \sum_i f_i(x_i \ldots x_n)dx_i \tag{1.1}$$

(x_i-state variable) expressing the increase of order, structure or similar tasks, will not be integrable. This means that in every stage of evolution there is a well defined halfspace into which the forthgoing process is directed, but the weakness of the differential principle does not forbid circular or even chaotic processes.

2. Evolution as Hill-Climbing in Valuation Landscapes

The usual description of a biological population with different interacting species, homogeneously distributied on an area or in a space, is given by a set of ordinary differential equations

$$(d/dt)x_i(t) = x_i \cdot W_i(x_i \ldots x_n), \tag{2.1}$$

where x_i is the population density of species i and w_i its net rate of reproduction. Most common examples are

i) the Fisher - Eigen equation

$$w_i = E_i - \langle E \rangle; \quad \langle E \rangle = \frac{1}{c} \sum E_i x_i \tag{2.2}$$

i) the Lotka-Volterra equation

$$w_i = a_i + \sum_j b_{ij} x_j . \qquad (2.3)$$

The coefficients a_i, b_{ij} etc. describe the environmental conditions as well as the phenotypic properties and interactions of the species.

In this picture the formulation of evolution processes is difficult for mainly three reasons:

i) The dimension of the occupation number space is permanently changing (lowered by selection and extended by mutations).

ii) Velocity and general tendency of evolution is not described (since the question, what the values of the coefficients a_i etc. for advantageous mutations are, is not answered in the theory).

iii) For nonlinear functions w_i evolution may come to dead-ends ("once-forever" selection).

To overcome the third problem one can consider a stochastic theory; to deal with the first and second one we propose a functional formulation [3,4] of the equation 2.1:

$$\frac{\partial}{\partial t} x(q,t) = x(q,t) \cdot \mathbf{W}(q/x) + \frac{\partial}{\partial q} D(q) \frac{\partial}{\partial q} x(q,t). \qquad (2.4)$$

The vector q represents here the set of phenotypic properties belonging to a certain individual such that $x(q)$ is the number (density) of individuals exhibiting the property q. The additional diffusion term (acting not in the real, but in the phenotype space) expresses the appearance of error copies (varied phenotypes). Given a certain population distribution $x(q)$ the rate $w(q)$ forms a real-valued landscape over the phenotype space. Following Fisher's law, evolution favours phenotypic change along the gradient $\partial w/\partial q$ of this landscape: evolution exhibits a general hillclimbing tendency. This change of the population distribution $x(q)$ acts back on the hyperface $w(q)$; in the case of Fisher-Eigen the landscape

$$w(q) = E(q) + \frac{1}{c}\int E(q')\, x(q')\, dq' \qquad (2.5)$$

is only rigidly shifted, but for Lotka-Volterra systems the landscape

$$w(q) = a(q) + \int b(qq')\, x(q')dq' \qquad (2.6)$$

may be deformed as a sea-view or a "seascape" if $x(q)$ is changing. Let us now look for solutions $x(q,t)$ in the case of Fisher-Eigen dynamics with $D(q)=D$. After a transformation

$$y(q,t) = x(q,t) \exp \{\int \langle E\rangle dt\} \qquad (2.7)$$

we arrive at the one-particle Schrödinger problem in the potential $-E(q)$

$$\frac{\partial y}{\partial t} = E(q) \cdot y + D \frac{\partial^2 y_2}{\partial q} . \qquad (2.8)$$

Since little is known about the real shape of $E(q)$ let us assume that $E(q)$ is a stochastic function with random distribution valleys and hills like the potential in the theory of disordered solids [6].

Further let us assume that the operator has a discrete part of the spectrum. The discrete eigenvalues ε_n correspond to localized eigenfunctions ψ_n with the asymptotic behavior

$$\psi_n(q) \sim \exp \{- \frac{|q - q_n|}{l(\varepsilon_n)} \}. \qquad (2.9)$$

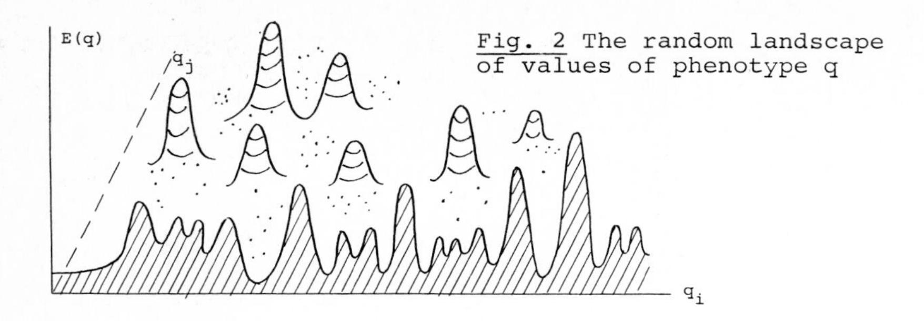

Fig. 2 The random landscape of values of phenotype q

We consider now the time evolution of an initial density concentrated in a region of volume l around a localized state m at q_m corresponding to a master species. The transition time to another localized state ψ_n (the new species) is given by

$$c_n e^{-\varepsilon_n t_{mn}} \simeq c_m e^{-\varepsilon_m t_{mn}}$$

$$t_{mn} \sim \frac{|q_n - q_m|}{l(\varepsilon_n)|\varepsilon_n - \varepsilon_m|} . \qquad (2.10)$$

The escape time is found to be [6]

$$t_m = \Gamma\left(\frac{d+1}{d}\right) \cdot \left[\frac{S_d}{d} \int_{\varepsilon_m}^{\infty} [l(\varepsilon) \cdot (\varepsilon - \varepsilon_m)]^d \cdot \rho(\varepsilon) d\varepsilon\right]^{-1/d} \qquad (2.11)$$

and the transition time from the value ε_m to ε_n [6]

$$t_{mn} = \Gamma\left(\frac{d+1}{d}\right)\left[l(\varepsilon_n) \cdot |\varepsilon_n - \varepsilon_m|\right]^{-1} \cdot \left[\frac{d}{S_d \rho(\varepsilon_n)}\right]^{1/d} . \qquad (2.12)$$

Since $\rho(\varepsilon_n)$ exponentially decreases at large ε_n the transition time will exponentially increase. For small differences in the value $|\varepsilon_n - \varepsilon_m| << |\varepsilon_m|$ the transition time diverges like $1/|\varepsilon_n - \varepsilon_m|$. Therefore transitions to phenotypic combination which show only little advantage $|\varepsilon_n - \varepsilon_m| << |\varepsilon_m|$ or such which show large advantage $|\varepsilon_n - \varepsilon_m| >> |\varepsilon_m|$ are equally seldom. Evolution prefers small but finite steps corresponding to the minimum of the t_{mn}. The average advantage per step $\delta\varepsilon$ corresponding to this minimum may be called the evolution quantum of value. In order to discuss the formulae given above let us assume that we are in a well-developed island-regime i.e. only high mountains are occupied. Let us assume for this regime

$$l(\varepsilon) \simeq (D/|\varepsilon|)^{1/2} \qquad (2.13)$$

and Gaussion distribution

$$\rho(\varepsilon) \approx C \exp\{-\varepsilon^2/2 \cdot B\}. \qquad (2.14)$$

We find now that the transition time has a minimum at

$$|\varepsilon_1| + \delta\varepsilon, \quad \delta\varepsilon = dB/|\varepsilon_1|. \qquad (2.15)$$

This means that transitions occur with the highest frequency to a mountain with a height by $\delta\varepsilon$ higher than the original one and the necessary time is

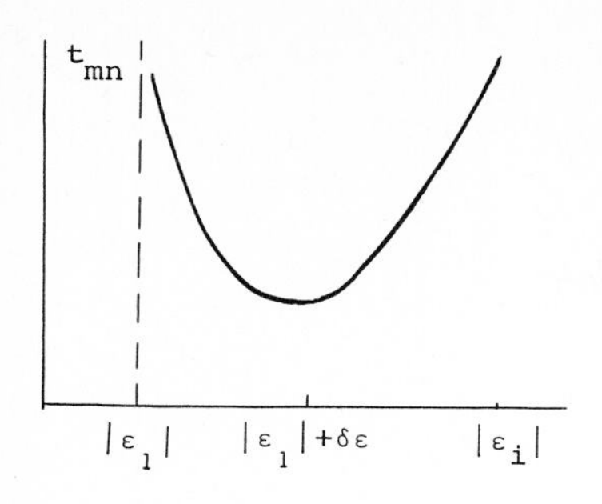

Fig. 3 The minimal transition time for evolutionary steps defines the evolution quantum

$$\delta t = t(\varepsilon_1,\varepsilon_1+\delta\varepsilon). \tag{2.16}$$

This expresses the hopping character of evolution. The mean evolution velocity is defined as

$$V_{ev} = \delta\varepsilon/\delta t, \tag{2.17}$$

The escape time of the original master state which is a measure of stability of the species can be estimated to be [6]

$$t_m \approx C_d\left(\frac{|\varepsilon_m|}{BD}\right)^{1/2} .\exp\left(\frac{\varepsilon_m^2}{2dB}\right). \tag{2.18}$$

The stability increases with the value already reached and decreases with the diffusion constant (mutation rate), the correlation strength (smoothness of the landscape) and with the dimension (number of developed properties). We should note here that all these calculations are restricted to a sufficiently smooth surface $E(q)$. If, on the other hand, $E(q)$ is very cleft (imagine a white noise random field), localized states do not exist at all if the dimension of the phenotype space is higher than four.

As pointed out by Conrad [7] this interesting picture with valuating landscapes should be treated with certain cautions for one never knows exactly the structure of the landscape. Nevertheless this picture allows several statements valid for wide classes of landscapes $w(q)$. The continuous density picture of evolution given above has many advantages however it is not able to describe the discrete characters of mutations. Remember that a new species appears always first in one exemplar only. The way to describe such processes will be the subject of the next chapter.

3. Evolution with Shot Noise

Evolution viewed as a continued sequence of self-reproduction processes (with a certain error) of discrete elements or individuals implies the presence of a special kind of noise - the shot noise - because the multiplication acts are stochastic events. Let us consider the Fisher-Eigen process (eg. 3.1) but replace the occupation density X_i by the entire number N_i of individuals of the species i. A master equation corresponding to the process (2.1) might be [2,8,9]

$$\frac{\partial}{\partial t}P(N_1\ldots N_n,t) = \frac{1}{N}\sum_{i\neq j} E_i\{(N_i-1)(N_j+1)\ P(N_i-1,\ N_j+1) - N_iN_jP\}. \tag{3.1}$$

Here P is the probability to find N_1 individuals of species 1 etc, and N is the total number of individuals.

Averaging eq. (3.1) yields

$$(d/dt)\langle N_i\rangle = \langle N_i\rangle E_i - \frac{1}{N}\sum_j E_j\langle N_i N_j\rangle, \qquad (3.2)$$

i.e. for small fluctuations one can neglect the correlations and we come back to eq. (2.1) again. Thus in many cases the difference between the stochastic and deterministic picture is not essential. But there are two situations where the stochastic aspect forms a quite new picture in evolution:

i) the survival of single individuals (mutants) [8],

ii) the break through evolution barriers [9].

When a new species begins its life, it appears the first time with the population number $N_n=1$, and the decision whether it remains and multiplies in the system or it dies out is a dice game. Let the master species be present initially with $N_m=N-1$ individuals, then we can solve eq. (4.1) and get the survival probability [8]

$$P_{surv} = \delta/[1 + \delta - \exp(-k\delta)]$$

after k generations of the new species and with its selection advantage $\delta=(E_n-E_m)/E_m$ (Fig.4).

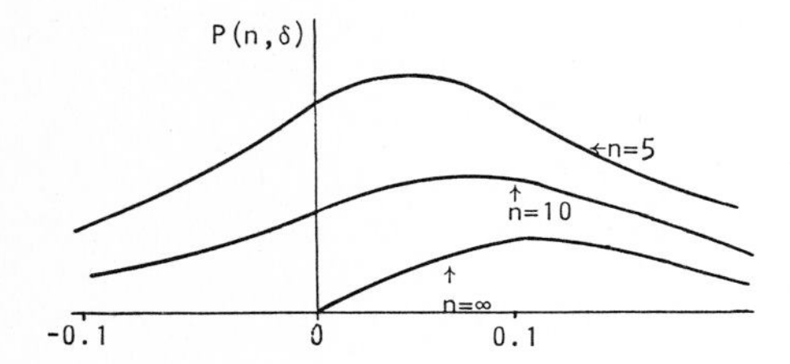

Fig. 4 Probability of appearance of mutants and survival (n generations). The maximum denotes the evolutionary quantum [8]

4. Conclusion

The main aim of this paper was to show in the framework of a simplifying physical picture, that evolution by self-reproduction and mutation may be considered as an autowave process in an appropriate space.

The two basic processes which drive the wave are

(a) amplification by exact reproduction, i.e. formation of identical copies

(b) propagation by error reproduction, i.e. by mutations

The most typical peculiarity of evolutionary autowaves is the hopping character of propagation. A specific difficulty is connected with the discrete character of mutations, i.e. a new species appears first only in one exemplar. Therefore in order to study the behaviour of new species, the discrete character of the number of individuals must be taken into account in the stochastic theory.

Acknowledgement: The authors want to thank A.Engel, B.Esser and F.Reglin for collaboration.

References

1.G.Nicolis, I.Prigogine. Thermodynamics of Structure, Stability, and Fluctuation, (Interscience, New York 1971)

H.Haken. Synergetics - An Introduction, 2nd Ed., (Springer, Berlin, Heidelberg, New York 1978) Springer Series in Synergetics, Vol.2-17, ed. by H.Haken (Springer, Berlin, Heidelberg, New York 1977)
G.R.Ivanizky, V.I.Krinsky, E.E.Selkov. Mathematical Biophysics of the Cell (in Russian) (Nauka, Moscow 1978)
2. W.Ebeling, R.Feistel. Physik der Selbstorganization und Evolution, Akademie-Verlag Berlin(1982)
3. W.Ebeling, R.Feistel. Stochastic Models of Evolutionary Processes, in A.I. Zotin (Ed.): Thermodynamics and Regulation in Biological Processes. Nauka, Moscow(1984)
4. R.Feistel, W.Ebeling. Models of Darwinian Processes and Evolutionary Principles. BioSystems 15 (1982) 291
5. R.Feistel. Conservation Quantities in Selection Processes. Studia biophysica 95 (1983) 107; Extremal Principles in Selection Processes. Studia biophysica 96 (1983) 133
6. W.Ebeling, A.Engel, B.Esser, R.Feistel. Diffusion and Reaction in Random Media and Models of Evolutionary Processes. Sumbitted J. Stat. Phys.
A.Engel. Selection and Diffusion in Stochastic Fields.Thesis. Humboldt-University (1983)
7. U.Conrad. Bootstrapping on the Adaptive Landscape, BioSystems 11 (1979) 167
8. P.M.Allen, W.Ebeling. On the Stochastic Description of a Predator - Prey Ecology, BioSystems 16 (1983)
9. W.Ebeling, I.Sonntag, L.Schimansky-Geier. On the Evolution of Biological Macromolecules. II Catalytic Networks. Studia biophysica 84 (1981) 87

Experimental and Theoretical Studies of the Regulatory Hierarchy in Glycolysis

A. Boiteux and B. Hess

Max-Planck-Institut für Ernährungsphysiologie, D-4600 Dortmund, Fed. Rep. of Germany

E.E. Sel'kov

Institute of Biological Physics of the USSR Academy of Sciences
SU-142292 Pushchino, USSR

Introduction

Of all known complex biochemical networks the glycolytic system (GS) has received the most extensive study (cf. [1]). During decades it has played the role of an experimental proving ground to test and refine the theories on the regulation of cellular metabolism.

The GS* consists of a great number of reactions in which ATP is synthesized from ADP and inorganic phosphate (P_i) at the expense of anaerobic degradation of glucose or storage polysaccharides (glycogen, trehalose) to the end products (ethanol, CO_2, glycerol, etc.). The relatively simple stoichiometric structure of the GS formed by the individual reactions is entangled in a complex network of allosteric regulations [1]. These regulatory influences are exercised by intermediates of the GS (and species from other metabolic pathways) capable of inhibiting or activating the key enzymes when binding to the specific regulatory sites of the enzyme molecules.

The GS shows a great variety of behaviour patterns due to its strong nonlinearity and multiloop circuitry. It combines incompatible, at first sight, properties such as the ability to generate self--oscillations in metabolite concentrations [2-11], to exhibit trigger behaviour [12,13] and to stabilize the relative concentration of its major product, ATP [11,14-18].

This latter property manifests itself as stability of the so called adenylate energy charge [19,20]. It is determined as

$$\varphi = \frac{[ATP]}{A_\Sigma} + \frac{1}{2}\frac{[ADP]}{A_\Sigma} \qquad (1)$$

*ABBREVIATIONS: ACA, acetaldehyde, ADH, alcohol dehydrogenase; ADP, adenosine monophosphate; AK, adenylate kinase; ALDH, aldehyde dehydrogenase; AMP, adenosine monophosphate; ATP, adenosine triphosphate; ATPase, adenosine triphosphatase; DAP, dihydroxyacetone phosphate; dATP, desoxyadenosine triphosphate; DPG, 1,3-diphosphoglycerate; ENO, enolase; ETOH, ethanol; Fru-6-P, fructose 6-phosphate; Fru-1,6-P_2, fructose 1,6-bisphosphate; Fru-2,6-P_2, fructose 2,6-bisphosphate; GAP, glyceraldehyde 3-phosphate; GAPDH, glyceraldehyde-phosphate dehydrogenase; Glc, glucose; Glc-1-P, glucose 1-phosphate; Glc-6-P, glucose 6-phosphate; GLG, glycogen; GLOH, glycerol; GL3PDH, glycerol 3-phosphate dehydrogenase; GS, glycolytic system; MAL, malate; NAD, nicotinamide dinucleotide (oxydized); NADH, nicotinamide dinucleotide (reduced); P_i, inorganic phosphate; PDC, pyruvate decarboxylase; PEP, phosphoenolpyruvate; PFK, 6-phosphofructokinase; 2-PG, 2-phosphoglycerate; 3-PG, 3-phosphoglycerate; PK, pyruvate kinase; PYR, pyruvate; UTP, uridine triphosphate; TIM, triosephosphate isomerase.

where $[ATP]/A_\Sigma$ and $[ADP]/A_\Sigma$ are the relative concentrations of ATP and ADP and $A_\Sigma = [ATP] + [ADP] + [AMP]$ is the total adenylate concentration.

Since the discovery of the mechanisms of allosteric regulation [20,21] and of their widespread occurrence in the metabolism of cells [23,24], it has become natural to consider the numerous manifestations of autoregulation in the cellular metabolism in terms of specific allosteric interactions [20,23-26]. However, theoretical analysis of a variety of stoichiometric schemes of energy metabolism, in particular for the GS, demonstrates the possibility of a stabilization of $[ATP]/A_\Sigma$ and φ without any allosteric interactions [27-30].

The data presented in this paper are the first experimental confirmation of the theory [27-30]. Moreover, they indicate conclusively that the homeostatic properties of the GS are the result of many regulatory mechanisms which provide a stable energy supply for the cell under drastic changes in (a) substrate availability, (b) metabolic load and (c) the very regulatory and stoichiometric structure of the GS.

Methods

The experiments were performed with the cell-free cytosolic system of yeast *Saccharomyces carlsbergensis* strain ATCC 9080 grown on glucose in aerobic or anaerobic 5 liter batch cultures. The cells were harvested in the late logarithmic growth phase at optical density of suspension $OD_{546}<1.5$. After being washed three times with 0.1 M phosphate buffer pH 6.5, the cells were used for preparation of extracts. The average yield was 60g of wet cells per batch culture. For depletion of endogenous substrates, the cells were resuspended in 0.1 M, phosphate buffer pH 6.5 and aerated vigorously for at least three hours. After washing and centrifugation 20g of wet cells were mixed with 3.3 ml of 0.1 M phosphate buffer pH 6.5 and carefully sonicated keeping the temperature always below 20°C. When microscopic control showed only rare unbroken cells the sonicate was subjected to high speed centrifugation (100.000g for 2 hrs) to remove unbroken cells as well as cell debris and membrane fractions. The resulting supernatant contains all soluble cytosolic components, including the complete GS. It is clear, slightly yellow and opalescent and contains about 50 mg protein per ml.

To carry out experiments, the extract was thermostated at 25°, supplemented with the required effectors or adenine nucleotides (in the form of MgATP) and adjusted to pH 6.5. At time zero specific substrates for glycolysis were added, either glucose to make the final concentration 40 mM, or glycogen to yield a final concentration of about 40 mM glucose units. After charging the endogenous polysaccharide pools of the cytosolic system for 45 minutes, various amounts of highly purified mitochondrial F_1-ATPase prepared from baker's yeast were added. At different times during charging and decharging samples were taken to be analyzed subsequently for metabolic concentrations according to micro analytical techniques described elsewhere [31].

Results

Fig.1 illustrates the results of a typical experiment with addition of glucose to the extract. It shows changes with time of the concentration of glycolytic intermediates (A), adenine nucleotides (B), glucose,glycogen, ethanol and glycerol (C), and also of the variables

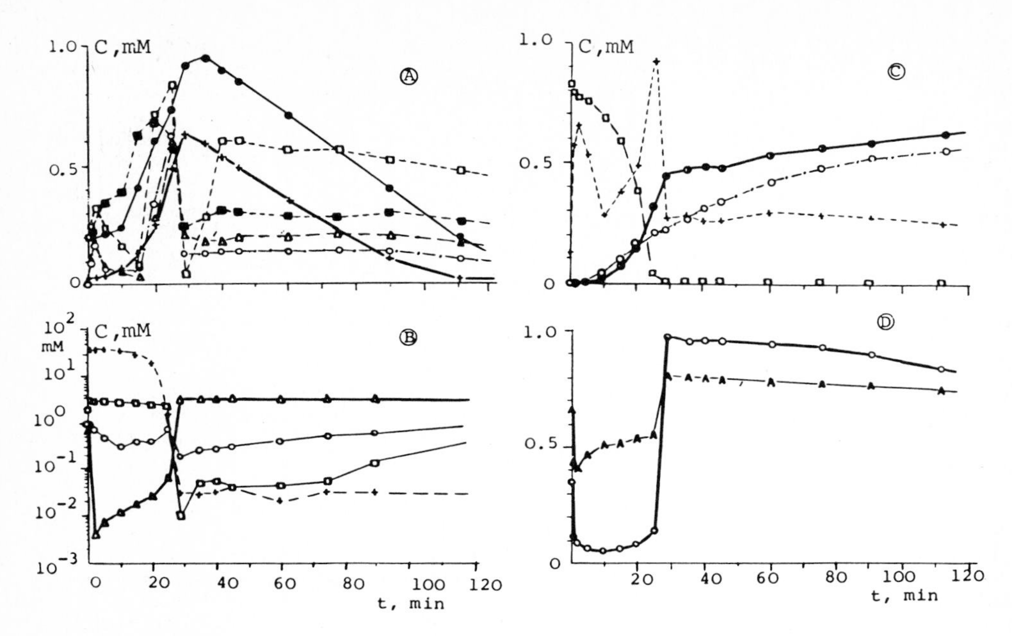

Fig. 1 Changes in time t in the yeast cytosolic cell-free system after addition of glucose (41 mM) at t=0.
A - concentrations of glycolytic intermediates: o-·-o, [Glc-6-P]/10; □ -- □, [Glc-1-P]x4; Δ--Δ, [Fru-6-P]/2; +——+, [Fru-1,6-P_2]/20; ●——●, [DAP]/2; ■ —— ■, [PYR]/2.
B - concentrations of adenine nucleotides (Δ——Δ, [ATP]; o——o, [ADP]; □ —— □, [AMP]) and glucose (+-+).
C - concentrations of glucose (□ -- □, [Glc]/50), glycogen (+-+, [GLG]/50), ethanol (◑——◑, [ETOH]/100 and glycerol (o-·-o, [GLOH]/5).
D - the variables characterizing the energy status; o——o, φ; A——A, A_{ATPase}/50 kJ/mol.

characterizing the energy status of the GS: the energy charge φ and the chemical affinity of the ATPase reaction, $A_{ATPase} = \Delta G^{o}_{ATPase}$ (D). The experiment demonstrates the existence of a long-lasting (30'<t<110') quasisteady state during which stable values of φ and A_{ATPase} are maintained at the expense of the endogenous polysaccharides accumulated for 30 min after glucose addition. In this state the concentration of glucose is vanishingly small.

It is of interest to note that during charging of the extract (0'<t<30') free glucose is very rapidly phosphorylated by the highly active hexokinase (HK) at the expense of ATP (Glc + ATP $\xrightarrow{HK}$ Glc-6-P+ +ADP). This results in a dramatic fall of the ATP concentration such that both the deposition of glucose into glycogen, which requires two molecules of ATP per glycosyl bond formed, and the glycolytic phosphorylation of ADP, which is preceded by a two-step hexose phosphorylation, may seem impossible because of limitation by the low ATP level.

However, this is far from being so: the rapid accumulation of storage polysaccharides and ethanol (Fig.1C) give clear-cut evidence that both of the ATP-dependent functions are fulfilled even at such low concentration of ATP.

The surprisingly high flux rates driven by a very low ATP level can be explained by the fact that the chemical affinity of the ATPase reaction A_{ATPase} changes only slightly during the phase of charging (from 40 to 20 kJ/mol, Fig.1D), whereas ATP drops from 3.0 to $4\cdot 10^{-3}$mM. The high level of A_{ATPase} is maintained by the activities of ADP-phosphorylating reactions in the lower part of the GS, which keep the net ATPase reaction far from thermodynamical equilibrium.

In order to establish the relative role of allosteric and stoichiometric mechanisms in the stabilization of $[ATP]/A_{\Sigma}$, φ and A_{ATPase} in the quasisteady state, several independent series of experiments were carried out measuring a number of glycolytic substrates, intermediates and products. In these experiments additions of different effectors were made blocking the allosteric properties of the key enzymes and changing the structure of the GS to a partially non-allosteric network. The major results of the experiments are summarized in Fig.2.

Fig.2A illustrates the case where the ATP pool is substituted by the pool of dATP. This results in elimination of the allosteric inhibition by ATP of phosphofructokinase (PFK) and the corresponding accumulation of its product fructose 1,6-bisphosphate (Fru-1,6-P_2) at high levels (up to 20 mM). It is well known that Fru-1,6-P_2 is a specific allosteric activator (the activation constant $K_a \simeq 2$ mM) of pyruvate kinase (PK), which is the key allosteric enzyme of the lower part of the GS. At high Fru-1,6-P_2 levels PK is completely activated

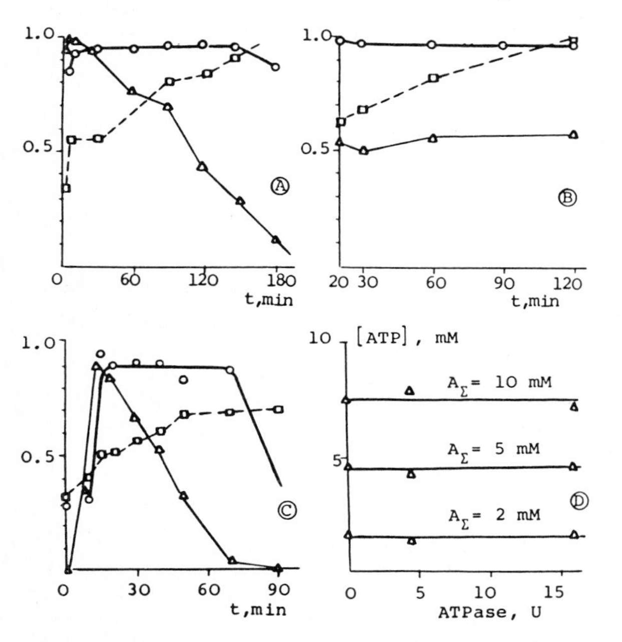

Fig. 2 Stabilization of the energy charge φ (A,B,C) and absolute concentrations of ATP (D) in the extracts deprived of the allosteric regulations of their key glycolytic enzymes. Allostericity was suppressed in different ways: A - substitution of adenylates by deoxyadenylates (the extract with a very low initial adenylate pool was loaded with 7.0 mM dATP, 4 units of F_1-ATPase and 5.3 mM $MgCl_2$). o——o, φ ; Δ——Δ, [Fru-1,6-P_2]/20 mM; □--□ , [ETOH]/200 mM. B - addition of 100 mM NH_4Cl and 1.5 mM MgATP. The energy charge $\varphi \simeq 0,96$ in the time interval from 20 to 120 min characterized by a nearly constant concentration of Fru-1,6-P_2([Fru-1,6-P_2]/10, Δ——Δ) and a linear accumulation of ethanol ([ETOH]/200, □--□). C - addition of 1.0 mM Fru-2,6-P_2 to the extract precharged with 44 mM Glc. A_{Σ}=1,6 mM and $P_i \simeq 80$ mM (o——o, φ; Δ-- Δ, Fru-1,6-P_2/20; □--□, [ETOH]/200). D - the quasisteady state load characteristics (Δ——Δ) of the GS loaded with 40 mM GLG at three different levels of the adenylate pool A_{Σ}.

and hence insensitive to Fru-1,6-P itself and other allosteric effectors. The linear fall of Fru-1,6-P_2 and the conjugated stoichiometric accumulation of ethanol at a practically constant value of $\varphi \approx 1$ prove the fact that the lower part of the **GS** does stabilize φ stoichiometrically. This stabilization cannot be mediated by the key allosteric enzymes of the upper part of the GS (the phosphorylase complex and PFK) because of the significant nonstationarity of Fru-1,6-P_2 and hexose monophosphates (not shown).

Fig.2B presents the results of the experiment on activation of both PFK and PK by high concentrations of NH_4^+, a powerful effector for these enzymes. In this experiment the stability of φ is maintained unaltered even with a non-allosterically operating PFK and an only partially allosteric PK.

Fig.2C demonstrates the outcome of experiments with a third method to block the allosteric properties of PFK and PK. A highly effective activator of PFK, fructose 2,6-bisphosphate (Fru-2,6-P_2) was used at a concentration of 1 mM to obtain completely activated PFK. Evidence for the strong activation of PFK in this condition are the rapid decrease of the hexose monophosphate pool to a very low level (not shown) and the accumulation of high concentrations of Fru-1,6-P_2 (up to 20 mM).

Fig.2D plots the absolute concentration of ATP as a function of the ATPase load with three different sizes of the adenylate pool, A_Σ. In these experiments, the initial substrate for glycolysis was glycogen added to the extracts. It is remarkable that the breakdown of exogenous glycogen proceeds in uncontrolled reactions [11], possibly, due to lack of the specific spatial organization of the respective enzymes which exists in the case of endogenously synthesized glycogen. In any case, the rate of glycogen breakdown was independent of ATPase activity. In the absence of such feedback, the upper part of the GS becomes overloaded with metabolites and PFK is activated by high concentrations of Fru-6-P and Fru-1,6-P_2. PK is also activated by a high level of Fru-1,6-P_2. Nevertheless, $[ATP]/A_\Sigma$ is stabilized in spite of the change in the activity of added ATPase, as seen from the load characteristics.

It is important to note that only the relative concentration ($[ATP]/A_\Sigma$) and not the absolute concentration $[ATP]$ is stabilized by the GS, contrary to what might be expected from the allosteric principle of regulation. The principle of stoichiometric regulation is very simple: the phosphorylating pathway of the GS tends to maintain the least possible concentration of ADP. Actually this principle is realized in glycolysis as is evident from the insignificant decrease in A_{ATPase} at a very strong overloading (Fig.1D) and from the dramatic decrease in the rate of glycolysis V_{ETOH} for $\varphi \to 1$ (Fig.3).

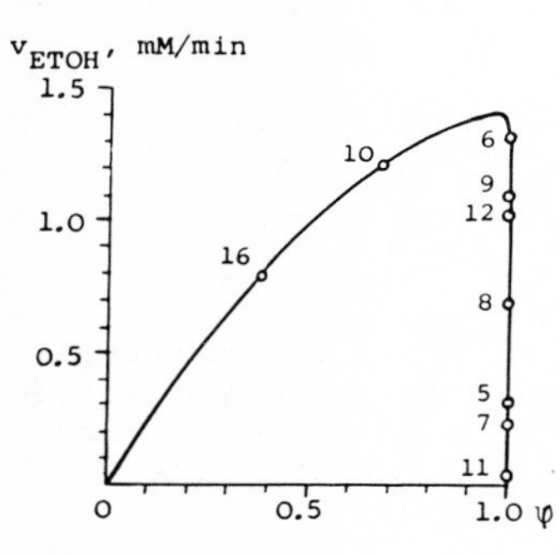

Fig. 3 Quasisteady state output characteristics of the GS. Circles are the quasisteady state values of the rate of ethanol production (V_{ETOH}) averaged for the time during which $[Fru\text{-}1,6\text{-}P_2] \approx const$. Different amounts of ATPase were added to the untreated extract charged with 40 mM glucose and containing ≈ 3.0 mM MgATP

Fig.3 plots V_{ETOH} as a function of φ for experiments with extract which underwent no changes other than the addition of ATPase. The values of φ and V_{ETOH} correspond to the average quasisteady state values of these veriables at $[\text{Fru-1,6-P}_2] \approx$ const.

The quasisteady state (Fig.4) is maintained by the slow expenditure of metabolic Fru-1,6-P_2 precursors, which counterbalances the expenditure of Fru-1,6-P_2 in the lower part of the GS ($v_{PFK} \approx v_{ALD}$). Depletion of the precursors makes the formation of Fru-1,6-P_2 in the PFK reaction impossible and $[\text{Fru-1,6-P}_2]$ decreases at a rate $v_{ALD} \approx 4v_{ATPase}$. As seen from Fig.5, φ remains constant even when Fru-1,6-P_2 decreases steadily. This proves the existence of at least one more quasisteady state in the lower part of the GS. The energy to maintain this state is derived from the Fru-1,6-P_2 breakdown.

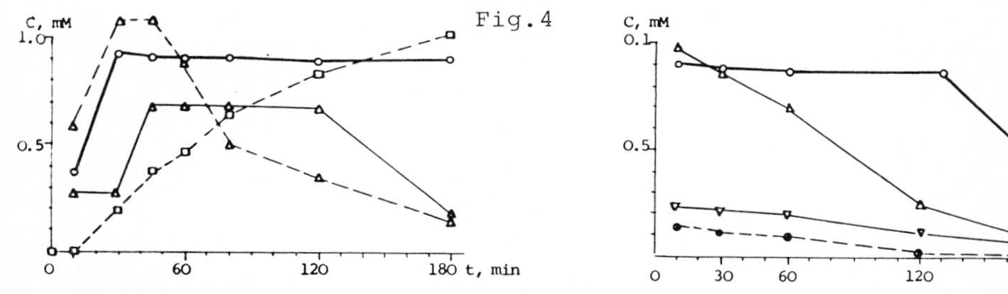

Fig. 4 Quasisteady state in Fru-1,6-P_2 concentration (45'<t<120') in the untreated extract with the initial pool of endogenously synthesized polysaccharides equal to about 38.5 mM glucose equivalents. Note that φ remains stable even during the phase of Fru-1,6-P_2 breakdown (t>120 min). △—△, $[\text{Fru-1,6-P}_2]/10$; △--△, [Fru-6-P]; □--□ [ETOH]/200; ○—○, φ

Fig. 5 Stabilization of φ by the lower part of the GS at the expense of the Fru-1,6-P_2 + GAP pool in conditions of $v_{PFK} \approx 0$. The extract was precharged with polysaccharides synthesized endogenously from 40 mM glucose and was then loaded with 4 units of ATPase at t=0. Total adenylate cocentration A_Σ=6.0 mM. △—△, $[\text{Fru-1,6-P}_2]/10$; ▽—▽, [DAP]/10; ●—●, [Fru-6-p]; ○—○, φ

In the upper part of the GS the quasisteady state may become transient and even non-existent because of high load imposed on the system by the added ATPase, while in the lower part it persists long enough to provide the stabilization of φ for an extended period of time

$$t = 2(2[\text{Fru-1,6-P}_2]_o + [\text{DAP}]_o)/(L \cdot [\text{ATP}]_{st}) \qquad (2)$$

determined by the initial concentrations of Fru-1,6-P_2 and DAP ($[\text{Fru-1,6-P}_2]_o$ and $[\text{DAP}]_o$), the rate constant of the net ATPase (endogenous + added), L, and the concentration of ATP being stabilized, $[\text{ATP}]_{st}$.

Discussion

The experiments demonstrate unequivocally that allosteric regulation is not necessary to make the GS a stable energy source. Even when the enzymes PFK and PK are deprived of their allosteric properties, the GS stabilizes φ and $[\text{ATP}]/A_\Sigma$ (Fig.2) as effectively as does the untreated GS (Fig.1).

This result is consistent with the theory [27-30], according to which the stabilization of φ is accomplished stoichiometrically by the lower part of the GS (below the PFK reaction).

The idea that the lower part might be the only stabilizer in glycolysis is confirmed by the experiments in which Fru-1,6-P_2 is produced in great excess by different ways. The resulting prolonged transient state (Fig.2A,C; Fig.5) is characterized by a linear fall of [Fru-1,6-P_2] coupled stoichiometrically with the accumulation of ethanol and with the quasisteady state in the lower part, in which φ and metabolite concentrations between GAPDH and alcohol dehydrogenase (ADH) are constant.

Jonnalagadda and co-authors [11] tried to evoke such a state adding to yeast extracts either glycogen or large amounts of Fru-1,6-P_2 (up to 60 mM). In these conditions a linear decrease in the concentrations of (Fru-1,6-P_2) and DAP and the conjugated linear accumulation of ethanol were observed, suggesting the existence of a rapidly attainable quasisteady state in the lower part of the GS. Based on these findings, these authors [11] first proposed that the stabilization of φ should be accomplished stoichiometrically by the lower part of glycolysis. However, in their experiments the value of φ did not remain constant on addition of excess Fru-1,6-P_2. And when the mean value of $\varphi \approx 0.84$ was maintained (linear accumulation of ethanol at the expense of added glycogen) oscillations of [Glc-6-P] and [Fru-1,6-P_2] were observed with a phase shift of 180°. This indicated that the PFK reaction continued to be allosterically regulated.

In our experiments the stabilization of φ was next to ideal in all conditions in which the contribution of PFK was excluded either because of the exhaustion of the metabolic precursors of Fru-1,6-P_2 ($v_{PFK} \ll v_{ALD}$) or because of the complete activation of PFK (fig.2).

Though the theory of the stoichiometric stabilization of φ by the cellular energy metabolism, in particular by the GS, has been primarily developed for the case of steady state [27-30], it can be easily extended to the nonstationary case [32]. Fig.6 presents a kinetic model of the lower part of the GS. If we take the simplest approximations for the rates

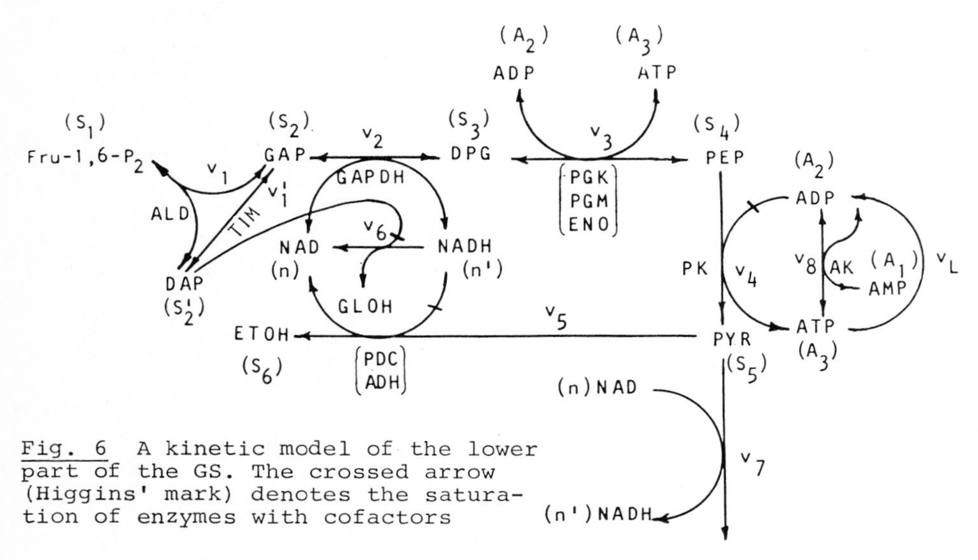

Fig. 6 A kinetic model of the lower part of the GS. The crossed arrow (Higgins' mark) denotes the saturation of enzymes with cofactors

$$v_1 = k_{+1}S_1 - k_{-1}S_2S_2', \quad v_1' = k_{+1}'S_2 - k_{-1}'S_2',$$
$$v_2 = k_{+2}S_2N - k_{-2}S_3N', \quad v_3 = k_{+3}S_3A_2 - k_{-3}S_4A_3, \tag{3}$$
$$v_4 = k_4S_4A_2/(M_4 + A_2), \quad v_5 = k_5S_5N'/(M_5 + N'),$$
$$v_6 = k_6N'S_2'/(M_6 + S_2'), \quad v_7 = k_7S_5N,$$
$$v_8 = k_{+8}A_1A_3 - k_{-8}A_2^2 \ ,$$

$v_L = LA_3$ (for numbering of reactions, see the kinetic scheme in Fig.6), then with constant nucleotide pools

$$A_\Sigma = A_1 + A_2 + A_3, \quad N_\Sigma = N + N', \tag{4}$$

the behaviour of the lower part of the GS in time t can be described by the following equation system:

$$\frac{dS_1}{dt} = -v_1, \quad \frac{dS_2}{dt} = v_1 - v_1' - v_2, \quad \frac{dS_2'}{dt} = v_1 + v_1' - v_6,$$
$$\frac{dS_3}{dt} = v_2 - v_3, \quad \frac{dS_4}{dt} = (v_3 - v_4)e_4, \quad \frac{dS_5}{dt} = v_4 - v_5 - v_7, \tag{5}$$
$$\frac{dN'}{dt} = v_2 - v_5 - v_6 + v_7, \quad \frac{dA_3}{dt} = v_3 + v_4 - v_8 - v_L,$$
$$\frac{dA_2}{dt} = -v_3 - v_4 + 2v_8 + v_L .$$

In this system the following notation is used:

A_1 = [AMP], A_2 = [ADP], A_3 = [ATP], N = [NAD], N' = [NADH];
S_1 = [Fru-1,6-P_2], S_2 = [GAP], S_2' = [DAP], S_3 = [DPG], S_4 = [PEP], S_5 = [PYR];

$v_1 = v_{ALD}$, $v_1' = v_{TIM}$, $v_2 = v_{GAPDH}$, $v_3 = v_{PGK} = v_{PGM} = v_{ENO}$, $v_4 = v_{PK}$, $v_5 = v_{PDC} = v_{ADH}$, $v_6 = v_{GL3PDH}$, v_7 stands for the net NAD-dependent leakage of PYR or ACA from the GS, $v_8 = v_{AK}$, $v_L = v_{ATPase}$;

$$e_4 = \{1 + K_{PGM}(1 + K_{ENO})\}/(K_{ENO}K_{PGM}) \approx 3 \tag{6}$$

where K_{PGK}, K_{ENO} and K_{PGM} are the equilibrium constants for the corresponding reaction; the K's are the rate constants and the M's are the Michaelis constants.

With consideration for the typical enzyme activity distribution in the extract [4], and also for the initial conditions $S_{10} \sim 2\cdot10^{-2}$M, $A_\Sigma = 2\cdot10^{-3}$M, $N_\Sigma = 1\cdot10^{-3}$M (experiment of the type shown in Fig.2D), model (5) can be reduced to a simpler one

$$\dot S_1 = -v_1, \quad \dot S_2 = v_1 - v_1' - \tilde v_2, \quad \dot S_2' = v_1 + v_1' - v_6,$$
$$\dot S_4 = (\tilde v_2 - \tilde v_4)e_4, \quad \dot S_5 = \tilde v_4 - v_5 - v_7, \quad \dot A_3 = \tilde v_2 + \tilde v_4 - v_L, \tag{7}$$

where

$$\tilde{v}_2 = k_{+2}\frac{N\tilde{A}_2S_2 - N'A_3S_4Q_2}{m_2N' + A_2}, \quad \tilde{v}_4 = k_4S_4\frac{\tilde{A}_2}{M_4 + \tilde{A}_2} \tag{8,9}$$

$$\tilde{A}_2 = 0.5K_{AK}A_3\left\{\sqrt{1 + \frac{4}{K_{AK}}\left(\frac{A_\Sigma}{A_3} - 1\right)} - 1\right\} \tag{10}$$

$$\tilde{A}_1 = A_\Sigma - \tilde{A}_2 - A_3, \quad \tilde{\varphi} = (A_3 + 0.5\tilde{A}_2)/A_\Sigma \tag{11,12}$$

are the quasistationary variable values attainable at

$$v_2 = v_3, \; v_8 = 0 \tag{13,14}$$

Computer fitting of the solution of model (7) (Fig.7) to the behaviour patterns of intermediates and nucleotides observed in the experiment with an almost linearly decreasing Fru-1,6-P_2 (Fig.2C) yields the following parameter values:

$$\begin{aligned} &A_\Sigma = 1.66 \text{ mM}, N_\Sigma = 1.0 \text{ mM}, L = 0.8, k_{+1}=0.13, k_{-1}=3.85, \\ &k'_{+1}=3.34, k'_{-1}=0.45, k_{+2}=300, m_2=0.04, Q_2=0.2, \\ &k_4=5.83, M_4=0.2, k_5=1.15, M_5=0.01, k_6=0.12, M_6=0.1, \\ &k_7 = 0.019, K_{AK} = k_{+8}/k_{-8} = 5. \end{aligned} \tag{15}$$

With these parameter values, model (7) describes quantitatively (within the experimental error) the change with time in the concentrations of Fru-1,6-P_2, DAP, GAP (Fig.7) and also of PYR, ETOH, GLOH, and NADH. The model predicts, however, a somewhat higher value of φ as compared to the experiment (Fig.7).. The discrepancy may be explained by the fact that the kinetic model (Fig.6) takes no account of inhibition of GAPDH by DPG and NADH and the inhibition of PK by ATP.

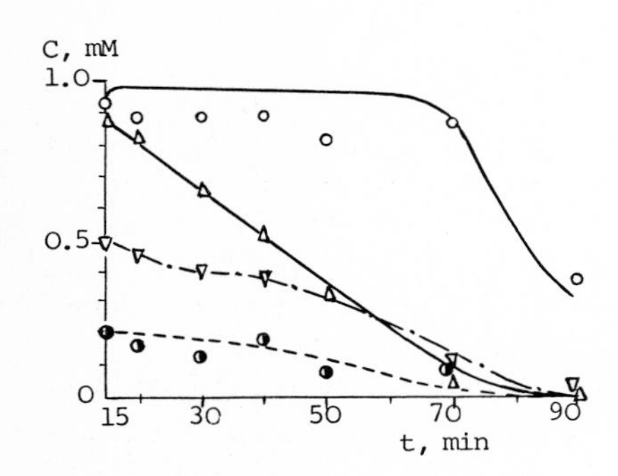

Fig. 7 Solution of the mathematical model (7) (curves) fitted to the experimental data (points); △——△, [Fru-1,6-P_2]/20; ▽-·-▽, [DAP]/5; ◑---◑, [GAP]; ○——○, φ

Fig.8 presents the solution to model (7) for the numerical data (15), which demonstrates the ability of the lower part of the GS to stabilize φ during the phase of Fru-1,6-P_2 breakdown. The high stability of φ in this phase (Fig.8B) is due to strong apparent saturation of GAPDH with GAP. Because of the saturation, the decrease in [Fru-1,6-P_2] and the resulting decrease in the quasisteady state value of [GAP] do not affect the reactions below GAPDH, unless [Fru-1,6-P_2] drops to extremely low values (Fig.8A). It should be noted that the high degree of saturation and the high apparent affi-

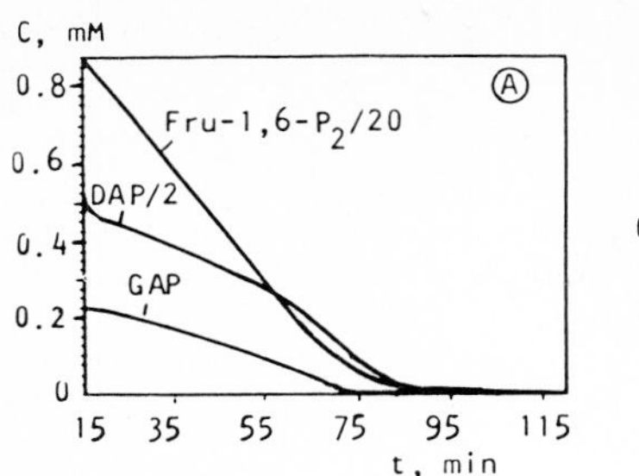

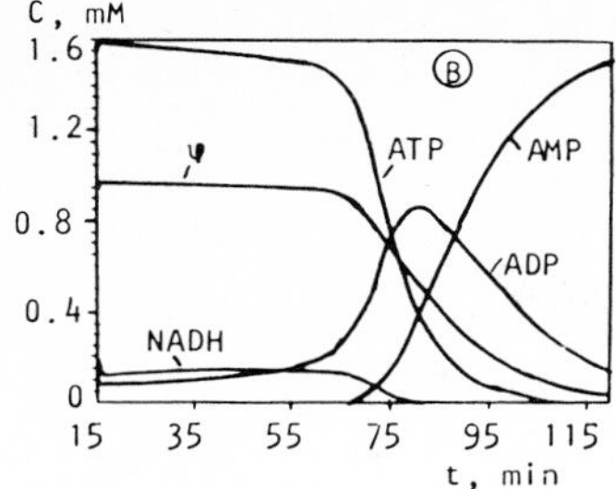

Fig. 8 Extended solution of model (7)

nity of GAPDH for GAP are purely system effects which manifest themselves only in the quasisteady state when $\dot{S}_3=\dot{S}_4=\dot{S}_5=\dot{A}_i=\dot{N}_1\approx 0$. Indeed, in our model v_2 is a linear function of S_2 at constant values of N and S_3.

As follows from [33], the lower part of the GS can be decomposed to a number of less effective stabilizers of ψ, interacting synergistically with each other. Some of them are very simple, such as

$$\text{[PEP]} \xrightarrow[\text{PK}]{\text{ADP} \to \text{ATP}} \text{PYR}, \qquad \text{[2-PG]} \underset{\text{ENO}}{\rightleftharpoons} \text{PEP} \xrightarrow[\text{PK}]{\text{ADP} \to \text{ATP}} \text{PYR}, \qquad (16,17)$$

(with v_L: ATP $\to$ ADP in each cycle)

others are much more complicated

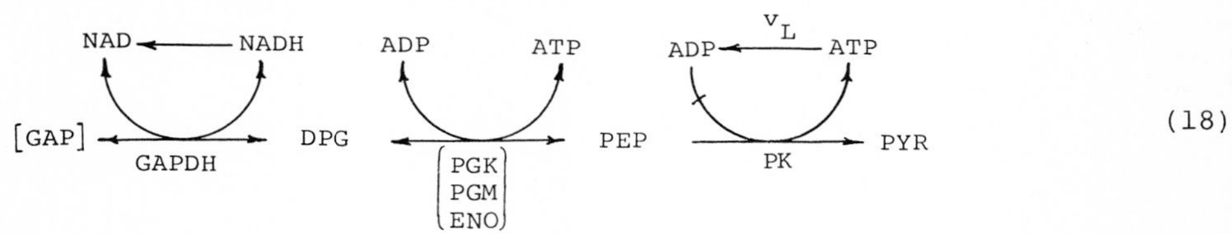

(18)

Saturation of the enzymes is shown in the schemes (16-18) by Higgins' mark. Concentrations of the initial substrate (square brackets) must be either excessive or kept constant.

Because of its consequences it seems noteworthy to stress the high stability of A_{ATPase}. Even in the strongly overloaded GS which shows a dramatic decrease of [ATP] (Fig.1), the affinity of the ATPase reaction, A_{ATPase}, changes very little and is able to drive effectively glycogenesis at a high flux rate. In fact, this stability serves as the basis for a hierarchical priority among the many different ATP consumers in the cell.

When the concentration of ATP is low, it becomes accessible only to those consumers which have significantly greater affinity for ATP than others. This hierarchy immediately results in the temporal organization of the consumers. This statement is illustrated by Fig.1: after the addition of glucose, it is first deposited by the "starved" GS in the form of storage polysaccharides, and it is not until the free glucose is completely exhausted that the stable supply of the consumers with ATP from the endogenous energy sources begins.

This intrinsic selection of metabolic priority is obviously the consequence of the very design [1] of the whole metabolic network: those pathways which are absolutely necessary for survival of the cell (in this case the requisition of food by phosphorylation) do have the highest affinity for substrates, i.e. the lowest values for half saturation of the respective enzymes. In due order follow all less important pathways according to their priority rating which automatically translates to the respective temporal sequence.

In our case not only a quick requisition of the offered glucose and the temporally delayed expenditure of the gained energy takes place but a major part of the phosphorylated hexose is reversibly converted to storage polysaccharides. This adds a new dimension of regulatory complexity to the system: the balance between katabolic and anabolic reactions. In either pathway the GS operates with different enzymes and coenzymes. In yeast the deposition of storage polysaccharides proceeds via uridine diphosphoglucose pyrophosphorylase, cofactor is UTP, whereas the opposing katabolic reactions are catalyzed by phosphorolytic or hydrolytic enzymes with the general use of adenine nucleotide coenzymes throughout the glycolytic reaction sequence.

Such diversity of enzymes, coenzymes (and eventually effectors) constitutes a higher order of regulatory hierarchy [1] than the allosteric mechanisms of enzyme reaction or the basic stoichiometric regulatory principles of the metabolic network outlined in this study. This class of the regulatory hierarchy is designed to switch on or off major reaction pathways avoiding effectively futile cycles by temporal separation of opposite or competitive reaction sequences [33,34].

The basis, however, for all regulatory principles of higher order is a stable energy source. For the glycolytic system it has been shown here that a stable energy output is maintained by a number of equivalent regulatory mechanisms. Their common principle is that pools of various glycolytic intermediates separately or in concerted action do serve as buffering reservoirs to maintain a quasisteady state in the ATP generating part of the system. Concomitant with the stability of this state, the energy charge, φ , is very stable and well protected against experimental perturbations which impose on the GS drastic changes in a) metabolic load, b) substrate availability, c) the length of its operating reaction sequence, and d) the very regulatory structure of higher hierarchical order. The simple stoichiometric nature of these surprisingly effective mechanisms of energy stabilization suggest that they are a very early product of evolution, much older than the allosteric regulation of enzymes or the sophisticated switching of reaction pathways.

Acknowledgements

We are very grateful to Mrs. B.Klein, Mrs. U.Schacknies and Miss W.Siebers for the skilful enzymatic determinations of metabolites as well as for expert technical help in data handling and to Mr. K.H.Müller and Mrs. L.Ja.Lesnetchuk for computer facilities.

References

1. A.Boiteux and B.Hess. Design of Glycolysis, in: The Enzymes of Glycolysis: Structure, Activity and Evolution; The Royal Society of London, Cambridge, At the University Press (1981) pp.5-22
2. B.Chance, K.Pye, J.J.Higgins, IEEE Spectrum, 4, (1967) pp.79-86
3. B.Hess, A.Boiteux, Hoppe-Seyler's Z. Physical. Chem. 349 (1968) pp.1567-1574

4. B.Hess, A.Boiteux, J.Krueger, Adv.Enzyme Reg., 7 (1969) pp.149-167
5. E.K.Pye, Canad. J. Bot, 47, (1969) pp.271-285
6. L. von Klitzing, A.Betz, Arch. Mikrobiol., 71 (1970) pp.220-225
7. B.Hess, A.Boiteux, Ann. Rev. Biochem., 40 (1971) pp.237-258
8. B.Chance, E.K.Pye, A.K.Ghosh, B.Hess (Eds.). Biological and Biochemical Oscillators, Acad. Press, New York (1973)
9. A.Goldbeter, R.Caplan, Ann. Rev. Biophys. Bioeng., 5 (1976) pp. 449-475
10. K.Tornheim, J.Theoret. Biol., 79 (1979) pp.491-541
11. S.B.Jonnalagadda, J.U.Becker, E.E.Sel'kov, A.Betz, Biosystems, 15 (1982) pp.49-58
12. L.von Klitzing, Arch. Mikrobiol., 72 (1970) pp.106-110
13. P.M.Greenwald, Med. Hypothesis, 3 (1977) pp.138-143
14. E.Racker, Mechanisms in Bioenergetics, Acad. Press, New York, (1965)
15. T.A.Rapoport, R.Heinrich, S.M.Rapoport, Biochem. J. 154 (1976) pp. 449-469
16. F.I.Ataullakhanov, V.M.Vitvitsky, A.M.Zhabotinsky, A.V.Pichugin, B.N.Kholodenko, L.I.Ehrlich, Acta biol. med. germ., 40 (1981) pp. 991-997
17. F.I.Ataullkhanov, V.M.Vitvitsky, A.M.Zhabotinsky, B.N.Kholodenko, L.I.Erlich, Biofizika, 22 (1977) pp.483-488
18. R.Heinrich, S.M.Rapoport, T.A.Rapoport, Progr. Biophys. Molec. Biol., 32 (1977) pp.1-82
19. D.E.Atkinson, Biochem., 7 (1968) 4030-4034
20. D.E.Atkinson, Cellular Energy Metabolism and its Regulation. Acad. Press, New York (1977)
21. Cold Spring Harbor Symposia on Quantitative Biology, vol. XXVI, Cellular Regulatory Mechanisms. The Biological Laboratories, Cold Spring Harbor (1961)
22. J.Monod, J.Wyman, J.-P. Changeux, J. Mol. Biol., 12 (1965) pp. 88-118
23. H.E.Umbarger, Ann. Rev. Biochem., 38 (1969) pp.323-370
24. E.R.Stadtman, in: The Enzymes (P.D.Bayer, Ed.), 3rd ed., vol. 1, Acad. Press, New York (1970) pp.397-459
25. E.A.Newsholme, C.Start, Regulation in Metabolism, John Wiley and Sons, London (1973)
26. P.W.Hochachka, G.N.Somero, Strategy of Biochemical Adaptation, W.B.Sounders Co., Philadelphia (1973)
27. E.E.Sel'kov, Eur. J. Biochem., 59 (1975) pp.151-157
28. J.G.Reich, E.E.Sel'kov, Th.Geier, V.Dronova, studia biophys., 54 (1976) pp.57-76
29. E.E.Sel'kov, in: Pattern Formation by Dynamical Systems and Pattern Recognition (E.H.Haken, Ed.), Springer Verl., Berlin, (1979) pp.166-174
30. J.G.Reich, E.E.Sel'kov, Energy Metabolism of the Cell. A Theoretical Treatise. Acad. Press, London (1981)
31. A.Boiteux, E.E.Sel'kov, BioSystems (in the press)
32. E.E.Sel'kov, BioSystems (in the press)
33. E.E.Sel'kov, Ber. Bunsenges. Phys. Chem., 84 (1980) pp.399-402
34. A.Boiteux, B.Hess, E.E.Sel'kov, Cur. Topics Cell. Reg., 17 (1980) pp.171-203

Evolution and Value of Information

M.V. Volkenstein and D.S. Chernavskii

Institute of Molecular Biology and P.N. Lebedev Physical Institute
Academy of Sciences of the USSR, SU-Moscow, USSR

1. The problem of origination of valuable information in biology is of great importance. But first of all we have to define the meaning of this concept and to find a proper quantitative measure.

Of special interest is to consider the possibility of a gradual increase of the value of information in the course of evolution. If this is possible, then is solved one of the main problems of evolution, the problem of small probability of spontaneous (accidental) generation of a great amount of valuable information.

The case is that the probability of a process, in which non-valuable information is initially generated (it is rather probable), and then the value of this information increases, is not small.

We shall try to discuss this problem in terms of dynamic system theory, where the typical synergetic phenomena, such as bifurcations, instability, multistationarity, etc., play an important role [1]. We should emphasize that the apparatus of dynamic systems has not yet been applied to the theory of information (but for [2]).

The quality of information (or its value) depends on the goal pursued. If the goal is clearly formulated, the value of information can be expressed as

$$V = \log_2 P_f/P_{in}, \tag{1}$$

where P_{in} and P_f are the probabilities to attain the goal before and after the information being received. The expression (1) is somewhat similar to that suggested by Kharkevich [3]. It is convenient in the case where the probability of the goal attainment is small. In the book by Stratanovich [4] the value of information is connected with the decrease of "penalty" or "expenditures". The definition by Stratanovich is suitable when the goal attainment is probable and there are many pathways leading to the goal. The information, which allows to choose the optimal pathway with minimal expenditures, can be considered as valuable.

In biological evolution the probability of the goal attainment is small, so we shall use the measure (1). In the particular, rather important case, where the goal is attained with near-unity probability after information being received,the value coincides with the maximal amount of information. According to (1) the value can be positive, negative (desinformation), or zero. The amount of zero value information in biology is never small compared with the amounts of positive and/or negative information.

In this connection we have to pose the problem to define the total amount of positive and negative information, or to separate the information with zero value [2,5] . The amount of information with a non-zero value depends on the broadness of the goal formulation, it is the greater the more clear and exact is the definition of the goal. The problem to measure the broadness of the goal has not yet been discussed. It is, however, important in biology since in many cases the goal can be defined only approximately, e.g. "the purpose of life is livimg".

In the case of a well formulated goal, the value of information depends on the interactions inside a system and on the state of this system. Thereby, the value can vary in time, in particular it can grow from zero to the maximal magnitude.

2.Let us consider the properties of the phase space of a dynamic system in which the above mentioned processes can occur. We call such a system informational. According to Quastler [6], we consider information as a choice of one option (or n options) from N possible (N>n) and memorizing the choice. The choice being made under the action of non-random external factors means reception of information, the random choice means generation of new information.

Therefore, an informational system must possess the following properties [2]:

1) Multistationarity is necessary for a choice among N options to be made.

2) In the phase space there must be a region of instability, it is necessary for generation of new information.

3) The system must be dissipative. This is needed for memorizing information, i.e. for the absolute stability of a finally chosen state.

Some remarks concerning these properties.

i. Not all dynamic systems possess such properties. In particular, the equilibrium thermodynamic system is not informational, since its macroscopic state is unique and microscopic states are unstable and cannot be memorized. Hence the condition of memorization is very important. Without it the choice of any microscopic state can be considered as information. We use the term microinformation for non-memorizable and the term macroinformation (or simply information) for memorizable information.

It is clear that in any system the amounts of micro- and macro-information differ strongly, $I_{mic} >> I_{mac}$. The values also differ, microinformation possesses no value. In practical problems of biological evolution, we always deal with memorizable information.

Let us emphasize that in the relation between information and entropy

$$I_{mic} + S = S_o = \text{const} \qquad (2)$$

microinformation is meant. Macroinformation is not connected directly with entropy, we can only say that generation and reception of information is accompanied by a change of entropy, which is much greater than the amount of information, $[\Delta S] >> \Delta I_{mac}$.

ii. Instability is the necessary condition for the generation of new information. The physical examples of systems satisfying the conditions (1)-(3) are chinese billiards, roulette, and other game machines. The phase space here is four-dimensional, a ball on the surface has two coordinates and two moments. The initial conditions are defined in a small region of the phase space, practically at a point. There exists, however, an intermediate region where all trajectories are unstable, as in Sinai's billiards [7]. In the case of chinese billiards this region corresponds to a layer of the phase space where the kinetic energy of the ball is higher than the portential barriers between the holes. Lower, there is a non-stochastic layer of the phase space, divided into regions of attraction of stationary states.

New information is generated at the moment t_c when the system makes a transition from the stochastic region to the attraction region of a stationary state. Just at this moment the system makes a choice and memorizes it. The probability to choose a given j-th stationary state depends on the position of the state point at the stochastic level at the moment just before t_c. The presence of an intermediate region of instability (stochasticity) is a non-trivial property of informational systems. This region has the properties of a strange attractor, though it is not an attractor in the strict sence of the word. The final stationary states are here attractors, in their vicinity the system behaves dynamically.

No mathematical model of a dynamic system with these properties (such as the Lorentz model of a strange attractor) has as yet been proposed. Four-dimensional phase space is, seemingly, necessary and sufficient for the basic model of this type.

3. Let us now discuss the concept of the goal in biology. The problem is quite clear as far as the higher nervous activity is involved. In an artificial system, such as a game machine, the goal is given from outside, the machine itself is not supposed to pursue its own goals. The question arises whether a dynamic system can exist, which (or a part of which) pursues its own formulated goals. This question is important for the problem of the origin of life. In this case we come across the transition from the non-living matter, which apparently has no purpose, to living organisms pursueing definite goals. Thus, the goal of an elementary evolutionary system, of a population, is to attain a state favourable for its life and reproduction.

The concept of the goal has the physical sense of transition from a relatively unstable state to a chosen relatively stable state. The population achieves higher stability by adaptation to the corresponding ecological niche. In fact, the choice of a niche is not determined unambiguously and can be considered as random.

The goal of DNA is to synthesize protein to provide the existence of the cell and organism. The biosynthetic system attains a more stable state when producing proteins. However, in this case there is no choice, since the synthesis of a given protein in the cells of an individual of the given species cannot be considered as creation of new information. This synthesis is determined. New information arises as a result of mutations.

The general goal of a biological system at every level of consideration is life, that is the optimal conditions of existence of individuals, population, species and higher taxons. We see that the notion of the goal in biology has a real but somewhat diffuse sense.

New information in biology originates in every formation of a new individual, population, species, etc. The appearance of a new genome, as a result of recombination of paternal ones in sexual reproduction, is memorizing of a random choice, because no law of nature foresees the progeny of just this pair. (An exception is the offsprings of Eve, since Adam had no choice, hence Abel and Cain had not carried new information). The same concerns the creation of new species in the course of microevolution. In [8] it has been shown that speciation is described by equations similar to those for phase transitions. Biological development proceeds in a discrete way.

Thus, discussing the value of biological, particularly, genetic information, we can use the concept of the goal.

Let us consider the properties of a dynamic system for which the concept of genuine goals (the goals not thrust from outside) can be introduced. It is clear that conditions (1)-(3) must be necessarily satisfied. Of particular importance is the presence of an unstable intermediate state and of a set of stable final states. The transition from an unstable state into a stable one means attainment of the goal [9]. However, to establish a goal the conditions (1)-(3) are necessary but not sufficient. In order to formulate one more condition, let us consider a system satisfying the following equations:

$$dx_i/dt = F_i(x_1, x_2, \ldots, x_m). \quad (3)$$

Here x_i is a vector in an m-dimensional phase space ($i=1,2,\ldots,m$). The functions F_i are such that conditions (1)-(3) are satisfied and there are N stationary states. Before the moment t_c the system is in the stochastic layer, and at t_c it finds itself in the region of attraction of one of the N stationary states. Thereby the system makes the choice.

Let us divide the system into ν subsystems (elements) and consider the state of each subsystem at $t<t_c$, assuming it to be isolated.

It is possible that at the moment preceding t_c every subsystem is already in the region of attraction of the j-th stationary state ($j=1,2,\ldots,N$), j being different for different subsystems. Just this situation is realized in concrete systems. Every subsystem can be said to possess information already, but this does not hold true for the whole system (3). Under these conditions the goal for every subsystem can be formulated as necessity to conserve the produced information. The goal of the subsystem which has chosen the j-th state is that the whole system takes the same j-th state. As follows from conditions (1)-(3), the goals of different subsystems, possessing different information, are antagonistic, since the choice of a given stationary state by the system (3) excludes a different choice.

In biology such a goal is equivalent to population survival. After the goal being formulated, the value of information originated in a given subsystem can be defined. Clearly, the value depends on the state of the whole system at a given moment (i.e. on the states of other subsystems) and it changes in time. Thus the fourth condition can be formulated as follows.

4) The possibility to divide a system into subsystems, which already possess information at the time moment when the whole system does not.

4. Let us illustrate these ideas by a model of selection of a unique code [10-12].

$$dx_i/dt = ax_i - \gamma\sum_{j\neq i}^{N} x_i x_j; \qquad i,j = 1,2,\ldots,N. \quad (4)$$

Here N is the number of various codes (N is of order 10^{15}), x_i is the specific number of individuals carrying the i-th code. The linear term describes autoreproduction, the square term describes antagonistic interactions. This model is similar to the Eigen hypercycle models [13], but it allows one to select a unique code even if all variants are equivalent, i.e. if no code has a *priori* advantages. The system (4) has the following properties:

i. There are N stationary states corresponding to survival of pure populations ($x_i \to \infty$, $x_{j \neq i} \to 0$; $j,i=1,2,\ldots,N$).

ii. The system is dissipative and stationary states are stable.

iii. There are two unstable points $x_i=0$ and

$$x_1 = x_2 = \ldots = x_N = \bar{x} = a/\gamma .$$

The region of instability, which is necessary according to (2), is absent here. However, if external noise is taken into account, the region adjacent to $x_i=\bar{x}$ plays the role of a stochastic layer. The elements of system (4) are individuals, their goal is to survive after encounters with other elements and, therefore, to preserve the code information they possess.

Let the number of each species' representatives be initially equal, i.e. at t=0 $x_1=x_2=\ldots=\bar{x}$. The a *priori* probability of survival in an encounter at this moment is

$$P_{in}(t=0) = \sum_{i=1}^{N} p_i^o W_i^o = \sum_{i=1}^{N} N^{-1}\bar{x}/X^o = N^{-1}. \qquad (5)$$

Here $W_i^o = N^{-1}$ is the probability to choose the i-th code, $p_i^o=\bar{x}/X^o$ is the probability to meet an organism of the i-th species, $X^o=\Sigma x_i=N\bar{x}$ is the total number of individuals at t=0.

The *a posteriori* probability to survive after the choice of the k-th code is obtained by taking $W_i=\delta_{ik}$ in (4). Hence,

$$P_f(t=0) = P_k^o = N^{-1}. \qquad (6)$$

The value of information is

$$V_k(t=0) = \log_2 P_f(t=0)/P_{in}(t=0) = 0. \qquad (7)$$

It means that reception of information does not affect attainment of the goal. This is quite natural under symmetrical conditions.

The initial symmetric state is not stable, it is destroyed in the course of time and the system tends to a stable state, where one (say, the k-th) version of the code predominates. At the moment t (such that at>>1) the number of individuals of the k-th type is

$$x_k(t) \simeq \bar{x}\exp(at). \qquad (8)$$

The number of individuals of the type i is

$$x_i(t) = \bar{x}\exp(-e^{at}) = \bar{x}(t) \ll x_k(t); \qquad i \neq k. \qquad (9)$$

The a *priori* probability to survive at this moment is

$$P_{in}(t) = \sum_{i\neq k}^{N} P_i(t)W_i + P_k(t)W_k = N^{-1}x_k(t)/X(t) + \bar{x}/X(t). \qquad (10)$$

The a *posteriori* probability, after the k-th code being chosen is

$$P_f(t) = P_k(t) = x_k(t)/X(t) \approx [1 + N\varepsilon(t)]^{-1}.$$

Here

$$X(t)=x_k(t) + (N-1)\bar{x}(t); \qquad \varepsilon = \bar{x}(t)/X(t)<<1 .$$

The value of this information is

$$V_k(t) = \log_2 N \; [1 + N\varepsilon \;(t)]^{-1}. \tag{11}$$

ε tends to zero with time and the value $V_k(t)$ increases to the maximal amount of information about the code

$$V_k(t) \rightarrow \log_2 N - N\varepsilon \;(t) = I_{max} - N\varepsilon(t). \tag{12}$$

This example shows that the value of information grows from zero to the maximal possible value, and this growth is determined by interactions inside the system, rather than by external causes.

Spontaneous increase of the value of information is of great importance in evolutionary processes. This concerns biology as well as linguistics, economics, and other sciences, considering developing systems. Indeed, the accidental generation of valuable information has a very small probability, but this information, nevertheless, is produced during development. The problem of small probability vanishes if the information is not valuable at the moment of generation but becomes progressively valuable as a result of interactions inside the system.

5. The amount of zero-value information can be estimated even in the case of a "broad" formulation of the goal.

Redundant information is an example of non-valuable information. On the contrary, non-redundancy, irreplaceability can serve as a measure of valuable information.

Redundant information does not change the probability of the goal attainment. Therefore, in accordance with formula (1), the value of redundant information is zero. Usually, a repeated message is redundant, as well as spare characters, which can be ommited without damage of understanding. Natural languages possess high redundancy, exceeding 50 per cent. An example is decoding of the message,damaged by the sea water in "Les enfants du captaine Grant" by J.Verne. Redundancy is used in technologies to eliminate noises, to increase the probability of adequate reception of a message.

The broadness of biological goals produces some ambiguity in the definition of redundancy of biological information. The genes of eucaryotes are multiple, the genomes containing plenty of repeats. Are they redundant? If the goal is synthesis of a definite protein, the answer is positive. But it is repeated genes that provide the definite amount of protein which is necessary for development of an individual. In this sense multiplicity of genes is not redundant.

The problem of the broadness of a goal needs a more detailed study with the help of dynamic systems theory. We hope to do this in our next works.

The genetic code is strongly degenerate. Does it mean redundancy? It would be so if the codons, degenerated together, were identical in an evolutionary sense. But mutations of these codons are different [5]. They are not identical both for ontogenesis and evolution [14]. The degenerate codons catalyze the inclusion of an amino acid residue into protein at different rates. For biological development not only the structure of a given protein is of importance, but also its amount, the place and time of its action. That is, the questions "how much, where and when?" must be exactly answered.

Nevertheless, irreplaceability can be considered as a tentative measure of value, of non-redundancy. Conventional scales of such values of amino acid residues in proteins and of codons in mRNA have been proposed in [5] (see also [15-17]). These scales are based on experimental data about mutational replacements of residues in proteins. Tryptophenyl has the lowest replaceability and the highest value. Alanyl has the lowest value.

In biological development, both individual and evolutionary, the irreplaceability, i.e., the value of information, increases. Let us consider some examples.

In classical experiments by Spemann at the beginning of our century, the totipotency of a rudiment in embryogenesis was shown to be replaced by unipotency, i.e. presumptive rudiments become determined and their irreplaceability increases. The formation of new species from common ancestors in Darwinian evolution occurs as divergence, as bifurcation. This means that new information is created by a phase transition and, therefore, irreplaceability increases. The species cannot substitute each other. That is why there are no reproductive hybrids.

Using the tentative scale of values of amino acid residues, it was shown that the total value of residues in the ancient universal protein, cytochrome C, increases corresponding to the evolutionary trees of both mammals and birds [5,15-17]. However, this regularity is not valid for relatively new and specialized proteins, hemoglobins. Evolutionary substitutions in this case are random.

These results correspond to the neutralist theory by Kimura. According to this theory evolution at the molecular level is, mainly, of random character.The replacements of nucleotides in DNA and of corresponding amino acid residues in proteins are not directly affected by natural selection, acting at the phenotype level [18]. The level of neutral mutability for hemoglobin is much higher than that for cytochrome C.

The neutralist theory is well grounded. Its physical meaning is that the correspondence between the primary genetically coded structure of a protein and the spatial structure, responsible for biological functionality, is degenerate and ambiguous [19].

There are additional arguments in favour of neutralism. It has been shown that the genetic code is built in a specific way [20,21, 26]. Single replacements of nucleotides in codons bring to substitution of an amino acid residue by another one with analogous properties and structure. In this sense the code possesses high stability against disturbances and the corresponding mutations are mainly neutral. The second argument is provided by the work by Ptitsyn [22] who has shown that a globular protein can be treated as an "edited static copolymer". Mutational replacements in the greater part of the protein are, therefore, neutral. It was indicated [23] that for an evolutionary increase of the protein values the supply of low-value residues is necessary.

6. The principle of evolutionary growth of biological systems complexity has been formulated (see, e.g. [24]). We have to define the notion of complexity. The definition has been given by Kolmogorov [25] (see also [26]). Complexity of an object is the minimal number of binary digits, containing information about the object, sufficient for its reproduction (decoding). Complexity is the length in bits of the shortest program needed to build a description of the object.

To establish complexity of a given sequence of numbers we have to prove that a shorter program generating this sequence does not exist. According to Gödel's theorem, such a statement cannot be proved, since the proof requires a system of higher complexity [26].

In this connection we have to emphasize that the aim of science is to find a minimal program generating (explaining) some aggregation of facts. This is the principle of Okkam's blade. Thus, the Newton gravitation law explains both the fall of an apple and motion of planets. But Gödel's theorem fails to prove the minimality of the program in a logical way. Therefore science is impossible without intuition.

The most complicated systems in nature are individual living organisms, humans being among them. Every personality is unique, it cannot be represented by a shortened program. Nobody is replaceable in this sense. This relates also to the human creative activity, to pieces of art and literature.

However, every organism is not purely individual. It is a representative of a kingdom, type, class, etc. down to a species. The discovery of this real hierarchy is one of the greatest events in the history of sciences. Evidently, the complexity increases from a kingdom towards a species. In every taxon there are no irreplaceable organisms. All representatives of a given species are described by the same minimal program.

The notion of complexity is a relative one [24]. For biologists, the brain of an ox is a very complicated system, requiring thousands of bits for its description. But for a butcher, this description is as short as five bits, because the brain is one of some thirty parts of the ox's body used as food. We mean here different levels of reception, the relative value of information. Complexity is seen to be equivalent to irreplaceability, to non-redundance at a given level of reception. Impossibility to minimize the program, generating a complex message, implies irreplaceability. Both the value of information and complexity increase when proceeding from a kingdom to a species, becoming maximal in individuals.

At the same time the notion of value is richer than that of complexity. Complexity concerns an object as a whole, while separate elements of the object are also of value. Complexity characterizes the structure, the value expresses also the function.

In biological evolution a system may become not only more complex, but also it may become simpler. This concerns, in particular, the transition of some biological species from free to parasitic existence. Complexity drops down, but irreplaceability (value) increases in accordance with ecological conditions.

The ability to estimate information and to select valuable information also increases in evolution. Such ability of selection is the basis of the human creative activity. It does not require additional expenditures, since expenditures per one bit of information does not depend on the information value [16].

Natural selection means comparative estimation of phenotypes in a given ecosystem, i.e. search for the maximal information value. Here the analogy with the game of chess is instructive. Game models are generally useful in theoretical biology (see [27]).

A move in chess plays the role of mutation. The game is irreversible, moves cannot be taken back. The game, developing along a certain pathway, cannot be switched over to another one. This is an analogy with evolution which has a directed, channelized character.

As Steinitz and Lasker have shown, chess must be played in a positional way, gradually collecting small advantages. If advantages are sufficient, you have to look for a combinational, decisive way towards winning. If such a way is not found, advantages dissipate. Collecting small advantages is similar to microevolution, combinations correspond to a phase transition, to speciation. There are two types of constraints in playing chess: definite rules of moves and a definite pathway in a given game. In biology an obligatory functionality of proteins and nucleic acids corresponds to the rules of moves. Channeling of evolution is determined by the structure of already formed organism and by possible changes of this organism. The problem of evolution is inseparable from the problem of ontogenesis (see [28]).

Creation of new information and increase of its value in evolution are possible only in an open system of biosphere, because of entropy export into environment [29]. It is obvious that the corresponding physical problems require further investigation.

References

1. H.Haken. Synergetics, Springer-Verlag, Berlin, Heidelberg, New York, (1978)
2. M.V.Volkenstein, D.S.Chernavskii. J. Soc. Biol. Struct., 1, 95 (1978)
3. A.A.Kharkevich. Information Theory, Nauka, Moscow (1973)
4. R.L.Stratanovich. Information Theory, Sovjetskoe Radio, Moscow, (1975)
5. M.V.Volkenstein. J. Theor. Biol., 80, 155 (1979)
6. H.Quastler. The Emergence of Biological Organization, Yale Univ. Press, New Havn, L. (1964)
7. Ya.G.Sinai. Doklady Akad.Nauk SSSR, 158, 1261 (1963)
8. B.N.Belintsev, M.V.Volkenstein. Doklady Akad. Nauk SSSR, 235, 205 (1977)
9. M.V.Volkenstein. J. Soc. Biol. Struct., 3, 67 (1980)
10. N.M.Chernavskaya. D.S.Chernavskii, J. Theor. Biol., 35, 13 (1975)
11. Yu.M.Romanovskii. N.V.Stepanova, D.S.Chernavskii. Mathematical Modeling in Biophysics, Nauka, Moscow, (1975)
12. D.S.Chernavskii. J. of the Mendeleev Chemical Soc., 25, 362, (1980)
13. M.Eigen, P.Schuster. The Hypercycle, Springer-Verlag, Berlin, Heidelberg, New York (1979)
14. M.Conrad, C.Friedlander, M.Goodman. Biosystems, in press.
15. M.V.Volkenstein. General Biophysics, Vol.2, Academic Press, N.Y. (1983)
16. M.V.Volkenstein. Biophysics, Mir Publishing House, Moscow, (1983)
17. M.V.Volkenstein. Physics and Biology, Academic Press, N.Y. (1982)
18. M.Kimura Sci. American, 241, No 5, 94, Nov. (1979)
19. M.V.Volkenstein. J. Gen. Biol (Moscow), 42, 680 (1981)

20. M.V.Volkenstein. Biochim.Biophys.Acta, 119, 421 (1966).
21. M.V.Volkenstein. Molecular Biophysics, Academic Press, N.Y.,197
22. O.B.Ptitsyn In: Conformation in Biology, Eds. R.Srinivasan, R.Sarma, Academic Press, N.Y., 1983
23. M.Conrad, M.V.Volkenstein. J.Theor.Biol., 92, 293 (1981)
24. P.Saunders, M.Ho. J.Theor.Biol., 63, 375 (1976); 90,515(1981)
25. A.N.Kolmogorov. Problems of Information Transfer, 1, 3.(1965); 5, 3, (1969)
26. G.Chaitin. Sci.American, 232, No. 5, 47 (1975)
27. M.Eigen, R.Winkler. Das Spiel, R.Piper Verlag, München, Zürich, 1975
28. Evolution and Development, J.Bonner ed., Springer-Verlag, Berlin, Heidelberg, New York, 1982
29. W.Ebeling, R.Feistel. Physik der Selbstorganisation und Evolution, Academie-Verlag, D., 1982

Index of Contributors